WASTEWATER SYSTEMS ENGINEERING

WASTEWATER SYSTEMS ENGINEERING

Dr. Homer W. Parker, P.E.

Consulting Engineer
Lake Jackson, Texas

Prentice-Hall, Inc. Englewood Cliffs, New Jersey

Library of Congress Cataloging in Publication Data

PARKER, HOMER W   1921–
   Wastewater systems engineering.

   Includes bibliographical references.
   1. Sewerage.   2. Sewage—Purification.   I. Title.
TD645.P37   628′.3   74-28429
ISBN 0-13-945758-5

10 9 8 7 6 5 4 3 2 1

Printed in the United States of America

PRENTICE-HALL INTERNATIONAL, INC., *London*
PRENTICE-HALL OF AUSTRALIA, PTY. LTD., *Sydney*
PRENTICE-HALL OF CANADA, LTD., *Toronto*
PRENTICE-HALL OF INDIA PRIVATE LIMITED, *New Delhi*
PRENTICE-HALL OF JAPAN, INC., *Tokyo*

Dedicated to
CONSUELO
Wife and Partner

Contents

6 SEWAGE TREATMENT—*PART I*   117

Preface

This book is a presentation of information useful in the calculation and design of wastewater treatment systems including so-called primary, secondary, tertiary, advance and non-biological plants. Most of the equations can be used with data readily obtainable by the average designer. The book is written in a form that is very easy to use by any designer having a B.S. degree or its equivalent. Techniques of system's design are included to provide any degree of domestic wastewater treatment desired including production of potable water.

Designers sometimes waste many manhours making plans that are subsequently not approved by regulatory agencies. The Recommended Standards for Sewage Works by the Great Lakes—Upper Mississippi River Board of Sanitary Engineers has been incorporated into the text in addition to design suggestions that have been accepted by the majority of the state regulatory agencies. Virtually all states have considerable latitude on what they will accept, but the designer must show rational calculations and present a basis for his design that satisfies the regulations for public health. The content should be of assistance in this regard.

While the book was written for the practicing engineer, it can readily be used as a graduate level textbook. This book contains information that has been found necessary for engineers if they are going to be efficient in a consulting office, industry, or in certain types of government positions. The problems at the end of chapters force the making of assumptions, the exercise of engineering judgment, and a search for data that is readily available in any accredited university library. Many of the problems are actual field problems that have been encountered except that the names of cities have all been changed.

HOMER W. PARKER

WASTEWATER SYSTEMS ENGINEERING

Introduction 1

Sewage consists of any wastewater and solids that find their way into the sewers and are transported through the sewer system. Domestic sewage is all the sewage that an average town produces, excluding industrial wastes but including water from factories that is of nonmanufacturing origin. Chemical or biological waste resulting from a manufacturing process is classified as industrial wastewater. Some industrial wastewaters can be processed in a municipal treatment plant without any special provisions. Some industrial wastes require considerable pretreatment before going into city sewers. Still other types of industrial wastes must be treated by the manufacturer before discharge into a receiving stream.

Fresh domestic sewage is gray-colored, that is, dirty-looking, water. Solids represent an insignificant percentage (approximately 0.1%) of total volume. Hydraulically, all calculations for sewage flows and mechanical handling are computed for water and exclude the solid content. The organic material in sewage is in the process of decomposition at all times unless the waste is heavily chlorinated or otherwise disinfected. This decomposition process is aerobic, or nonseptic, as long as the water that transports the waste has dissolved oxygen present. If the putrefaction process consumes all the available dissolved oxygen before reaching the treatment plant, then the process turns anaerobic and putrefaction continues in the absence of oxygen. In both cases, the decomposition is caused by bacteria and other microorganisms. The essential difference is that different species of bacteria predominate under aerobic (in presence of oxygen) and anaerobic (in absence of oxygen) conditions. Anaerobic bacteria produce foul-smelling gases and the sewage turns black. If either air containing oxygen or oxygen gas is transmitted into the water in adequate quantities, septic sewage turns dark brown or gray and loses its foul smell. Thus, by artificially supplying adequate air in sewage, a nonseptic state can be maintained at all times.

While the terms "sewage" and "wastewater" are used interchangeably in this book, the term sewage is becoming obsolete. Wastewater, the preferred term, includes both organic and mineral matter. The mineral matter may combine with the water to form dissolved solids, such as chlorides, or it may be present as undissolved suspended matter. The mineral matter may contribute to eutrophication of lakes due to the nutrients present in the sewage, it may form unsightly sludge deposits, and it may contribute to the hardness of the water in the effluent. The organic portion of the sewage is putrescible and is undergoing biological decomposition. This decomposition process consumes available dissolved oxygen in accordance with the quantity of degradable waste present. In the second chapter some of the applicable procedures and parameters are explained. Included in these parameters are explanations of BOD (biochemical oxygen demand), COD (chemical oxygen demand), and other measures of the condition of a waste and the demand that it places on a receiving stream for dissolved oxygen. If the dissolved oxygen content of a stream falls below certain levels, the wastewater will kill the aquatic inhabitants. Each oxygen-dependent aquatic dweller has a critical level of oxygen requirement under a given set of accompanying conditions such as water temperature.

In engineering design practice, regulatory agencies are tending to classify each stream in the state as well as segments of these streams. As an example, a developer decides to purchase a large parcel of land outside the range of a municipal treatment plant. Before a sewage treatment effluent can be discharged into a nearby stream, authorization is required from the concerned regulatory agencies. The regulatory agency may not permit any effluent to be discharged into the stream. Thus, the classification of a stream and what regulatory agencies will accept is one of the first things that should be considered before plans are made.

It is sometimes necessary to make a rather detailed analysis of the organic waste load on a stream. One of the best presentations available on this subject is found in Chapter 4 of N. L. Nemerow's *Theories and Practices of Industrial Waste Treatment*, Addison Wesley Publishing Co., Inc., Reading, Mass., 1963.

1-1. Sewage Treatment Processes. Sewage treatment is a broad term that applies to any process, or combination of processes, that can reduce the objectionable properties of water-carried waste, that is, wastewater, and render it less dangerous and repulsive to man. A series of graduated screens that remove a large portion of the solids would be a form of so-called primary sewage treatment. In its usual senses, primary sewage treatment consists of coarse screens, grit chambers, skimming tanks, and sedimentation basins where the settleable solids could settle out. Secondary treatment usually follows the primary process and involves establishing an environment favorable for rapid degradation of organic material by biological processes. Tertiary treatment usually consists of a more refined treatment of the effluent of a secondary process to achieve an even greater degree of treatment of the effluent. While all of these terms are still used, we shall approach the subject more from a systems-oriented viewpoint.

A number of processes or methods are employed to treat sewage. The nomenclature of these processes will be preserved in most cases, but the reader should not assume it is necessary to retain the traditional component grouping contained in individual processes. A system is a series of interacting component parts, groups of items, or subsystems forming a unified whole to achieve a specified vital function. In wastewater treatment it may be necessary to use mechanical (physical) devices, biological processes, and chemical treatment in the same treatment plant to establish a system appropriate to meet a given set of conditions.

Table 1-1 lists some of the traditional sewage treatment plant combinations and indicates the percentage of BOD reduction that each process combination can be expected to achieve. The BOD of a waste throughout this book represents the five-day BOD demand of a waste unless otherwise indi-

TABLE 1-1. Approximate Percent Reduction Achieved by Various Types of Sewage Treatment

Type of Treatment	Percent BOD Reduction
Septic tank	15–25
Imhoff tank	25–40
Sedimentation	25–40
Sedimentation + trickling filter	80–95
Sedimentation + conventional activated sludge	85–95 +
Sedimentation + sand filters	90–95
Extended aeration*	75–85
Tapered aeration*	90–95
Step aeration*	90–95
High rate*	60–75
Modified aeration*	60–75
Contact stabilization* (biosorption)	85–90
Hatfield process*	85–90
Kraus process*	85–90
Rapid bloc*	90–95
Supra activation*	55–65
Activated aeration*	80–85
Waste stabilization lagoon	85–95
Oxidation ditch	90–95 +

*Modification of activated sludge process.

cated. It is a measurement of the oxygen that would be required to stabilize a given waste.

No table or group of tables, charts, or graphs can replace the engineer's investigation, judgment, and experience in resolving a treatment problem. The engineering investigation should, among other things, look at the future operation of the plant. What level of operating personnel does the owner have available or will he be able to realistically afford? Also, it is necessary to consider the current state of the art on treatment processes and the degree of enforcement that will be applied several years hence. Can the plant meet the needs forthcoming? Will it lend itself to automation in lieu of skilled operators?

Table 1-2 is a guide to the first and second most likely solutions to a particular treatment problem. The selections offered are based on recent,

TABLE 1-2. Sewage Treatment Process Selection

Activity[1]	Unit	Units/Day Maximum	Units/Day Minimum	Recommended[2] Process	Alternate[3] Process
Airport	Capita	*	1,500	OD	AL
Apartment house	Capita	*	150	OD	TP
Assembly hall	Seat	*	1,500	OD	TP
Bar or tavern	Capita	*	1,500	OD	TP
Bowling alley	Lane	100	2	TP	ST
Cafeteria	Capita	*	160	OD	TP
Camp, construction	Capita	*	10	TP	TP
Camp, day, central bath	Capita	*	270	OD	TP
Camp, resort, luxury	Capita	*	150	OD	TP
Camp, summer	Capita	*	170	OD	TP, WSP
Church, large	Capita	*	900	OD	TP, WSP
Church, small	Capita	900	0	ST	TP, WSP
Country club	Capita	*	280	OD	TP
Dance hall	Capita	*	1,000	OD	TP
Factory[4]	Capita	*	300	OD	TP
Hotel	Capita	*	150	OD	TP
Hospital	Capita	*	75	OD	TP
Institution, resident	Capita	*	150	OD	TP
Marina	Slip	*	15	OD-SP	AL-SP
Municipality, large	Capita	*	500,000	AS	TF
Municipality	Capita	500,000	250,000	AS	OD, TF
Municipality	Capita	250,000	5,000	OD	AS, TF
Municipality, Small	Capita	5,000	150	OD	AL ,TP
Motel	Capita	*	150	OD	TP
Mobil home park	Mobil home	150	10	OD	TP
Offices	Capita	*	450	OD	TP
Picnic parks	User	*	300	OD	TP
Recreation area	Capita	*	450	OD	TP
Restaurant	Seat	*	150	OD	TP
Rest area, roadside	Capita	*	600	OD	TP
Rest home	Capita	*	150	OD	TP
School, boarding	Capita	*	450	OD	TP

TABLE 1-2. (Continued)

Activity	Unit	Units/Day Maximum	Units/Day Minimum	Recommended Process	Alternate Process
School, day or elem.	Capita	*	600	OD	TP
School, high	Capita	*	450	OD	TP
Subdivision, luxury	Capita	*	140	OD	TP
Subdivision	Capita	*	150	OD	TP
Service station-vehicular	Bay	*	1	TP	ST
Shopping centers	Sq Ft	*	25,000	OD	TP
Stadium	Seat	*	1,500	OD	TP
Swimming pool	Capita	*	300	OD	TP
Theater, drive-in	Stall	*	750	OD	TP
Theater, movie	Seat	*	750	OD	TP
Trailer dump station	Trailer	*	15	OD-SP	AL-SP

[1] Assumed that local conditions require a separate treatment plant for activity.
[2] Probably best all-around solution provided local conditions permit usage.
[3] Probably second-best choice if local conditions favorable to its use.
[4] The nonindustrial waste portion of factory sewage.
* Recommended process can be designed for largest facility likely to be built.
OD = Oxidation ditch configuration described in Chapter 9.
TP = Package treatment plant of appropriate type. See Chapter 8.
AS = Custom-designed activated sludge plant. TF = Custom designed trickling filter.
AL = Aerated lagoon. SP = Special provision in addition to basic plant required.
ST = Septic tank and drain field. WSP = Waste stabilization pond.

previously unpublished, field investigation and research by the author. Thus, some of the selections offered are several years ahead of common practice. The study considered the problem as a whole, looking not only at treatment efficiency but also labor, future automation, changes expected to be imposed on the owners as tougher and tougher enforcement takes place, plant modification, aesthetics, etc. It should be emphasized that the oxidation ditch, recommended in a number of the applications, is based on the configuration indicated in Chapter 9 and it includes provision to vary the water level of the ditch to regulate the amount of oxygen transfer. This is a practical, manually operated plant in its present state of development, and it is considered a superior configuration to the earlier plants in the United States and Europe. A properly designed, constructed, operated, and maintained plant may achieve over 97% settleable-solids removal and over 95% BOD removal for a significant percentage of the time. It also has the potential capability of being amenable to future conversion to automated operation. The oxidation ditch is the product of an engineer's work and it is a custom-designed plant. A municipal oxidation ditch wastewater treatment facility is shown in Fig. 1-1.

Package treatment plants are sold in a fiercely competitive market. These plants are usually capable of producing a good effluent if operated continuously under the supervision of a sufficiently skilled person. The quality of design and construction, the degree of operating skill required, the amount of attention necessary, plant flexibility, and maintenance necessary may vary significantly among different manufacturers' products. It is a buyer-beware

Figure 1-1. Municipal Oxidation Ditch Wastewater Treatment Plant. (Courtesy Lakeside Equipment Corp.)

Figure 1-2
Typical Package Treatment Plants. (Courtesy Can-Tex Industries)

Figure 1-2
Typical Package Treatment
Plants. (Courtesy Can-Tex
Industries) (Continued)

market, and Chapter 8 should be carefully studied. There is a need for package plants and they serve a definite role. In Table 1-2, many of the applications below the minimum units use per day indicated would be served by package plants. Typical package plants are shown in Fig. 1-2.

Figure 1-3 is a flow diagram of a typical trickling filter sewage treatment plant. Not all of the treatment processes described in this text will be found economical under a given set of wage rates, physical location with respect to equipment availability, skill of available labor supply, and similar engineering

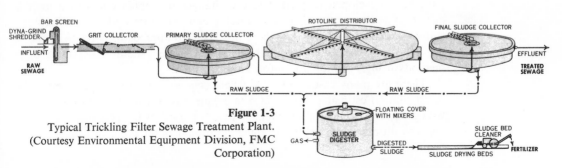

Figure 1-3
Typical Trickling Filter Sewage Treatment Plant.
(Courtesy Environmental Equipment Division, FMC
Corporation)

considerations. Each design must consider all the parameters, and the engineer must become acquainted with the costs of different alternatives.

1-2. Ten-State Standards. The so-called Ten-State Standards are published under the title "Recommended Standards for Sewage Works" by the Health Education Service, P.O. Box 7283, Albany, N.Y. 12224. The cost is low. These standards were the outgrowth of a joint meeting of the Upper Mississippi and Great Lakes Boards of Public Health Engineers (now Great Lakes-Upper Mississippi River Board of State Sanitary Engineers) in Chicago on Feb. 6, 1947. At this meeting the Committee on Development of Uniform Standards for Sewage Works was created.

The standards are intended as a guide in the design and preparation of plans and specifications for sewage works, to list and suggest limiting values for items upon which an evaluation of such plans and specifications will be made by the reviewing authority, and to establish, as far as practical, uniformity for practice among the several states. Statutory requirements among the states are not uniform, and use of the standards must adjust itself to these variations (1).

The member states are Illinois, Indiana, Iowa, Michigan, Minnesota, Missouri, New York, Ohio, Pennsylvania, and Wisconsin. It will be found, however, that Alaska, Arkansas, Connecticut, Delaware, Georgia, Hawaii, Kentucky, Maine, Montana, Nebraska, Nevada, New Mexico, North Carolina, South Dakota, Virginia, Washington, West Virginia, Wyoming, and Guam generally subscribe to the Ten-State Standards, and some of the other states readily accept plans that meet or exceed those standards. These standards, then, are the most widely used of any standard of their type in existence. The standards are revised at periodic intervals.

1-3. Population Estimation. The Ten-State Standards (1) are the most observed for sewage treatment plant design, and they indicate that engineering services are performed in three steps: an engineering report; preparation of construction plans, specifications, and contract documents; and construction compliance, inspection, administration, and acceptance. The standards amplify what is expected in the report, and the first item is the development of the predicted population. Unfortunately, no exact methods or formulas have been developed to predict future changes in population with a high degree of accuracy (2). With the development of more and more reliable statistical reporting programs, there has been an appreciable decline in the use of purely mathematical methods and a more general acceptance of methods using symptomatic information.

There are many cases on record where engineers have made large errors in population estimates. However, the engineer is still required to come up with a population prediction over the design period of the proposed facility and the ultimate population at some additional point more distant in the future.

It is recommended that raw data inputs be obtained from several of the following sources (listed in order of preference): local planning and estimating reports (where various components such as births, deaths, and net migration and significant local developments that will alter migration have been considered); U.S. census data; Chamber of Commerce; school census; public utilities; health district or authorities; city directory; building permits; freight shipments; Federal Post Office; bank transactions; employment, unemployment, and welfare records, are all potentially useful. There is material in the literature (3, 4, 5, 6, 7) that provides a broader base in census prediction that will be feasible in this text. Unfortunately, in engineering sufficient data are often not available for doing something the best way. This is where good judgment becomes necessary instead of blindly following a method.

The arithmetic progression method consists simply of examining the census data from several census reports. If the town has been growing an average of 1,000 persons per year for several census reports, then this increment of projection is added to the first tabulated census data to estimate the population at each year for the desired number of years.

$$a, \quad a + g, \quad a + 2g, \quad a + 3g, \quad a + 4g \qquad (1\text{-}1)$$

and

$$S = \frac{n}{2}[2a + (n-1)g] \qquad (1\text{-}2)$$

where

$$S = \text{sum of first } n \text{ terms}$$

If the growth rate has been favored by a constant percentage of growth for each time increment of census tabulations, then a geometric progression method may be applicable. A geometric progression is a series of terms, each of which except the first is derived from the term preceding by multiplication by a constant called the *ratio* (8). For ratio $= 3$,

$$3, 9, 27, 81$$

$$a, \quad ar, \quad ar^2, \quad ar^3 \qquad (1\text{-}3)$$

$$S = a\left(\frac{r^n - 1}{r - 1}\right) \qquad (1\text{-}4)$$

where

$$S = \text{sum of the first } n \text{ terms}$$

When the population is decreasing by an outward migration, it is advisable to use the last-known census figure even though the population may be less than this value. If it is mandatory to estimate the population, then an infinite geometric progression (7) can be used, where

$$S = \frac{a - ar^n}{1 - r} \qquad (1\text{-}5)$$

Another approach is to simply study the available data and, using either statistical methods or engineering judgment, graph the known data and project it for the desired period of time. For one of the best methods for free-hand curve fitting, and for general information on the interpretation of statistical calculations, refer to Ezekiel's *Method of Correlation Analysis*, 2nd ed., John Wiley & Sons, Inc., New York, 1941.

Another method consists of selecting three or four cities that have similar characteristics but larger populations. All the cities are selected and arranged so that the population curve of each city will pass through the same graphical point that coincides with the present population of the town on which the report is being made. The growth curves of the other cities are studied, and from this data a prediction curve is established.

Verhulst (9) established a theory which Pearl (10) put into a mathematical expression in a logistic curve of the form

$$y = \frac{k}{1 + me^{ax}} \tag{1-6}$$

where y = population at any time x from an assumed orgin,
k = saturation population,
m = a constant,
a = a constant,
p = percentage of saturation.

McLean (11) speaks rather highly of Pearl's equation. The constants are determined by selecting three points uniformly spaced along the x-axis, $(0, y_0)$, (x_1, y_1), and $(2x_1, y_2)$, and substituting in the following equations

$$k = \frac{2y_0y_1y_2 - y_1^2 (y_0 + y_2)}{y_0y_2 - y_1^2} \tag{1-7}$$

$$m = \frac{k - y_0}{y_0} \tag{1-8}$$

$$a = \frac{1}{x} \log e \frac{y_0(k - y_1)}{y_1(k - y_0)} \tag{1-9}$$

If it is necessary to use school enrollment, utility data, or telephone service connections, it is advisable to check the latest census ratios. In general, where this has been used in the past, a ratio of five persons for every child in school, or three persons for every water, gas, or electric service connection, or four persons for every telephone has been applicable.

1-4. Design Period. When it is economically feasible, sewer systems should be designed for the estimated ultimate tributary population, except in considering parts of the system that can readily be increased in capacity. Outfalls, interceptors, and main sewers may have to be limited to 40- to 50-year design projects by economics. However, all laterals and submains less than 18 inches in diameter should be designed for full development. Sewage treatment plants are designed for populations 15 to 25 years hence. In general, if the growth and interest rates are low (on the order of 3 % or less), a 20- to 25-year design period is recommended. If growth and interest rates are very high, then a 10- to 15-year design period may be warranted. In any wastewater design, it is good engineering practice to consider how the facility can be altered or expanded to meet future growth and/or industrialization.

It is advisable to check on regulatory agency regulations *before* starting any design. It may be found that local regulations (written or unwritten) may dictate what will be approved.

1-5. Hydraulic Flow Estimates. The best method for estimating sewage flows for design purposes is to actually measure the flows in existing sewage systems and make the necessary corrections to predict future requirements. Another method is to estimate and total the various components of flow. Tables 1-3 and 1-4 will be useful in the estimating method.

The least desirable method involves the use of some arbitrary value as the basis of design. The Ten-State Standards offer the following rules (1):

New sewer systems shall be designed on the basis of an average daily per capita flow of sewage of not less than 100 gallons per day. This figure is assumed to cover normal infiltration, but an additional allowance should be made where conditions are unfavorable. Generally, the sewers should be designed to carry, when running full, not less than the following daily per capita contributions of sewage, exclusive of sewage or other waste flow from industrial plants. Laterals and sub-main sewers, 400 gallons. Main, trunk, and outfall sewers, 250 gallons. Intercepting sewers, in the case of combined sewer systems, should fulfill the above requirements for trunk sewers and have sufficient additional capacity to care for the necessary increment of storm water. Normally, no interceptor shall be designed for less than 350% of the gauged or estimated average dry-weather flow. When deviations from the foregoing per capita rates are demonstrated, a description of the procedure used for sewer design shall be included.

Industrial wastewater flow should be determined by contacting the industries involved. Another rule of thumb in estimating domestic flow is to take 0.7 or 0.8 times the value recorded by the water works for domestic consumption. In some communities, considerable engineering judgment may be involved in deciding whether to use the latter method.

The average daily flow of sewage and ground water is the average 24-hr discharge during a period of one year. This figure is valuable in estimating the cost of operation of sewage treatment works and pump stations. The engineer must estimate the minimum and maximum 24-hr discharge for initial operation of a new plant and project what these values will be at the end of the projected design period of the plant.

1-6. Sewage Composition. In domestic waste treatment, the composition of sewage varies from location to location. It also varies with the time of day or season at a given location, depending upon the number and type of facilities being served by the treatment plant. The magnitude of the dissolved and total solids is a function of the hardness and other mineral substances in the water supply and the infiltering ground water in the sewers. The BOD in

TABLE 1-3. Sewage Data Guide[1]

Type of Establishment	Unit	Flow (gal/unit/day)	BOD[5] (lb/unit/day)
Airport	Employee	15	0.05
Airport	Passenger	5	0.02
Apartment house	Capita	75	0.17g*
Assembly halls	Seat	2	0.02
Bars and taverns[1]	Employee	15	0.05
Bars and taverns[1]	Customer	2	0.01
Bowling alley[1]	Lane	75–160	0.13–0.3
Camp, construction[2]	Capita	50	0.15
Camp, day, central bath	Capita	35	0.10
Camp, resort, luxury[2]	Capita	100	0.17
Camp, summer[2]	Capita	50	0.15
Church, large[2]	Seat	6	0.03
Church, small	Seat	4	0.02
Cottage, summer	Capita	50	0.17
Country club resident	Capita	100	0.17
Country club member pres.	Capita	40	0.10
Dance halls	Capita	2	0.02
Factories[1, 3, 12]	Employee	30	0.07
Factory cafeteria	Employee	5	0.02
Hotel, luxury	Capita	70	0.15
Hotel, average	Capita	50	0.15
Hospitals[2]	Capita	200	0.30
Institutions, resident[2]	Capita	100	0.17
Laundry (coin-operated)	Machine	300	
Multiple-family residence[6]	Capita	50	0.17
Municipalities	Capita	100	0.17
Motel, luxury[4]	Capita	70	0.15
Motel, average[4]	Capita	60	0.15
Mobile home parks	Trailer space	125	0.15
Offices[1, 8]	Employee	20	0.05
Picnic parks[7]	Capita	15	0.08
Picnic parks[8]	Capita	5	0.05
Restaurant, average	Seat	80	0.15
Restaurant, 24-hour	Seat	110	0.17
Restaurant, curb service	Car space	100	0.17
Rest areas, roadside	Capita	5	0.04
Rest homes	Capita	100	0.17
Schools, boarding[12]	Capita	75	0.04g
Schools, day[1, 8, 9]	Capita	10	0.04g
Schools, elementary[2]	Capita	15	0.04g
Schools, high[2, 12]	Capita	25	0.05g
Subdivision, luxury homes[11]	Capita	150	0.20g
Subdivision, average homes[11]	Capita	110	0.18g
Subdivision, economy homes[11]	Capita	75	0.17g
Service stations, vehicular	Bay	1,000	10.
Shopping centers[1]	Sq Ft Floor Space	0.1	0.001
Swimming pools, average	Capita	10	0.03
Swimming pools[14]	Capita	15	0.08

TABLE 1-3 (Continued).

Type of Establishment	Unit	Flow (gal/unit/day)	BOD[5] (lb/unit/day)
Theater, drive-in	Stall	5	0.03
Theater, movie	Seat	5	0.03
Vacation cottage	Capita	50	0.17

[1]Excluding restaurant or cafeteria.
[2]Food service wastes included.
[3]No industrial process waste included.
[4]Compute 1.5 persons/room (minimum).
[5]Average figure varies 150–300 gal/day/capita.
[6]Economy neighborhood.
[7]Includes bathhouse, shower, and toilets.
[8]Excluding showers.
[9]No. gym.
[10]Increase by factor 1.25 if garbage grinders are used.
[11]Assume 3.5 persons/house or 3 persons/apartment.
[12]If predominantly female, add 15%.
[13]No laundries or other high-waste concessions.
[14]Snack bar, luxury.
[*g] Various health departments may have local area standards.

TABLE 1-4. Sewer Infiltration Rates*

Unit	Minimum Flow	Average Flow	High Flow
GPD/acre sewered area	500	2,000	5,000
GPD/mile of sewer pipe	5,000	30,000	200,000
GPD/mile sewer pipe/inch diameter	500[T]	5,000	2,500
GPD/capita	25		200
GPD/manhole cover	75	100	150
GPD/mile 8-inch sewer†	1,600	4,000[T]	12,000
GPD/mile 24-inch sewer†	4,000	12,000[T]	36,000

*Typical values used in different engineering offices.
†Values indicated in some specifications on amount of infiltration permissible (new sewers).
[T]Ten-State Standards for new sewer construction (1).

terms of mg/liter (milligrams per liter) depends on the amount of water carriage. The amount of water used per person in a given treatment situation may range from 1 gal/day to more than 200 gal/day. Therefore a wide variation would naturally occur in terms of BOD mg/liter.

Hoskins (11) reported that human beings excrete coliform bacteria at the rate of 125 to 400 billion per day. Domestic wastewater contains on the order of 10^5 coliform organisms/ml. The ratio of coliform organisms to enteric pathogens in wastewater is normally on the order of 10^6. Camp (12) states that the total number of live bacteria in fresh feces is estimated at about 2×10^{12} per person per day, of which about 10% or 200 billion are coliform bacteria. It is also stated in Camp (12) that "The number of polio viruses in stools of patients is less than 1%, and the number of infectious hepatitis viruses in the stools of patients is less than 0.1% of the number of coliform bacteria."

Since nearly all feces contain coliform organisms and only a relatively small portion (2 to 20%) contribute pathogenic virus (13, 14, 15), domestic sewage normally contains approximately 10,000 times as many coliforms as viruses. Total average excreta amounts to approximately 83 g of feces and 970 g of urine per capita per day. About 20% of the feces and 2.5% of the urine is putrescible organic material.

The constituents of sewage of most concern and the variations of concentrations likely to be encountered in practice are discussed in Chapter 2.

1-7. Municipal Treatment Plants. The per capita cost of sewage treatment usually decreases as the size of the plant increases. Every piece of equipment in a large plant is usually selected on the basis of highest efficiency to reduce power cost (other factors being equal). While the skill level required may be high, the necessary trained personnel are characteristically employed in such plants on a 24-hr basis.

In most cases, the activated sludge process is capable of removing the most sludge from wastewater at the lowest cost. The most competitive alternate process in large plants is the trickling filter using new techniques of construction. Under a given set of treatment conditions, that is, where only BOD reduction is considered, high-rate trickling filters may cost less to construct and operate for performance up to a certain percentage of BOD reduction. However, in terms of total treatment where the requirements are severe, the activated sludge process is superior. As the size of the plant decreases, depending on local available terrain and many other factors, the point is usually reached where the long-term solution is an extended aeration-process variation of the activated sludge known as the oxidation ditch. This process produces a sludge that can be directly used for agricultural applications without the need for anaerobic digesters. Combinations of processes may be warranted to meet special needs.

Municipal plants usually employ mechanized screens, grit chambers, oil skimming, and grease removal provisions before starting any biological stabilization of the waste.

The cost of a more expensive initial plant and the use of processes where the cost of operating power per unit of waste treated is more expensive may be fully warranted after a total evaluation of all factors. The availability of suitably trained operation personnel must also be considered. The fact that courses are planned or being offered to train operators in the state does not remove the responsibility of the engineer to consider the reality of the case for the plant in question.

1-8. Suburban Sewage Systems. Many suburban sewage problems exist as a result of the lack of an adequate master plan and suitable ordinances to control development in a county area outside the city limits. A typical problem occurs when several developers build on the outskirts of a city but install sewers of the minimum size to handle the development. When land is later developed beyond these developments, there is no trunk sewer of suitable size to connect with, and the cost of cutting through an inhabited area to install such a sewer becomes prohibitive, particularly if the sewer lines inside the city itself are inadequate to handle the extra demand. Figure 1-4 shows a typical cost per capita curve that results in order to achieve adequate sewage treatment in suburban sections. The engineer must, in each case, conduct a study to determine the cost of sewers to the closest acceptable municipal sewer trunk, plus usage fees, etc., to the users, as compared with shorter sewers to a new treatment plant and any subsequent fees and maintenance charges. If a new treatment plant is warranted, then the necessary plans must be prepared and submitted to the proper regulatory agency for approval. The engineering investigation should include a study of the disposal of the plant effluent and the ultimate receiving stream.

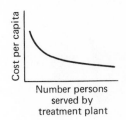

Figure 1-4
Cost of Construction Versus Size.

The oxidation ditch plant is substantially lower in cost than a conventional, custom-built activated sludge plant, yet it is capable of producing equal settleable solids and BOD reduction. This process has thus made it feasible to construct plants in locations that can serve suburban communities.

1-9. Town Sewage System. In designing any town treatment plant, care should be taken to determine any prospective industry being sought and the role of the town plant if this expansion takes place. It is also advisable to consider how any town or municipal plant can be expanded, unless it is in an area that has been static for a long time or has had a slow but steady out-migration.

If package treatment plants are used, it is desirable to select one of the plants that can be started on extended aeration. Then by opening and shutting valves and sliding gates, it can subsequently be converted to step aeration and finally contact stabilization. The plant should be built in modular units such that the first unit will be at least half-loaded at the start of operations. Each unit should be operated between half loading and full loading at all times.

1-10. Industrial Complexes. Cities, counties, developers, and other agencies are today building industrial parks and complexes with the hope of attracting industrialization in select areas. In some instances specific industries are in mind, and in other cases the ultimate makeup of the users may be one of substantial speculation. Each case must be handled on its own merits. However, the leasing contract established for the development should contain specific limitations on the strength and types of waste that any occupant of the park can discharge into the sewer system. It may be advisable to arrange the sewers in a special manner so that waste from a large user can be brought to the plant area, separated from the other users, and processed in a special pretreatment unit before combination with the rest of the park waste.

1-11. Recreation Areas. Recreation areas are often associated with water-use activities. An exceptionally high degree of treatment, including tertiary treatment, may be desirable in addition to a two-hour or longer chlorination contact time.

Work in recreation area sanitation is a multiple-discipline problem. Public health aspects assume particular importance and the potential transmission of pathogens is of substantial concern. Sewage treatment processes and chlorination reduce the number of viruses present, but large numbers of viruses will survive even a six-hour chlorination contact time and long after bacteria are of no concern. It is therefore desirable to prohibit the effluent of the treatment plant from reaching the recreation waters. This text presents applicable information in various chapters.

1-12. Roadside Rest Areas. The original method of waste treatment for roadside rest areas was that of the septic tank and its associated drainage field. The demand on such rest area facilities has brought substantial pressure in some areas to go to a secondary treatment process or even a tertiary system. The rest areas are in a class similar to that of recreation areas. The effluent from the rest area can be sprayed on the roadside in such a manner that it does not arrive at a receiving stream. It is feasible to recirculate part of the effluent in a roadside rest area treatment plant. This is particularly important in areas having a severe shortage of water.

Care must be taken that there is no possibility of cross connections. This danger must be considered also in terms of future plumbers who may be unaware that the effluent is being used in the toilet and urinal flush water. It is possible to treat the effluent to such a degree that it can be used as potable water. It can also be treated to a point that it can be used as recharge water of the underground aquifer either by irrigation, settling basins, or an injection well designed for the purpose.

1-13. Rural Restaurants. Rural restaurants seldom have properly trained personnel to operate a treatment plant. If adequate land area of suitable soil is available, the septic tank system should be considered for a small restaurant (50 seats or less). If suitable land is available at a lower elevation, a waste stabilization pond may be feasible for any size restaurant. For restaurants with over 150 seats, an oxidation ditch is probably one of the alternatives that should be explored. An aerated lagoon is another possibility for a large restaurant.

If a package treatment plant must be used, seek the simplest unit to operate, even if its initial cost is higher. The mechanical-type plants such as that offered by Lakeside Equipment Corporation or the Infilco Division of Degremont Inc. were being better used by the owners than some of the diffused aeration plants, according to the results of a field investigation by the author.

1-14. Marinas. A marina has three separate waste treatment problems. One problem is the boat sanitary holding tanks, the second is that of pumping of bilges which may be heavily contaminated with oil and hydrocarbons, and the third is the domestic waste of the customers using marina facilities. The third problem should be conveyed directly to the waste treatment plant. The sanitary holding tank waste should not be combined with the material pumped from bilges.

QUESTIONS

1. Why is untreated sewage always in a state of decomposition?

2. Why does anaerobic sewage smell bad whereas the same sewage in the aerobic state will have only a light odor?

3. Why is some sewage aerobic and other sewage anaerobic?

4. What are the undesirable aspects of mineral matter in sewage?

5. Give one reason to explain why dumping an excess of sewage in receiving streams kills the aquatic inhabitants.

6. What is sewage treatment?

7. Differentiate between primary, secondary, and tertiary sewage treatment processes.

8. What is a system and what processes might it include?

9. What type of treatment would occur in a septic tank and what percentage in BOD reduction can be expected?

10. What type of treatment is sedimentation only? What degree of treatment in terms of percentage of reduction of BOD is achieved?

11. Describe an oxidation ditch and explain why it might perform better than a package plant?

12. In making population predictions, why shouldn't an engineer always make his estimates the mathematically best way? If best way is not used, what must prevail? Why?

13. With an interest rate of 3%, what would be the design period for a sewage treatment plant?

14. What role would regulatory agency regulations play in the preparation of a preliminary plan for a sewage treatment plant?

15. What are the so-called Ten-State Standards?

16. Why is the engineer concerned with how much infiltration occurs in a sewer after it is completed?

17. How many coliform bacteria are excreted per day by a person? (Indicate the probable range.)

18. What ratio of coliform bacteria to viruses would you normally expect to encounter?

19. Podunk Junction just completed a new sewage treatment plant. The town has

a population of 1,000. The sewage treatment plant has 100% reserve capacity. Would it be a good idea to invite a large tannery to locate in Podunk Junction and use the local sewage treatment plant facility? Justify your answer.

20. A salesman from Hot Shot Super Package Treatment Plants offers the town council an exceptionally good buy on a contact stabilization package type of plant. The manufacturer builds good equipment. The town sewage is subject to slugs of industrial waste and variable composition. The loading cycle varies from day to day. As the town's consultant, would you recommend the purchase or advise against it? What would the basis for your position be one way or the other?

PROBLEMS

The following problems require consulting census data. They may require making assumptions, using engineering judgment, etc. Show all calculations and methods used in arriving at the solution. Justify your assumptions.

1. Beckley, West Virginia wants to build a sewage treatment plant. The interest rate is 2%. Establish the design population.

2. Savannah, Georgia plans to completely rebuild its sewage facilities. The interest rate is 3%. Compute the design population for a new plant and interceptor sewer system.

3. Poplarville, Mississippi decides to construct a new treatment facility. The interest rate is 8%. Compute the design population for the sewage facility.

4. Austin, Texas decides to build a new interceptor sewer system and a new treatment plant to handle the entire town's water carriage waste. Compute the design population for the sewers and the treatment plant.

5. Repeat Problem 4 using at least two different methods from that employed the first time.

6. Baton Rouge, Louisiana decides to build a completely new sewer system and new treatment plant. The interest rate is 1%. Use every method of population estimation presented in the text. Note the different answers that you obtain and explain the discrepancies. What answer should you use? Why?

REFERENCES

1. *Recommended Standards for Sewage Works, Great Lakes—Upper Mississippi River Board of State Sanitary Engineers*, 1971 rev. ed., Health Education Service, Albany, N.Y.

2. ZITTER, MEYER, "Population Projections for Local Areas." *Public Works*, 88, 6, 110 (1957).

3. STANBERY, VAN BEUREN, *Better Population Forecasts for Areas and Communities*. U.S. Department of Commerce (Sept., 1952).

4. SIEGEL, JACOB S., "Some Aspects of the Methodology of Population Forecasting for Geographic Subdivisions of Counties." *Proceedings of World Population Conference*, Rome, Italy (Aug. 31–Sept. 10, 1954).

5. JAFFE, A. J., *Handbook of Statistical Methods for Demographers*. U.S. Bureau of Census, Washington, D.C. (1951).

6. JAFFE, A. J., and HAMMER, CARL, *Estimating a Nation's Urban Growth*. (For the Human Resources Research Institute, Maxwell Air Force Base, Ala.) Technical Research Report No. 12 (Feb., 1953).

7. U.S. Bureau of the Census, *Current Population Reports*. Series P-25, No. 123. Revised Projection of the Population of the United States, by Age and Sex: 1955 to 1975 (Oct. 20, 1955).

8. DULL, RAYMOND W., *Mathematics for Engineers*, 2nd ed. McGraw-Hill Book Company, Inc., New York, 1941.

9. VERHULST, P. F., "Recherches mathématiques sur la loi d'acroissment de la population." *Mén. Acad. Roy.*, Bruxelles, Vol. 18 (1844), pp. 1–58.

10. PEARL, RAYMOND, *Studies in Human Biology*. The Williams and Wilkins Co., Baltimore, Md., 1924, pp. 523, 578–581, 590.

11. HOSKINS, J. K., "Quantitative Studies of Bacterial Pollution and Natural Purification in the Ohio and Illinois Rivers." *Trans. ASCE*, 89, 1365 (1925).

12. CAMP, THOMAS R., *Water and Its Impurities*. Reinhold Publishing Corp., New York, 1963.

13. BANCROFT, P. M., ENGELHARD, W. E., and EVANS, C. A., "Poliomyelitis in Huskeville (Lincoln) Nebraska—Studies Indicating a Relationship Between Clinically Severe Infection and Proximate Fecal Pollution of Water." *Jour. Amer. Med. Assn*, 164, 836 (June 22, 1957).

14. KELLY, S., SANDERSON, W. W., WINSSER, J., and WINKELSTEIN, W. JR., "Poliomyelitis and Other Enteric Viruses in Sewage." *Amer. Jour. Public Health*, 47, 72 (Jan., 1957).

15. MELNICK, JOSEPH, et al., "Seasonal Distribution of Coxsackie Viruses in Urban Sewage and Flies." *Amer. Jour. Hygiene*, 59, 164 (March, 1954).

2 *Pollution Parameters*

Wastewater consists of the summation of liquid-carried wastes that find their way into sewers and are transported by water carriage to the sewage treatment facility. To successfully design the treatment facility it is necessary to learn certain biological, chemical, hydraulic, civil, and mechanical engineering principles as well as segments of information from a number of other disciplines, including a knowledge of the most widely used parameters.

In order to facilitate the design study, a number of component parts are grouped together under so-called basic wastewater treatment processes. However, these component parts may be likened to an erector set toy. There are numerous combinations of components that can be put together to achieve a specified set of conditions for a given collection of parameters, but the manner must not violate regulations established by regulatory agencies having jurisdiction in any given locality. These agencies may include both the state water pollution control agency and the state health department. The design may also involve federal agencies or other state agencies.

The end objective is to establish the best engineering system design feasible considering the parameters involved including cost, economics of operation, performance, terrain, climatic conditions, and similar factors. If there is a conflict with provisions of the regulatory agency, the designer must have a well-documented case for his system design in order to secure the necessary approvals.

A comprehensive treatment of parameters would fill a book, so only the most significant parameters of immediate interest will be discussed. Table 2-1

TABLE 2-1. Parameters and Data Sources

Parameter	Domestic Waste Analysis	Industrial Waste Analysis	Supplemental Information
Acidity	1, 85, 86	1, 3, 85, 86	2, 10
Alkalinity	1, 9, 85, 86	1, 3, 8, 85, 86	1, 5, 10
Bioassays	1	1	1
BOD	1, 10, 86	1, 10, 85	11, 13, 14–36
Coagulation	11	11	10, 38–42
COD	1, 43	1, 85, 86	1, 43
Colloids			10, 38–42
Color	46	1	1, 46–50
DO	1, 45	1, 3, 45	10, 51–54
Fluoride	1, 85	1, 85	10
Hardness	1, 85	1, 3, 85	1, 5, 10
Iron		1, 3, 45	10
Manganese		1, 3, 45	10
MPN	1		1, 13, 55, 56
Nitrogen	1, 86	1, 85, 86	1
Odor	1, 3	1, 3	1, 3
Oil and grease	1, 85	1, 85	58, 10, 57, 75, 76
Organic pollution	1	1	10, 59, 60
Pesticides	64	64	68, 69, 64
pH	1	1	70, 10, 58
Phenols	1, 85, 86	1, 85, 86	1, 10, 58, 77
Phosphates	1, 10, 86, 85	1, 3, 10, 85, 86	1, 3, 10, 71–74
Solids	10, 85, 86	3, 85, 86	10
Sulfate	10, 86	3, 10, 85, 86	10
Sulfides	1, 85, 86	1, 85, 86	58
Surfactants	1, 85, 86	1, 85, 86	
Taste	1	1, 3	1, 3, 61–63
Temperature			10, 78–84
Turbidity	1, 10	1, 10	10
Volatile Matter	1, 10	10, 1	10
Enzymes			13, 10

Numbers in columns indicate references at the end of this chapter.

contains a more detailed listing of parameters and sources where additional information can be obtained on each. Biological factors are discussed separately in Chapter 3. While some of the parameters are comparatively simple, they are sometimes of a nature that a high order of repeatability is very difficult to achieve. Therefore it is often necessary to use a more complex method in field practice in order to improve accuracy.

The most comprehensively used standard in the United States for analysis of municipal and general usage is the book *Standard Methods for the Examination of Water and Wastewater*, American Public Health Association (APHA), Washington, D.C. Industrial waters have additional water quality analysis methods and standards in *Manual on Industrial Waste Water*, American Society for Testing Materials, Philadelphia. The British counterparts to these publications are the Ministry of Housing and Local Government,

Methods of Chemical Analysis as Applied to Sewage and Sewage Effluents, H.M. Stationery Office, London, and ABCM/SAC Joint Committee, *Recommended Methods for Analysis of Trade Effluents,* Cambridge.

Raw sewage is putrescent, odoriferous, unsightly, and a health hazard. If it is dumped into a stream in quantities too great for the stream to assimilate the organic demand of the sewage for oxygen, the stream will become septic and fish and other aquatic organisms will die. Domestic sewage is a term commonly used to describe the sewage exclusive of industrial wastes. This book is directed toward municipal and smaller waste treatment facilities handling domestic-type sewage, but it contains sufficient design information to cope with a number of industrial wastes.

PRIMARY PARAMETERS

2-1. Biochemical Oxygen Demand. The biochemical oxygen demand (BOD) determination is a measure of the oxygen required to oxidize the organic matter in a sample, through the action of microorganisms contained in the sample. The classical dilution method has been to determine the dissolved oxygen prior to and following a period of incubation at 20°C. The most used time period in the past has been five days. Other time intervals and temperatures can be used provided the temperature is held constant. Thus BOD_5 is used to designate an incubation period of five days and the temperature is understood to be 20°C unless otherwise designated. In the dilution method, if the oxygen demand of the sample is greater than the available dissolved oxygen, a dilution is made. The amount of dilution depends on the oxygen demand and must be such that an appreciable amount of dissolved oxygen (1.5 to 2.0 ppm minimum) remains after the incubation period (11). For wastes or sewage having an unknown demand, it is necessary to make up a number of dilutions to be sure that one will meet the requirements. There are a number of factors which influence the rate of oxidation of organic matter by bacteria and, hence, the five-day oxygen demand. The type of diluting water, the pH value, and the bacteria content are the most important.

The ultimate carbonaceous BOD of a liquid waste is the amount of oxygen necessary for the microorganisms in the sample to decompose the carbonaceous materials that are subject to microbial decomposition. The BOD value is not an absolute measure. It is actually a qualitative test with a narrow range of variation that permits it to assume semiquantitative proportions. There is substantial disagreement among different researchers on the causes of variations in a BOD curve.

This text shall use the viewpoint that two distinct phases of activity take place. In the first phase, the microorganisms in the sample use the organic matter in the wastewater for energy and growth until the organics present are completely utilized. During the second phase, the organisms continue to use oxygen for autooxidation or endogenous metabolism of their cell masses. The reaction is completed when the cell mass is completely oxidized, leaving a nonbiodegradable residue. The first phase takes place normally within 18 to 36 hours and total oxidation takes over 20 days. For a more rigorous treatment, consult references (15) through (26).

A number of sources use a first-order equation for fit of BOD data. Others raise the question of whether a second-order equation might be more applicable (26, 27, 28, 29, 30). Here we will use the classical first-order equation

$$y = L_0(1 - 10^{-kt}) \qquad (2\text{-}1)$$

where y = amount of oxygen consumed or BOD after any time t,
L_0 = ultimate BOD or total amount of oxygen consumed in the reaction,
k = average-rate constant,
t = time of incubation in days.

Methods to determine the value of k are available in the literature (11, 31, 32, 33, 34).

From Van't Hoff's law

$$k_t = k_{20}\theta^{(T-20)} \qquad (2\text{-}2)$$

where k_t = BOD reaction rate at temperature t, °C,
k_{20} = BOD reaction rate at temperature of 20°C,
θ = 1.056 (20–30°C) and 1.135 (4–20°C)

In Eq. (2-2) the values of θ are from the work of Schroepfer (35).

BOD is usually expressed in mg/l; however, conversion to parts per million (ppm) is very simple

$$\text{ppm by weight} = \frac{\text{mg/l}}{\text{specific gravity}} \qquad (2\text{-}3)$$

Other useful relations are:

$$\frac{\text{mg}}{\text{l}} = \frac{\text{lb/24 hr.}}{mgpd \times 8.34} \qquad (2\text{-}4)$$

$$\frac{\text{lb}}{24 \text{ hr}} = \frac{\text{mg}}{\text{l}} \times \text{cfs} \times 5.39$$

$$PE = \left(\frac{\text{mg}}{\text{l}} \text{ of BOD}_s\right)(mgpd)\frac{(8.34)}{(0.17)} \qquad (2\text{-}5)$$

where PE = population equivalent and $mgpd$ = millions of gallons per day. The figure 0.17 is based on pounds of BOD waste per person per day (see Table 1-3). The figure 8.34 is a conversion factor × gallons of water to yield pounds of water.

One of the best explanations of computation of organic waste loads on streams is contained in Nemerow (36). Clark and Viessman (37) also have an elementary presentation on this topic.

Theriault (32) tables can be used to simplify BOD calculations and eliminate the need for log tables. Sample calculations and derivations of various

BOD equations are contained in Clark and Viessman (37) based on the classical first-order equations.

2-2. Chemical Oxygen Demand. Chemical oxygen demand (COD) is defined as the amount of oxygen, in parts per million, consumed under specified conditions in the oxidation of organic and oxidizable inorganic matter in wastewater, corrected for the influence of chlorides (3). This method is recommended as a supplement to, but not as a replacement for, the biochemical oxygen demand (BOD) test which is the only test that indicates directly the quantity of oxygen that will be used by natural agencies in stabilizing organic matter (3). There is no inherent constant relationship between COD and BOD, but the COD test can be useful in evaluating the treatment and control of wastewaters.

There are a number of COD tests and all of them can be performed more rapidly than BOD tests. This provides a more rapid tool, but the result is sometimes confusing due to lack of inherent consistency with the BOD test and the variation of available COD tests. There is no correlation between the different COD methods. There is a COD method available where COD of an aqueous sample can be determined in about two minutes after homogenization or dilution (43).

2-3. Dissolved Oxygen. The dissolved oxygen (DO) is the amount of oxygen dissolved in a specified sample of water. It can be expressed in mg/l, ppm, or percent of saturation. Each species of aquatic animal, such as fish, has a minimum DO level at a given temperature for survival over an average specified time period.

Sewage places a dissolved oxygen demand on the water serving as a carriage for the waste. As long as there is adequate dissolved oxygen in the carriage water, the sewage will not turn septic and the aerobic microorganisms will predominate. Odor does not become a serious problem until a sewage turns septic. If adequate oxygen is continuously supplied artificially, sewage will remain aerobic throughout the treatment process.

2-4. Residue. Take a 2000-cc sample of well-mixed wastewater containing impurities and divide it into two 1000-cc portions. Evaporate one 1000-cc sample at low temperature. The material left in the container is the *total residue* per 1000 cc of sample. Filter the remaining sample. The portion that remains on the filter is the *nonfilterable residue.* Evaporate at low temperature the water from the filtered portion of the material remaining in the container. The solid remaining in the container after evaporation of the water is the *filterable residue.* The two values obtained in the latter case should equal the total residue of the first case. The minimum diameter of the solids on the filter should be approximately 1 micron or larger.

Many solids crystallize as hydrates which lose water at various temperatures and form decomposition products before all their water of hydration is given off. Volatile compounds may be lost during the process. Thus, this simple process may be subject to substantial errors unless necessary precautions are observed. The correct procedures and precautions to be observed can be found in *Standard Methods* (1).

$$\frac{mg}{l} \text{ residue on evaporation} = \frac{mg \text{ residue} \times 1000}{mg \text{ sample}}$$

2-5. Total Organic Carbon. One of the primary indicators of the degree of water "pollution" is the concentration of organic matter. One approach used to evaluate the degree of organic pollution is to determine the amount of organic carbon in the pollutants. There is a procedure in Standard Methods (1). Over the past several years, however, this measurement has been increasingly accomplished by the use of the combustion/infrared technique developed and patented by Dow Chemical Company, exclusively licensed to Beckman Instruments, Inc., and marketed as the Carbonaceous Analyzer. This analyzer provides a measure of total carbon content of an aqueous sample in approximately two minutes. Total organic carbon is commonly expressed as TOC.

2-6. Total Solids. It is customary in sanitary engineering to refer to residues as solids. The *total residue* in Section 2-4 is the *total solids* including both organic and inorganic materials. The residue remaining on the filter paper in the second sample represents most of the *suspended solids* in the waste. The filterable liquid contains the *dissolved solids* and the *colloids*.

The suspended solids, in turn, are made up of settleable and nonsettleable solids. The settleable solids are larger than 10^{-2} mm in diameter, and if placed in a nonagitated container, they should settle out within approximately two hours (Fig. 2-1). The nonsettleable solids are intermediate in size between colloidal particles and the settleable suspended solids (10^{-2} to 10^{-3} mm). They can be removed by coagulation. Both the settleable and nonsettleable solids contain organic and inorganic materials.

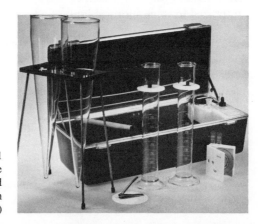

Figure 2-1
Typical Test Kit for Settleable
Solids, Sludge Settleability, DO and
Chlorine. (Courtesy of Hach
Chemical Co.)

The colloids and dissolved solids comprise the filterable solids portion remaining after evaporation. The colloids are 10^{-3} to 10^{-6} mm in size and can be separated from the dissolved solids by coagulation before the sample is evaporated. The dissolved solids are approximately 10^{-6} to 10^{-10} mm in diameter. Again, both the colloidal and dissolved solids contain both an organic and an inorganic portion.

As a general rule of thumb, computing the dissolved solids is a more reliable method than determining the residue on evaporation for solids concentrations above 1,000 ppm. The method for computing solids is available in the literature (89).

2-7. Volatile Solids. Assume that we weigh the residue in the first example of Section 2-4 and then ignite the residue at 600°C for 15 minutes. The loss of the residue in mg/l represents the volatile solids. The volatile portion of the solids represents the organic matter present, and the residue represents the inorganic material. If the material of the filter paper in the second example of Section 2-4 is weighed and then ignited at 600°C for 15 minutes, the loss in weight of the residue represents the volatile suspended solids in the sample.

SECONDARY PARAMETERS

2-8. Acidity. The Johannes Brönsted concept states that an acid is any compound which can furnish a proton or hydrogen ion. The concept of G. N. Lewis is that an acid is anything that can attach itself to something with an unshared pair of electrons. It is a substance that dissolves in water with the formation of hydrogen ions or a substance containing hydrogen which may be replaced by metals to form salts. NH_4^+ is an acid and NH_3 is a base. One of the primary concerns over acidity in wastewater is that it is a source of corrosiveness. It is also of concern biologically from the ecological viewpoint because it may destroy or damage flora and fauna or otherwise alter the ecosystem. Acidity is frequently expressed as mg/l $CaCO_3$. Acids turn blue litmus paper red, and they have a sour taste. They usually must be neutralized to some specified limit of acceptance in a particular stream.

2-9. Alkalinity. When carbonates, bicarbonates, and hydroxides and occasionally borates, silicates, and phosphates are present in water, they give the water the ability to neutralize acids. This property is called alkalinity. In reference (1), alkalinity is defined as the capacity of water to accept protons.

Nordell (5) states:

Alkalinity is determined by titration with a standard acid solution using phenolphthalein and methyl orange as indicators. The result of titration with the methyl orange indicator is expressed as *methyl orange alkalinity* or *total alkalinity*. The result of titration with the phenolphthalein indicator is expressed as *phenolphthalein alkalinity*. (Most natural water supplies contain some free carbon dioxide and show no phenolphthalein alkalinity.)

1. If no phenolphthalein alkalinity is present, all of the alkalinity is assumed to be bicarbonate alkalinity.
2. If phenolphthalein alkalinity is present, then twice the phenolphthalein alkalinity, if less than or equal to the methyl orange alkalinity, is assumed to be carbonate alkalinity.
3. If twice the phenolphthalein figure exceeds the methyl orange alkalinity, the excess is presumed to be *caustic* or hydroxide alkalinity.

Note: There are certain errors inherent in these determinations, and these assumptions are not strictly in accordance with physical chemistry concepts. But these methods are simple and very satisfactory for most of the calculations involved in water treatment.

The molecular weight of calcium carbonate was originally believed to be 100.00. One source (6) now indicates 100.09. Brown (7) used a value of 100.08 and computed a table of alkalinity equivalents. Copies of nomographs may be obtained from the American Water Works Association, 2 Park Ave., New York, N.Y., 10016. These can be used to determine the free carbon dioxide and the three forms of alkalinity.

2-10. Bioassays. A bioassay is a test which is used as the basis for judging whether or not a waste can be discharged at a given rate, and under a specified set of conditions such as temperature, without causing direct injury to fish and other organisms in the receiving water. *Standard Methods* (1) states:

The TL_m corresponds to the median lethal dose or 50% lethal dose (LD_{50}), which is frequently reported in toxicologic literature (commonly without specifying the duration of observation), and should not be confused with the minimum lethal dose (MLD). The expression *lethal dose* is not appropriate for designating a certain concentration in an external medium, since a dose, strictly speaking, is a measured quantity administered. Unlike lethal dose and lethal concentration (LC), the term *tolerance limit* is universally applicable in designating the effective level of any measurable lethal agent, including high and low temperatures, pH, etc. For this and other reasons, the term *median tolerance limit* and the symbol TL_m are recommended.

The details of how to conduct a bioassay are contained in *Standard Methods* (1).

2-11. Chlorides. Chlorides are the most important and widely distributed member of the halogen group to be found in nature. In dilute solutions they are present as dissociated chloride ions, but in concentrated brines, interionic effects and incomplete dissociation are to be expected. Chloride in domestic sewage may cause important increases in the chloride content in streams into which such wastes empty. In waste disposal recycled-water-use systems, the buildup of chlorides presents a serious problem and becomes one of the parameters that must be taken into consideration. This statement is applicable to other dissolved solids, but chlorides are usually the most predominant.

2-12. Coagulation. In any domestic treatment process there are particles present in the colloidal or near colloidal state that will either not settle out as sedimentation or will settle at such a slow rate that it is impractical to build a settling basin large enough for the settling time required. Coagulation involves the addition of chemicals, such as alum and lime, whose properties cause the formation of a jelly-like material (aluminum hydroxide in this case) which will settle and carry the impurities with it.

Mysels (39) emphasizes the physical chemical principles of colloids. Kruyt and Overbeck (40) have a fairly easy-to-understand physical chemistry approach. Adamson (41) presents an advanced treatment of both theory and applications concerning colloids. General treatment, in concise form, with emphasis on industrial waste treatment is given by Eckenfelder (38) and Rich (42).

2-13. Color. Color is not normally an important factor in domestic waste treatment. In industrial waste treatment, the suspended material, including pseudocolloidal particles, is removed and the color of the light transmitted by the waste is measured. The Hach high-range test for color is based on APHA platinum-cobalt standard and uses colored glass discs expressed 0 to 500. It reads apparent color which is created by dissolved substances (45). A spectrophotometric method and the Tristmulus filter method are both described in *Standard Methods* (1). Color and turbidity are aesthetically undesirable and impose extra loads on downstream water treatment plants.

An arbitrary standard for color is established which is assigned a number of 500. This standard is made up of 1 gram of cobalt chloride and 1.245 grams of potassium chloroplatinate, to which is added 100 milliliters of concentrated hydrochloric acid, and the whole diluted to one liter with distilled water.

2-14. Hardness. There is no completely satisfactory classification system for water hardness. Hardness is a term whose meaning has gone through transition. The term is one of the carry-overs from past methods. It is reported as the hardness that would be produced if a certain amount of $CaCO_3$ were dissolved in water. More than one ion contributes to water hardness. The *Glossary of Water and Wastewater Control Engineering* (4) defines hardness as: A characteristic of water, imparted by salts of calcium, magnesium, and ion, such as bicarbonates, carbonates, sulfates, chlorides, and nitrates, that causes curdling of soap, deposition of scale in boilers, damage in some industrial processes, and sometimes objectionable taste. Calcium and magnesium are the most significant constituents.

Standard Methods (1) indicates: When the hardness is numerically greater than the sum of the carbonate alkalinity, the amount of hardness which is equivalent to the total alkalinity is called *carbonate hardness*; the amount of hardness is numerically equal to, or less than, the sum of carbonate and bicarbonate alkalinities, all the hardness is *carbonate hardness*, and there is no *noncarbonate hardness*. The hardness may range from zero to hundreds of milligrams per liter in terms of calcium carbonate, depending on the source and treatment to which the water has been subjected.

Total hardness is sometimes referred to in industrial water treatment as the sum of the calcium hardness plus the magnesium hardness.

2-15. Nutrients. Nutrients are dissolved materials such as phosphorus, nitrogen, and carbon, which are stimulants to plant growth. Their most objectionable property is their contribution to entrophication of lakes and ponded areas.

2-16. Odor. The odor threshold (4) is the point at which, after successive dilutions with odorless water, the odor of a sample can just be detected. The threshold odor is expressed quantitatively by the number of times the sample is diluted with odorless water. There is a chart showing dilution of sample and reporting of results, as well as a more detailed explanation, in *Standard Methods* (1).

2-17. pH. The acidity of a solution is frequently expressed as the common logarithm of the reciprocal of the hydrogen ion concentration in moles per liter. This value is known as the pH of the solution. Since the molecular weight of the hydrogen ion is 1, a solution containing 1 gram of hydrogen ion per liter has a concentration of 1. The term *concentration of hydrogen ion* is written in the form $[H^+]$ where the brackets stand for "concentration" and the H^+ means "hydrogen ion." The most popular method of measurement uses one of the electrometric instruments on the market. Colorimetric methods can also be used. In the colorimetric method, dyes that change color within a specified pH range are used.

2-18. Refractory Materials. This term is sometimes applied to a material such as ABS detergent that causes foaming in streams.

2-19. Thermal Pollution. Thermal pollution is not often of concern in domestic wastewater. It is most often caused by cooling water discharged from industry. Heating water reduces its ability to hold oxygen in solution. This causes depletion of dissolved oxygen (by lowering the saturation value) and results in ecological disturbances of aquatic life, including fish kills.

2-20. Toxicity. Toxicity is the property or reaction of a substance, or a combination of substances reacting with each other, to deter or inhibit the metabolic process of cells (without completely altering or destroying them) in a particular species, under a given set of physical and biological environmental conditions for a specified concentration and time of exposure. To determine whether a substance is toxic under a given set of conditions, it is common practice to perform a bioassay on a test organism.

Acute toxicity is any direct lethal action of pollution to freshwater fishes that is demonstrable within 96 hours or less (ASTM D1345). The *median tolerance limit* (TL_m) is the concentration of pollutants at which 50% of the test animals are able to survive for a specified period of exposure (ASTM D1345). The *median inhibitory limit* (IL_m) is the concentration of test material which decreases the amount of growth to 50% of that in the controls, within a test period of seven days (ASTM D1129). It is the recommended measure or index of relative toxicity. Specifically, it is a concentration value derived by graphical interpolation and is based on the amount of growth made in seven days in the test flasks, as compared with that made in a control. Inhibitory toxicity is any direct inhibitory action of pollutants on the rate of reproduction of diatoms which is demonstrated within seven days of testing (ASTM D1129). The critical concentration range in the interval between the highest concentration at which all test animals survive for 48 hours and the lowest concentration at which all test animals die within 24 hours (ASTM 1345, ASTM 1129, and ASTM 1345) is contained in reference (3).

The maximum allowable limit is the concentration level (in mg/l) which may have adverse effects on the health of man when present in concentrations above this limit. The recommended limit is the concentration level which should not be exceeded whenever more suitable water supplies are, or can be made, available at reasonable cost. Substances in the latter category, when present in concentrations above this limit, are either objectionable to an appreciable number of people or exceed the levels required for good water control practices. The excess limit is a value that is tolerated by man under some circumstances in some geographic locations but are not necessarily an acceptable level that would be tolerated in a highly developed country. All the terms in this paragraph apply to the freeflowing outlet to the ultimate consumer and are either based directly on or derived from terminology used in the U.S. Public Health Service "Drinking Water Standards" of 1962.

QUESTIONS

1. How are sewage calculations handled hydraulically? Why?

2. Niagara Falls decides to construct a new sewage treatment plant. How many different regulatory agencies may the engineer have to consult in order to get proper approval for this project? Name them.

3. Define domestic sewage and explain what sources make up domestic sewage.

4. What is the ultimate carbonaceous BOD of a liquid waste?

5. Explain at least one method for determining the value of k in eq. (2-1).

6. Why would a COD test sometimes form a confusing result compared to the BOD test?

7. When does odor become a serious problem in sewage? Why?

8. What errors might occur in determining a residue? Why?

9. What is TOC and what is its value to the sanitary engineer in association with a waste?

10. How do the Brönsted and Lewis concepts compare?

11. What is phenolphthalein alkalinity? Is the molecular weight of calcium carbonate actually 100.00?

12. What can a bioassay be used to determine? Why?

13. Explain how to conduct a bioassay.

14. Explain the process of coagulation.

15. Exactly what does the number 500 in color measurement mean? How was it established? In industrial water treatment, how is industrial water hardness sometimes defined?

16. What is entrophication? What may cause it? How does this concern a sanitary engineer?

17. Why is thermal pollution of particular concern to sanitary engineers?

18. Why do most books avoid discussing the issue of toxicity? Is it something that can be precisely established and defended in a court of law? Why?

19. What are enzymes and what is their significance to the sanitary engineer?

20. How would you test for phenols?

PROBLEMS

1. A sewage has a strength of 224 mg/l in a plant that treats 5,000,000 gallons per day. What population equivalent is the plant treating?

2. Assume that k is 0.17 and the five-day BOD of a waste is 210 mg/l. What is the ultimate demand BOD or total amount of oxygen consumed in the reaction?

3. Indicate the rate at which a waste is being oxidized if the five-day BOD is 250 mg/l and the ultimate BOD is 360 mg/l?

4. If the value of the average rate constant is 0.17 at 20°C, what is its value at 30°C?

5. If the specific gravity is taken as 1.0 and the strength of a waste is 235 mg/l, what is its strength in ppm?

6. If the flow rate is 80 cfs how many pounds of waste in terms of BOD will be processed every 24 hours?

REFERENCES

1. APHA, AWWA, and WPCF, *Standard Methods for the Examination of Water and Wastewater*, 12th ed. American Public Health Association, New York, 1965.

2. DICKERSON, RICHARD E., et al., *Chemical Principles*. W. A. Benjamin, Inc., New York, 1970.

3. *Manual on Industrial Water and Industrial Waste Water*, 2nd ed. ASTM Special Technical Publication No. 148–1, American Society for Testing Materials, Philadelphia, Pa., 1966.

4. APHA, ASCE, AWWA, and WPCF, *Glossary of Water and Wastewater Control Engineering*. American Society of Civil Engineers, New York, 1969.

5. NORDELL, ESKEL, *Water Treatment for Industrial and Other Uses*, 1st ed. Reinhold Publishing Corporation, New York, 1951.

6. WEAST, ROBERT C., ed., *Handbook of Chemistry and Physics*, 50th ed. The Chemical Rubber Co., Cleveland, Ohio, 1969.

7. BROWN, C. ARTHUR, "Alkalinity Equivalents." *W. & S. W.*, Reference Numbers R-397 (1964) and R-442 (1962).

8. LARSON, T. E., and HENLEY, L. M., "Determination of Low Alkalinity or Acidity in Water." *Anal. Chem.*, 27, 85 (1955).

9. DONG, G., et al., "How to Make Alkalinity Measurements in Water." *W. & S. W.*, 104, 509 (1957).

10. SAWYER, CLAIR N., *Chemistry for Sanitary Engineers*. McGraw-Hill Book Company, Inc., New York, 1960.

11. THEROUX, FRANK R., et al., *Laboratory Manual for Chemical and Bacterial Analysis of Water and Sewage*, 3rd ed. McGraw-Hill Book Company, Inc., New York, 1943.

12. FAIR, GORDOWN MASKEW, and GEYER, JOHN CHARLES, *Water Supply and Waste-Water Disposal*. John Wiley & Sons, Inc., New York, 1966.

13. MCKINNEY, ROSS E., *Microbiology for Sanitary Engineers*. McGraw-Hill Book Company, Inc., New York, 1962.

14. ECKENFELDER, W. WESLEY, JR., *Water Quality Engineering for Practicing Engineers*. Barnes & Noble, New York, 1970.

15. COCHRANE, V. W., and GIBBS, M., "The Metabolism of Species of Streptomyces IV. The Effect of Substrate of Endogenous Respiration of Streptomyces Coelicolor." *Jour. of Bacteriology*, 71, 2, 126 (1964).

16. DAWES, E. A., and RIBBONS, D. W., "Some Aspects of the Endogenous Metabolism of Bacteria." *Bacteriological Review*, 28, 2, 126 (1964).

17. GRONLUND, A. E., "A Study of Endogenous Respiration in Pseudomonas Aeruginosa." *Dissertation Abstracts*, 25, 4, 2128 (1964).

18. BURLEIGH, I. G., and DAWES, E. A., "Studies on the Endogenous Metabolism and Senescence of Starved Scorcina Cutea." *Biochemical Jour.*, 102, 236 (1967).

19. LAMANNA, C., "Studies of Endogenous Metabolism in Bacteriology," *Annals of the New York Academy of Sciences,* 102, 3, 517 (1963).

20. MALLETTE, M. F., "Validity of the Concept of Energy of Maintenance." *Annals of the New York Academy of Sciences*, 102, 3, 521 (1963).

21. MIDWINTER, G. G., and BATT, R. D., "Endogenous Respiration and Oxidative Assimilation in Nocardia Corallina." *Jour. of Bacteriology*, 59, 1, 9 (1963).

22. MOSES, V., and SYRETT, P. J., "The Endogenous Respiration of Microorganisms." *Jour. of Bacteriology*, 70, 2, 517 (1964).

23. CAMPBELL, J. J. R., "Endogenous Metabolism of Pseudomonas." *Annals of the New York Academy of Sciences,* 102, 2, 669 (1963).

24. Dawes, E. A., and Ribbons, D. W., "Studies of the Endogenous Metabolism of Escherichia Coli." *Biochemical Jour.*, 95, 332 (1965).

25. Eaton, M. R., "Endogenous Metabolism of Reserve Deposits in Yeast." *Annals of the New York Academy of Sciences*, 102, 3, 678 (1963).

26. Gates, William E., and Ghosh, Sambhunath, "Biokinetic Evaluation of BOD Concepts and Data." *Proc. ASCE*, 97, SA3, 287 (1971).

27. Orford, H. E., and Ingram, W. T., "Deoxygenation of Sewage." *Sew. & Ind. Wastes*, 25, 4, 424 (1953).

28. Revelle, C. S., et al., "Biooxidation Kinetics and a Second-Order Equation Describing the BOD Reaction." *Jour. WPCF*, 37, 12, 679 (1965).

29. Woodward, R. L., "Deoxygenation of Sewage, A Discussion." *Sew. & Ind. Wastes*, 25, 948 (1953).

30. Young, J. C., and Clark, J. W., "Second-Order Equation for BOD." *Proc. ASCE*, 91, SA1, 43 (1965).

31. Thomas, H. A., Jr., "Graphical Determination of BOD Curve Constants." *W. & S. W.*, 97, 123 (1950).

32. Theriault, E. J., *The Oxygen Demand of Polluted Waters*. U. S. Public Health Service Bull. No. 173, 1927.

33. Thomas, H. A., Jr., "Analysis of the Biochemical Oxygen Demand Curve." *Sewage Works J.*, 12, 504 (1940).

34. Moore, E. W., et al., "Simplified Method for Analysis of BOD Data," *Sew. & Ind. Wastes*, 22, 1343 (1950).

35. Schroepfer, G. S., et al., in *Advances in Water Pollution Control*, Vol. 1. Pergamon Press, Oxford, 1964.

36. Nemerow, Nelson Leonard, *Theories and Practices of Industrial Waste Treatment*. Addison-Wesley Publishing Company, Inc., Reading, Mass., 1963.

37. Clark, John W., and Viessman, Warren, Jr., *Water Supply and Pollution Control*. International Textbook Company, Scranton, Pennsylvania, 1965.

38. Eckenfelder, W. Wesley, Jr., *Industrial Water Pollution Control*. McGraw-Hill Book Company, New York, 1966.

39. Mysels, K. J., *Introduction to Colloid Chemistry*. Interscience Publishers, Inc., New York, 1959.

40. Kruyt, H. R., and Overbeck, J. Th. G., *An Introduction to Physical Chemistry for Biologists and Medical Students with Special Reference to Colloid Chemistry*. Holt, Rinehart and Winston, New York, 1962.

41. Adamson, A. W. *Physical Chemistry of Surfaces*. Interscience Publishers, Inc., New York, 1960.

42. Rich, Linvil G., *Unit Processes of Sanitary Engineering*. John Wiley & Sons, Inc., New York, 1963.

43. Stenger, V. A., and Van Hall, C. E. "Rapid Method for Determination of Chemical Oxygen Demand." *Analytical Chemistry*, 39, 2, 206 (1967).

44. STENGER, V. A. and VAN HALL, C. E., Preprint 5.3-4-66, Instrument Society of America, 21st Annual Conference, New York, Oct. 24–27, 1966.

45. *Water Test Kits*. Hach Chemical Co., Ames, Iowa.

46. HARDY, A. C., *Handbook of Colorimetry*. Technology Press, Boston, Mass. 1936.

47. JUDD, D. D., *Color in Business, Science and Industry*. John Wiley & Sons, Inc., New York, 1952.

48. JONES, H., et al., *The Science of Color*. Thomas Y. Crowell Co., New York, 1952.

49. Committee Report, "The Concept of Color." *J. Opt. Soc. Am.*, 33, 544 (1943).

50. RUDOLFS, W., and HANLON, W. D., "Color In Industrial Wastes." "I. Determination by Spectraphotometric Method." *Sew. & Ind. Wastes*, 23, 1125 (1951).

51. RIDEAL, S., and STEWART, G. G., "The Determination of Dissolved Oxygen in Water in Presence of Nitrites and of Organic Matter." *Analyst*, 26, 141 (1901).

52. THERIAULT, E. J., *The Determination of Dissolved Oxygen by the Winkler Method*. U. S. Public Health Service Bull. No. 151, 1925.

53. RUCHHOFT, C. C., MOORE, W. A., and PLACAK, O. R., "Determination of Dissolved Oxygen by the Rideal-Stewart and Alsterberg Modification of the Winkler Method." *Ind. Eng. Chem.*, anal. ed., 10, 701 (1938).

54. POMEROY, R., and KIRSCHMAN, H. D., "Determination of Dissolved Oxygen. Proposed Modification of the Winkler Method." *Anal. Chem.*, 17, 715 (1945).

55. WOODWARD, R. L., "How Probable is the Most Probable Number?" *Jour. AWWA*, 49, 1060 (1957).

56. MCCARTHEY, J. A., et al., "Evaluation of Reliability of Coliform Density Tests." *Amer. Jour. Public Health*, 48, 12 (1958).

57. SWIFT, W. H., et al., "Oil Spillage Prevention, Control and Restoration—State of the Art and Research Needs." *Jour. WPCF*, 41, 8, Pt. 1, 392 (1969).

58. BEYCHOK, MILTON R., *Aqueous Wastes from Petroleum and Petrochemical Plants*. John Wiley & Sons, Ltd., New York, 1967.

59. MIDDLETON, F. M. and LICHTENBERG, J. J., "Measurement of Organic Contaminants in the Nation's Rivers." *Ind. Eng. Chem.*, 52, 99A (1960).

60. MIDDLETON, F. M., et al., "Manual for Recovery and Identification of Organic Chemicals in Water." Robert A. Taft San. Eng. Center, Cincinnati, Ohio (1959).

61. LAUGHLIN, H. F., "Influence of Temperature in Threshold *Odor Evaluation*." *Taste and Odor Control J.*, 28, 10 (1962).

62. BAKER, R. A., "Critical Evaluation of Olfactory Measurement." *Jour. WPCF*, 34, 582 (1962).

63. *Taste and Odor Control in Water Purification*, 2nd ed. Industrial Chemical Sales Div., West Virginia Pulp & Paper Co., New York (1959).

64. ZWEIG, G., *Analytical Methods for Pesticides, Plant Growth Regulators, and Food Additives. Principles, Methods and General Applications*, Vol. I. Academic Press, New York, 1955.

65. GUNTHER, F. A., and BLINN, R. C., *Analysis of Insecticides and Acaricides.* Interscience Publishers, New York, 1955.

66. METCALF, R. L., ed., *Advances in Pest Control Research*, Vols. I–V. Interscience Publishers, New York, 1957–1963.

67. GUNTHER, F. A., *Residue Reviews*. Academic Press, New York, Vol. I (1962), Vol. II (1963), Vol. III (1963).

68. *Handbook of Toxicity of Pesticides to Wildlife*. Cat. No. I-49.66:84 (1970). Supt. of Documents, U. S. Government Printing Office, Washington, D.C. 20402.

69. *Metabolism of Pesticides*. Cat. No. I49.15/3:127, Supt. of Documents, U. S. Government Printing Office, Washington D.C. 20402, 1969.

70. BATES, R. G., *Determination of pH*. John Wiley & Sons, Inc., New York, 1964.

71. SPIEGEL, MILTON, and FORREST, TOM H., "Phosphate Removal: Summary of Papers." *Proc. ASCE*, 95, SA5, 803 (1969).

72. DAVIS, E. M., and GLOYNA, E. F., "The Role of Algae in Degrading Detergent Surface Active Agents." *Jour. WPCF*, 41, 8, 1494 (1969).

73. ALBERTSON, O. E., and SHERWOOD, P. J., "Phosphate Extraction Process." *Jour. WPCF*, 41, 8, 1467 (1969).

74. NESBITT, JOHN B., "Phosphorus Removal—The State of the Art." *Jour. WPCF*, 41, 5, 701 (1969).

75. SMITH, J. E., ed., *Torrey Canyon, Pollution and Marine Life*. Cambridge Univ. Press., London, Eng., 1968.

76. CARTHY, J. D., and ARTHUR, D. R., *The Biological Effects of Oil Pollution on Littoral Communities*. Suppl. to Vol. 2 of *Field Studies*. Field Studies Council, London, England.

77. REICHENBACH-KLINKE, H. H., "Effect of Oil and Tar Products in Water on Fish." *Muencher Beitr. Abwasser. Fisch.-Flussbiol* (Germany), 9, 200 (1967).

78. PARKER, FRANK L., and KRENKEL, PETER A., *Thermal Pollution: Status of the Art*. Report No. 3, Vanderbilt University, School of Engineering, Nashville, Tenn., Dec., 1969.

79. *Biological Aspects of Thermal Pollution*. Proceedings of the National Symposium on Thermal Pollution, Sponsored by the Federal Water Pollution Control Administration and Vanderbilt University, Portland, Oregon, June 3–5, 1968. Vanderbilt University Press, Nashville, Tenn.

80. *Engineering Aspects of Thermal Pollution*. Proceedings of the National Symposium on Thermal Pollution, Sponsored by the Federal Water Pollution Control Administration and Vanderbilt University, Nashville, Tenn., Aug. 14–16, 1968.

81. SPEAKMAN, JAMES N., and KRENKEL, PETER A., *Quantification of the Effects of Rate of Temperature Change on Aquatic Biota.* Report No. 6, Vanderbilt University, School of Engineering, Nashville, Tenn., May, 1971.

82. MOTZ, LOUIS H., and BENEDICT, BARRY A., *Heater Surface Jet Discharged Into a Flowing Ambient Stream.* Report No. 4, Vanderbilt University, School of Engineering, Nashville, Tenn., Aug., 1970.

83. EDINGER, JOHN ERIC, *Initial Mixing of Thermal Discharges into a Uniform Current.* Report No. 1, Vanderbilt University, School of Engineering, Nashville, Tenn., Oct., 1969.

84. THACKSTON, EDWARD L., and PARKER, FRANK L., *Effect of Geographical Location on Cooling Pond Requirements and Performance.* Report No. 5, Vanderbilt University, School of Engineering, Nashville, Tenn.

85. ABCM/SAC Joint Committee, *Recommended Methods for Analysis of Trade Effluents.* W. Heffer, Cambridge, 1958.

86. Ministry of Housing and Local Government, *Methods of Chemical Analysis as Applied to Sewage and Sewage Effluents.* H. M. Stationery Office, London.

87. BRYAN, ARTHUR H., et al., *Bacteriology Principles and Practice*, 6th ed. Barnes & Noble, Inc., New York, 1962.

88. VAN HEYNINGEN, W. E., *Mechanisms of Microbial Pathogenicity.* Cambridge University Press, 1955.

89. HEM, JOHN D., *Study and Interpretation of the Chemical Characteristics of Natural Water.* Geological Survey Water-Supply Paper 1473, U. S. Government Printing Office, 1967.

Microbiology **3**

Microorganisms and other biological organisms are involved in many segments of the sewage treatment problem. The majority of microorganisms are beneficial to man. The major sewage treatment processes depend upon the controlled functioning of microorganisms. A microorganism that requires living tissue to grow is called a *pathogen* and is a parasite which must depend upon the host organism as a proper environment for growth and reproduction. Microorganisms can be either plants or animals. The plants are the viruses, rickettsiae, bacteria, fungi, and algae. The animals consist of protozoa, rotifers, and crustaceans.

This chapter is not intended to present a comprehensive treatment of microbiology, but it emphasizes the pathogenic microorganisms that are often not adequately considered in a conventional microbiology course for sanitary engineers. The tendency has been to divide microbiology into two general areas. The sanitary microbiologist and the sanitary engineer have been primarily concerned with the microorganisms to treat wastes. The water microbiologists and water supply design engineers have been most concerned with pathogenic microorganisms and their control. The trend confronting consultants on some types of projects, such as certain types of recreation areas, seems to be toward requiring the sanitary engineer to thoroughly remove pathogens from the waste.

In general, the information in print on the ability of different processes and combinations of processes to reduce pathogens has not been as complete as would be desired. This book does include pertinent fragments of information on the subject and provides methods in some cases for the engineers to at least make an educated guess of how to solve certain problems if encountered.

It can generally be assumed that if a large number of fecal coliform organisms, found in intestines of all warm-blooded animals, are present, a wide variety of pathogens may be present. If fecal coliforms are absent, it is an indication that a number of different types of pathogens are probably not present, but it is not necessarily proof that some such as viruses are not present. If any viruses are present in a waste sample, it may be assumed that other more difficult to detect viruses may be present.

Information is given on sizes of various microorganisms. Later in the book, information on the pore sizes of certain filter media is given. Also discussed is the period of time certain organisms may require to pass through a medium of irregular pore size. Effectiveness of chlorination and its deficiencies are discussed in Chapter 15. The periods of survival of various microorganisms combined with the retention time a given filter medium retards their passage provides one tool to the sanitary engineer in some applications for the elimination of pathogens or for inactivating them.

One of the purposes of this chapter is to provide information useful in the recreation area sanitary design problem. It is anticipated that the design of many recreation facilities will be confronting engineers in the near future. It was for this reason that examples of vector-borne disease transmission were included, as well as a few pathogens that have no relation to water carriage waste.

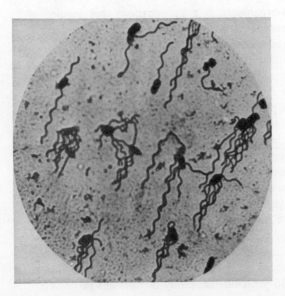

Figure 3-1
Bacillus Typhosa. (Courtesy Armed Forces Institute of Pathology)

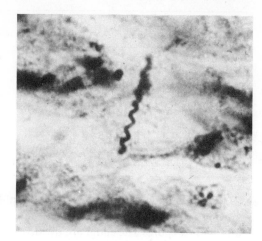

Figure 3-2
Borrelia Recurrentis Cause of
Relapsing Fever. (Courtesy Armed
Forces Institute of Pathology)

3-1. Bacteria. Bacteria are single-celled microorganisms that take in soluble food and convert it to new cells. The vast majority of bacteria are beneficial to man. Beneficial bacteria that obtain organic matter in solution from dead and decaying tissue in plants or animals are called *saprophytic.* Bacteria that require oxygen to live are called *aerobic.* Those that survive in the absence of oxygen are called *anaerobic.* Bacteria multiply most rapidly in warm water. According to the temperatures at which they flourish, bacteria are classified *psychrophilic* (10–20°C), *mesophilic* (20–40°C), and *thermophilic* (40–65°C). They can be of spherical or ovoid shape or in the form of curved or bent rods.

The rod-shaped bacteria are called *bacilli* and frequently have whip-like structures, called *flagella,* that enable them to move about. *Bacillus typhosa* is an example of this type (Fig. 3-1). Some bacilla have oval, egg-shaped, or spherical bodies in their cells, which are called *spores.* After the death of the bacterium, the spore may survive extremely adverse conditions of temperature or dehydration and still be capable of germination when the conditions become favorable.

The spherical or ovoid-shaped bacteria are called *cocci;* they do not form spores and cannot usually move about. When this type occurs singularly, it is called *micrococcus;* in chains, *streptococcus;* in pairs, *diplococcus;* in irregular bunches *staphylococcus;* and in cubical packets *sarcina.*

The curved- or bent-rod group does not form spores; however, some of them form a gelatinous covering that may provide protection from adverse environmental conditions. The genus *Vibrio,* the genus *Spirillum,* and the genus *Spirochaeta* (*spirochetes*) are found in this group, and all members of this group are mobile (Fig. 3-2).

Bacteria are technically fungi, but because of their importance, they are usually discussed separately. Bacteria are unicellular. Information on the size of bacteria is given in Table 3-1.

TABLE 3-1. Size of Bacteria

	Size in Microns	
Classification	Minimum	Maximum
Limits of cell sizes	0.3	50.0
Common bacteria	0.5	3.0
Average rod bacteria		
Width	0.5	1.0
Length	1.5	3.0
Average sphere bacterial	0.5	1.0
Average spiral bacteria		
Width	0.5	5.0
Length	6.0	15.0

3-2. Fungi. Fungi are multicellular, nonphotosynthetic plants capable of growing in low-moisture and low-pH environments. The reproductive stage of fungi is a spore. The fungi are divided into five classes:

1. Myxomycetes—slime fungi
2. Phycomycetes—algae fungi
3. Ascomycetes—sac fungi
4. Basidiomycetes—rusts, smuts, and mushrooms
5. Fungi imperfecti—miscellaneous

Fungi are 5 to 10 microns wide. They produce substantial numbers of spores which are transmitted long distances in the air by wind currents. Absence of suitable environment prevents most fungi spores from germinating.

The term "sewage fungus" usually refers to macroscopic growths of a collection of microorganisms, few of which are fungi, found in organically enriched waters. These organisms include the bacteria *sphaerotilus, zooglea ramigera, Beggiatoa* (the sulfur bacterium), and a microscopic animal (protozoan) which forms macroscopic colonies—*Carchesium* (1). True fungi such as *Leptomitus lacteus* and *Fusarium aqueductum* may be found associated with such growths.

3-3. Rickettsiae. Rickettsiae are substantially smaller than bacteria. They depend on living tissue for growth and cultivation, and thus they are pathogens. They have rod-like shapes with maximum dimensions of about 0.3 micron by 2 microns (2) (Fig. 3-3).

3-4. Viruses. Collectively, the viruses are infectious agents of both plant and animal cells (Fig. 3-4). They are ultramicroscopic, obligate introcellular parasites that manifest their presence by destruction or impairment of the host cells (3). Most viruses fall in the size range of 10 to 500 millimicrons. The genetic material of viruses may be either ribonucleic acid (RNA) or deoxyribonucleic acid (DNA)(4). A virus has the ability to remain virulent outside

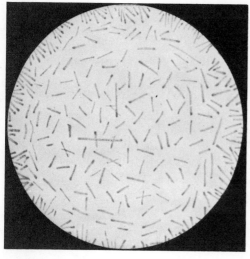

Figure 3-3
Rocky Mountain Spotted Fever.
(Courtesy Armed Forces Institute
of Pathology)

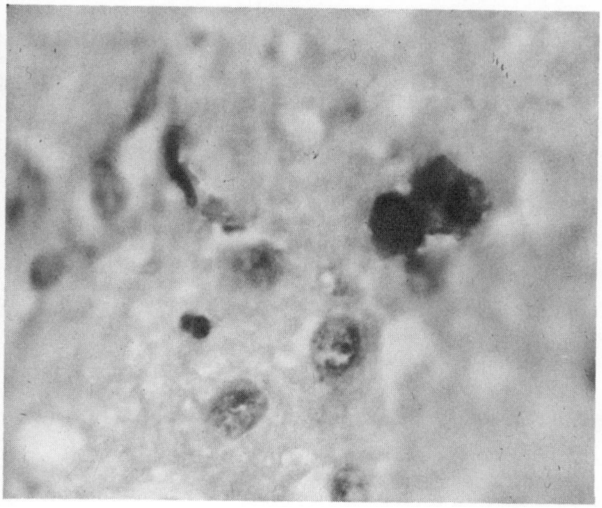

Figure 3-4
Tamilnad Virus Encephalitis.
(Courtesy Armed Forces Institute
of Pathology)

of the host even though it is not growing (5). It may pass through a series of secondary hosts with no reaction because the environment is not suitable. The adenoviruses are associated with upper respiratory infections, particularly in children (3). Enteroviruses are found in the gastrointestinal tract and feces of man and many lower animals. Enteric viruses include Coxsackie viruses, infectious hepatitis viruses, polioviruses, ECHO (enteric-cytopathogenic-human-orphan), and the reoviruses.

3-5. *Bacteriophages.* A bacteriophage is a virus that is parasitic to bacteria. It initiates infection by attaching itself by its tail to the wall of a bacterial cell (6). Through enzyme action, the wall is perforated and bacteriophage DNA passes through into the bacterial cell whose synthetic machinery it organizes to make more bacteriophages. These are released by lysis (rupture) of the host cell. The bacteriophages are highly host-specific, and this property may be used to differentiate between various strains of the same bacterial species. The bacteriophage counts cannot be correlated with water potability (7).

3-6. *Protozoa.* In sanitary engineering, protozoa may be considered as single-celled animals that reproduce by binary fission (5). They differ from bacteria by having at least one well-defined nucleus (6). They are not easily defined organisms, and there is much variation in what different authorities say about them. Aiba et al. (4) state: "Protozoa may be unicellular or multicellular and exhibit a wide variety of forms; some are nonmotile, some are motile with flagella, some are motile with cilia, some are amoeboid, some are parasitic, others are symbiotic." Hawkes (1) states: "These are microscopic animals

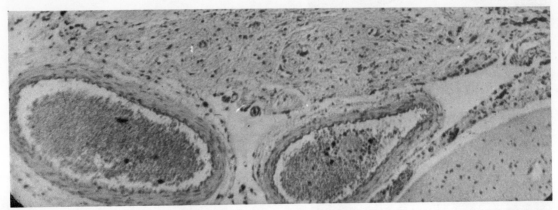

Figure 3-5. Protozoa (Dysautonomia). (Courtesy Armed Forces Institute of Pathology)

whose bodies are not made of several cells but are differentiated into different organelles which carry out the many functions of the many-celled organs of higher forms. Previously they were defined as unicellular, but are now considered noncellular." Thus, how protozoa are classed and whether algae are also included in the classification depend on whether the viewpoint of the phytologists or the protozoologists is accepted (Fig. 3-5).

The protozoa are bacteria eaters and survive in dilute organic wastes by eating bacteria. When the concentration of organic waste is sufficiently great, the protozoa can utilize the soluble organic compounds for their food.

3-7. Rotifer. The rotifer is the simplest of the multicell animals (5). It has the ability of pulling in tremendous quantities of liquid in search of food. Most species live in freshwater and are able to withstand being dried up by forming cysts, in which form they are probably dispersed by the wind (1).

3-8. Algae. In this book, algae are considered as simple, photosynthetic plants with unicellular organs of reproduction. They can utilize the energy of light and do not have to depend upon the oxidation of matter to survive. Algae produce oxygen during their growth when adequate light is available. The freshwater forms are generally microscopic in size, but saltwater forms, such as the giant kelps, may be several hundred feet in length. Some algae grow in colonies. Runaway growths of algae, in colonies, are known as *water blooms*.

3-9. Crustaceans. The class *Crustacea* covers nearly all aquatic organisms including crabs, lobsters, and shrimp. A copepod is any of a large subclass (*Copepoda*) of usually minute freshwater and marine crustaceans. Microbiology is concerned with the more primitive orders such as the copepods *Daphnia* and *Cyclops*. The *Cyclops* is less than 2 millimeters long.

3-10. Worms. Worms can be classified as *roundworms* and *flatworms*. Under this system, the roundworms are called *nematodes*. The most common forms of nematodes found in aerobic sewage treatment processes are the *Microphagus rhabditids* and *Diplogasterids* (whose burrowing and feeding keep biological beds porous and accessible to oxygen and also help to prevent clogging by encouraging the necessary sloughing of biological films) (8). Altogether, four groups (families) are commonly encountered. The other two (*Mononchs* and *Dorylaimids*) are much less common and are reported to be animal predators attacking rotifers, oligochaetes, and other nematodes. The flatworms contain two main groups, the tapeworms (*Cestodes*) and the flukes (*Trematodes*). The tapeworm is a segmented worm consisting of a common head, called a *scolex*, and a series of segments, one behind the other, called *proglottids*. Each segment of a tapeworm is capable of developing another worm.

Worms may also be classed as parasitic and free-living forms. The parasites include the ascarids, hookworms, flukes, and tapeworms. The worms that have host specificity are called *Enterobius*. Those that require multiple hosts are referred to as *schistosomes*, and those with durable ova are called *Ascaris*. Tapeworms vary in size from *Echinoccus* (forming large cyst-like tumors containing thousands of tiny larva worms) (Fig. 3-6) to the fish tapeworm (*Diphyllobothrium*), which may produce worms 60 feet in length.

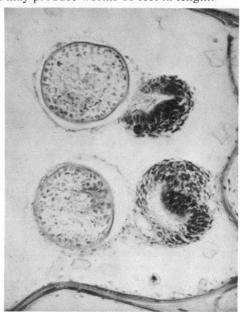

Figure 3-6
Echinococcosis of the Liver.
(Courtesy Armed Force Institute of
Pathoolgy)

3-11. Pathogens. There are 41 types of bacteria, 10 types of fungi, 10 types of protozoa, 5 rickettsia, 11 nematodes, 3 cestodes, 3 trematodes, and approximately 100 viruses that are of concern to the sanitary design engineer. Certain ones of each classification are more hardy than others, and if our prevention and control measures are adequate to control them, then the other pathogens within a given classification are eliminated. *Salmonella*, particularly *B. typhosa*, *S. typhosa*, *E. typhosa*, *Genus Shigella*, and *Mycobacterium Tuberculosis*, are of particular concern. Viruses are probably more of a problem than heretofore recognized, but we simply do not have adequate data from which to draw desired conclusions.

Within the *Genus Salmonella*, the typhoid organism *S. typhosa* has been studied most extensively since it is the longest lived, most resistant to bactericides, and is fairly representative of the other intestinal pathogens as a group (9).

Some of the other organisms noted for their survival ability include the tubercle bacilli in polluted waters (10, 11), Shigellae in sewage (12), *Leptospira australis* in soil and water (13), Coxsackie virus in water and sewage (14), *Ascaris* eggs and infectious hepatitis in water (15, 16).

3-12. Pathogen Survival. There are nonpathogenic bacteria that can survive indefinitely at the surface or at great depths in the soil. This section is limited to pathogenic microorganisms of most concern to the sanitary design engineer.

Granger and Deschamps (17) claimed to have recovered *Eberthella typhosa* from the soil after $5\frac{1}{2}$ months. Beard (18) found that the typhoid bacillus survived for two years in moist soil. Kligler (19) reported that moist, slightly alkaline soils were most favorable for the survival of *E. typhosa*. In moist soils the survival was up to 80 days, but only 20 days in dry soil and 10 days in acid soil. The longevity of the typhoid bacillus has been stated (18, 20, 21) to depend on the water-holding capacity of a soil, the temperature, rainfall, sunlight, organic material in the soil, and the hydrogen ion concentration. According to Firth and Horrocks (22), typhoid organisms do not multiply or move about in the soil, but can be washed by water even from closely packed soil. They also found that some of the organisms can exist up to 74 days in ordinary or polluted soil, depending on the soil moisture. The survival time is reduced to 25 days in dry sand. In the surface soil layer, the organisms were viable after 122 hours of exposure to sunlight distributed over a period of 22 days. Demster (23) considered that, if the soil itself does not have a destructive action on the bacillus, then the viability is chiefly dependent on the soil moisture. Rudolfs et al. (9) concluded from a review of the literature that the viability of Salmonella types was variable—having been reported ranging from less than 24 hours in peat to more than 2 years in freezing, moist soils—but generally less than 100 days. Colder temperatures increased the survival of *E. typhosa*.

Table 3-2 contains a summation of survival times of the typhoid bacillus in water. Jordan, Russell, and Zeit (24) found that the survival time of the typhoid bacillus was in inverse relation to the degree of contamination. The period of survival in raw river water was one to four days, and in a drainage canal the survival was two days. Wheeler (25) reported results opposite to

TABLE 3-2. Survival of Typhoid Bacillus in Water

Type of Water	Survival Time in Days	Reference
Fresh	4	20
Polluted	4	20
Sea	32 +	20
Muddy waste	180	31
Swimming pool	7–10	29
Groundwater	28	29
Septic tanks	27 +	32, 20
Water 0°C	63	33
Water 5°C	49	33
Water 10°C	35	33
Water 18°C	28	33
Cold water or sewage	months	26
Unspecified	490	34
0.85% NaCl	240	35

those of Jordan et al. He found that the greater the pollution or dissolved material, the longer the survival. Clark (27) found survival of *S. typhosa* to be several days in water or sewage in warm weather and a period of months in cold weather. Ruediger (28) found that typhoid and colon bacilli disappeared about five or six times more rapidly from polluted river water in the summer than in winter when the river was covered with ice and snow. The sun's rays were considered of value, but more important was the increase during the summer of microscopic plants, protozoa, and saprophytic bacteria. Wibaut and Moens (28, 29) found that *E. typhosa* disappeared from tap, rain, and swimming pool water in seven to ten days and corresponded with increases of certain bacteria-eating protozoa. However, in groundwater the survival was four weeks, in the presence of the same protozoa. They concluded that, although the disappearance of *E. typhosus* from water was influenced by protozoan population, other forces were also at work.

Table 3-3 shows the survival of *Genus Shigella* in water. This genus is the primary cause of bacillary dysentery in man. Table 3-4 shows the survival of *Mycobacterium Tuberculosis* in various environments. Table 3-5 shows the distances virulent tubercles have been found from point of introduction into a stream. Table 3-6 indicates the duration of survival of *Vibrio Cholarae* in water. Mukerjee et al. (30) found no relation between the survival of *Vibrio Cholarae* and temperature, humidity, rainfall, or the pH value of water of various types and temperatures.

3-13. Helminth Parasites. *Ascaris* eggs are extremely resistant to many chemicals ordinarily lethal to worms or microorganisms (59). Galli-Valerio (60) found that *Ascaris* eggs could develop the embryo stage in 50% sulfuric, hydrochloric, nitric, or acetic acid; saturated solutions of copper sulfate, ferrous sulfate, and copper acetate; and in 50% commercial formalin or 50%

TABLE 3-3. Survival of Genus Shigella in Water

Species	Type of Water	Survival Time in Days	Reference
S. flexneri	Water + Humus	391	36
S. shigae	Water + Humus	197	36
S. shigae	Water + Humus	7	36
S. flexneri	Saline	1–2	39
S. flexneri	Unidentified	52	38
S. boydi	Unidentified	52	38
S. sonnei	Unidentified	105	38
S. boydi	Synthetic well	469	37
S. sonnei	pH 7 or 9*	77	37
S. shigae	pH 7 or 9*	77	37
S. flexneri	pH 7 or 9*	77	37
S. sonnei	pH 8*	196	37
S. shigae	pH 8*	196	37
S. flexneri	pH 8*	196	37

*5° and 21°C.

TABLE 3-4. Survival of *Mycobacterium Tuberculosis* in Various Environments

Environment	Approximate Survival Time in Days	Reference
Waste water	90	31
Soil	180	31
Stagnant pools with manure	365 +	40
Sludge drying beds	330–500 +	41
Growing grass	49	43
Freshwater	6 +	11
Saline water	1 +	11
Sludge	365 +	42
0.85% NaCl	405	44
Distilled	652	44

TABLE 3-5. Distances Virulent Tubercles Have Been Found From Point of Introduction Into a Stream

Source of Entry to Stream (All from Sanatoriums)	Distance	Reference
Sewage outfall	3.5 miles	45
Sewage outfall*	250 meters	46
Sewage outfall	1,500 meters	47
Sewage outfall	2,000 meters	47
Imhoff tank effluent	1,250 meters	53
Sewage plant effluent	1,000 meters	51
Sewage outfall	2,000 meters	48
Septic tank effluent	1,750 meters	52
Sewage outfall	1 mile	49
Sewage outfall	10 miles	50

*In dilutions of 1 to 1000.

TABLE 3-6. Duration of Survival of *Vibrio Cholaeae* in Water

Type of Water	Survival Time in Days	Reference
Sea*	11	54
Sea and river mixed*	13	54
River water*	8	54
Tap water*	8	54
Well water	8	54
Reservoirs	2–5	58
Septic tanks	2–5	57
Sea	10†	56
River	1–3	55
Surface tank	1–3	55
Tap, Calcutta	1–3	55
Spring	1–3	55

*Winter conditions.
†Average.

antiformin. Caldwell and Caldwell (61, 62) indicate that desiccation is the greatest lethal factor to *Ascaris* eggs. Temperature and humidity conditions which either hasten or retard drying have an effect on the development of *Ascaris* eggs. Orenstein (63) indicated that in infected communities *Ascaris* ova are found in sewage effluents, unless filtered, and in digested sludges. Sand filters remove them mechanically without affecting their viability. Vassikova (64) found that, if feces are buried in pits, 13 months are required for 100% kill of helminth eggs. A 97% kill may result within 6 to 8 months.

A live *Ascaris* worm is shown in Fig. 3-7. The life cycle of the *Ascaris* is

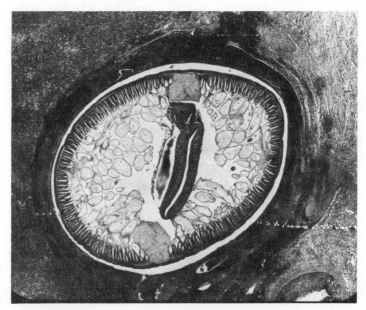

Figure 3-7
Live *Ascaris*. (Courtesy Armed Forces Institute of Pathology)

indicated in Fig. 3-8. Another helminth parasite, the hookworm, is shown in Fig. 3-9 consuming human body tissue.

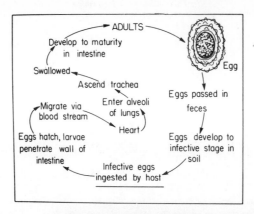

Figure 3-8
Nematode Cycle—Ascaris Type.
(Courtesy Armed Forces Institute of Pathology)

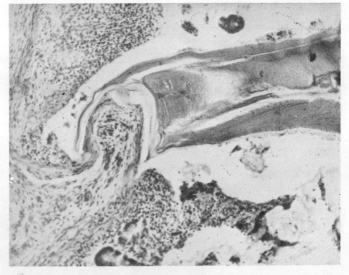

Figure 3-9. Hookworm. (Courtesy Armed Forces Institute of Pathology)

REFERENCES

1. HAWKES, H. A., *The Ecology of Waste Water Treatment*. Pergamon Press, New York, 1963.

2. CLARK, RANDOLPH LEE, JR., and CUMLEY, RUSSELL W., *The Book of Health*, 1st ed. Elsevier Press, Inc., New York.

3. MCLEAN, D. M., BROWN, J. R., and LAAK, R., "Virus Dispersal by Water." *Jour. AWWA*, 58, 7, 920 (July, 1966).

4. AIBA, SHUICHI, et al., *Biochemical Engineering*. Academic Press, New York, N.Y., 1965.

5. MCKINNEY, ROSS E., *Microbiology for Sanitary Engineers*. McGraw-Hill Book Co., Inc., New York, 1962.

6. ABERCROMBIE, M., et al., *A Dictionary of Biology*, 5th ed. Penguin Books, Baltimore, Md., 1966.

7. COUTURE, E., "The Value of the Presence of Bacteriophage in Waters as a Measurement of Their Quality." *Rev. Hyg. Med. Prevent.*, 58, 371 (1936).

8. CALAWAY, WILSON T., "Nematodes in Wastewater Treatment." *Jour. WPCF*, 35, 8, 1006 (Aug., 1963).

9. RUDOLFS, W., et al., "Literature Review on the Occurrence and Survival of Enteric Pathogenic and Relative Organisms in Soil, Water, Sewerage and Sludges, and on Vegetation. I. Bacterial and Virus Diseases." *Sew. & Ind. Wastes*, 22, 1261 (1950).

10. HEUKELEKIAN, H., and ALBANESE, M., "Enumeration and Survival of Human Tubercle Bacilli in Polluted Water. I. Development of Cultural Method of Enumeration." *Sew. & Ind. Wastes*, 28, 8, 955 (Aug., 1956).

11. HEUKELEKIAN, H. and ALBANESE, M. "Enumeration and Survival of Human Tubercle Bacilli in Polluted Waters. II. Effect of Sewage Treatment and Natural Purification." *Sew. & Ind. Wastes*, 28, 9, 1094 (Sept., 1956).

12. WANG, W. L., et al., "The Survival of *Shigella* in Sewage. An Effect of Sewage and Fecal Suspensions on Shigella Flexneri." *Appl. Microbiol.*, 4, 34 (1956).

13. SMITH, D. J. W., and SELF, H. R. M. "Observations on the Survival of *Leptospira Australis*. A. In Soil and Water." *Jour. Hygiene* (Brit.) 53, 436 (1955).

14. CLARKE, N. A., et al., "Survival of Coxsackie Virus in Water and Sewage." *Jour. AWWA*, 48, 677 (1956).

15. BERG, G., *Transmission of Viruses by the Water Route*. Interscience Publishers, New York, 1965.

16. MOSLEY, JAMES W., "Epidemiologic Aspects of Viral Agents in Relation to Waterborne Diseases." *Public Health Reports*, 78, 328 (1963).

17. GRANGER, J., and DESCHAMPS, E., *Arch. de Med. Exper. et D'Ant. Path.*, 1, 33 (1889).

18. BEARD, PAUL J., "Longevity of *Eberthella Typhosus* in Various Soils." *Amer. Jour. Public Health*, 30, 1077 (Sept., 1940)

19. KLIGLER, I. J., *Investigation of Soil Pollution and Relation of Various Types of Privies to Spread of Intestinal Infections*. Monograph 15, Rockefeller Inst. of Medical Research, New York, Oct. 10, 1921.

20. BEARD, PAUL J., "The Survival of Typhoid in Nature." *Jour. AWWA*, 30, 124 (1938).

21. BEARD, PAUL J., et al., "The Survival of *Eberthella Typhosa* in Soil." *Jour. of Bact.*, 33, 1, 74 (Jan., 1937).

22. FIRTH, R. H., and HORROCKS, W. H., "An Inquiry into the Influence of Soil, Fabrics and Flies in the Dissemination of Enteric Infection." *Brit. Med. Jour.*, 2, 936 (1902).

23. DEMSTER. *Brit. Med. Jour.*, 1, 1126 (1894).

24. JORDAN, E. O., et al., "The Longevity of the Typhoid Bacillus in Water." *Jour. of Infect. Diseases.* 1, 641 (1904).

25. WHEELER, J. M., "The Viability of *B. Typhosus* under Various Conditions." *Jour. Med. Res.*, 15, 269 (1906).

26. CLARK, R. N., "The Transmission of Disease by Sewage." *Sew. Works Jour.*, 18, 1139 (1946).

27. RUEDIGER, G. F., "Studies on the Self-Purification of Streams." *Amer. Jour. Public Health*, 1, 411 (1911).

28. WIBAUT, N. L., and MOENS, I., "The Disappearance of Typhoid Bacteria in Water." *Abs. Zentbl. Gesam. Hyg.*, 15, 586 (1927).

29. WIBAUT, N. L., and MOENS, I., "The Disappearance of Typhoid Bacteria in Water." *Verslag. Naturk. Konink Akad. Watensch*, 36, 1, 129 (1927).

30. MUKERJEE, S., et al., "Observations of Cholera Endemicity in Calcutta and Survival of *Vibrio Cholerae* in the Water Sources." *Ann. Biochem.*, 21, 31 (1961).

31. DEDIE, K., "Organisms in Sewage Pathogenic to Animals." *Stadtehygiene*, 6, 177 (1955).

32. GREEN, CARL E., and BEARD, PAUL J., "Survival of *E. Typhi* in Sewage Treatment Plant Processes." *Amer. Jour. Public Health*, 28, 762 (June, 1938).

33. WEBER, G., "Salmonella in Water." *Zentr. Bakteriol* (Germany), 163, 312 (1957).

34. KONRADI, D., "Ueber die Lebensdauer Pathogener Bakterien in Wasser." *Centralbl. f. Bakt. I. Abt. Orig.*, 36, 203 (1904).

35. BALLANTYNE, E. N., "On Certain Factors Influencing the Survival of Bacteria in Water and in Saline Solutions." *Jour. of Bact.*, 19, 5, 303 (1930).

36. SHREWSBURY, J. F. D., and BARSON, G. J., "A Note on the Absolute Viability in Water of *S. Typhi* and the Dysentery Bacilli." *British Med. Jour.*, 954 (1952).

37. SHREWSBURY, J. F. D., and BARSON, G. J., "On the Absolute Viability of Certain Pathogenic Bacteria in a Synthetic Well Water." *Jour. Pathol. Bacteriol* (Brit.), 74, 215 (1957).

38. KRATOCHVIL, J., and TARBCAK, M., "The Water Factor in the Epidemiology of Dysentery." *Caskoslov Epidemiol., Mikrobiol., Imunol.* (Prague), 7, 271 (1958).

39. MEITERT, T., et al., "Investigations on the Part Played by Sea and Lagoon-Water (Black Sea, Lake of Tekirghiol) in the Transmission of Enteric Infections." *Arch. Roumaines Path. Exper. et Microbiol.*, 17, 3/4, 513 (1958).

40. STEFFERUD, A., "Animal Diseases." *The Yearbook of Agriculture*, U.S.D.A. (1956).

41. JENSEN, K. E., "Presence and Destruction of Tubercle Bacilli in Sewage." *Bull. World Health Organ.*, 10, 171 (1954).

42. Jensen, K. A., and Jensen, K. E., "Occurrence of Tubercle Bacilli in Sewage and Experiments on Sterilization of Tubercle Bacilli—Containing Sewage with Chlorine." *Acta. Tub. Scand.*, 16, 217 (1942).

43. McCarthy, J. A., "A Laboratory Study of Sulfamic Acid as a Vehicle for Chlorine." *Jour. New England Water Works Assn.*, 74, 2, 166 (1960).

44. Ballantyne, E. N., "On Certain Factors Influencing the Survival of Bacteria in Water and in Saline Solutions." *Jour. of Bact.*, 19, 5, 303 (1930).

45. Brown, L., et al., "The Occurrence of Living Tubercle Bacilli in River Water Contaminated by Sewage from a Health Resort." *Amer. Jour. Public Health*, 6, 1148 (1916).

46. Gabe, E., "Desinfection bei Tuberkulose." *Beitr. Klin. Tuberc.*, 88, 304 (1936).

47. Barons, K., "Zur Frage der Sanatorien und Krandenhauserkanalisation in Lettland." Nosokomeion (*Quarterly Hospital Review*), 8, 243 (1937).

48. Bergsman, A., and Vahlne, G., "Tuberkulos som Vattenburen Smitta." *Nord. Hyg. Tidskr.*, 29, 92 (1948).

49. Emili, H., and Tomisic, P. S., "The Health Aspects of Polluted Water with Special Reference to the Epidemiology of Water-Borne Infections." *Water Pollution in Europe*, Fourth European Seminar of Sanitary Engineers, 22 (1954).

50. Kraus, E., "Studies of the Spreading of Tubercle Bacilli from the Sewage of a Sanatorium." *Arch. Hyg. u. Bakt.*, 128, 112 (1942).

51. Larmola, E., "Investigations of the Number, Form, Growth and Pathogeninicity of the Acid-Fast Microorganisms in the Sewage of a Sanatorium." *Act Tuberc. Scand.*, 18, 209 (1944).

52. Wagener, E., et al., "Detection of Tuberculosis Bacteria in Sewage From Tuberculosis Sanatoriums." *Gesundh. Ing.*, 74, 154 (1953).

53. Stenius, R., "Eiwirkung einer tuberkuloseheilstatte auf die Gesundheit von Rindern." *Finn. Vet. Zschr.*, 47, 105 (1941).

54. Yasuhara, S., "How Long Does *Cholera Vibrio* Live in the Water in Winter?" *Jour. Public Health Assn.* (Japan), 2, 9, 1 (1926).

55. Lahiri, M. N., et al., "The Viability of *Vibrio Cholarae* in Natural Waters." *Indian Med. Gazette*, 74, 742 (Dec., 1939).

56. Kiribayashi, S., and Aida, T., "A Study of the Fate of *Cholera Vibrios* in the Sea Water of Keeling Port, Formosa." *Jour. Public Health Assn.* (Japan), 9, 11, 1 (1933).

57. Flu, P. C., "Investigations of the Duration of Life of Cholera Vibriones and Typhoid Bacteria in Septic Tanks at Batavia." *Med. Burg. Geneesk. Dienst.*, 179 (1921).

58. Flu, P. C., "Investigations of the Auto-Purification of Water in Large Reservoirs Exposed to the Direct Rays of the Sun." *Med. Burg. Geneesk. Dienst.*, 299 (1921).

59. Spector, B. K., and Buky, F., "Viability of Endamoeba Histolytica and Endamoeba Coli—Effect of Drying." *Public Health Reports*, 49, 379 (1934).

60. GALLI-VALERIO, B., "Notes de Parasitologic et de Technique Parasitologique." Centr. f. Bakt., Parasit. u. Infekt., *I. ABT.*, 75, 46 (1915).

61. CALDWELL, F. C., and CALDWELL, E. L., "Are *Ascaris Lumbricoides* and *Ascaris Suilla* Identical?" *Jour. Parasit.*, 13, 141 (1937).

62. CALDWELL, F. C., and CALDWELL, E. L., "Preliminary Report on Observations on the Development of Ova of Pig and Human Ascaris Under Natural Conditions and Studies of Factors Influencing Development." *Jour. Parasit.*, 14, 254 (1928).

63. ORENSTEIN, A. J., "A Contribution to the Discussion on the Hazard of Ascaris Infestation from Sewage." *Jour. and Proc. Inst. Sewage Purif.*, Part 4, 481 (1949).

64. VASSILKOVA, Z. G., "Viability of Helminth and Parasitic Disease." Moscow, 9, 231 (1940); *Trop. Disease Bull.* 40, 318 (1943).

Sources
and Transmission
of Pollution

4

In the preceding chapter a very condensed presentation of general microbiology applicable to all areas of sanitary engineering was presented. This chapter continues the treatment of microorganisms with emphasis on their hazards and migration from specific sources of pollution. This chapter also begins to consider the geologic, hydrogeologic, hydraulic, mechanical, and other areas of science that are involved in the transmission of pollution.

POLLUTION TRAVEL IN GROUNDWATER

4-1. Basic Principles. Sanitation engineering requires an understanding of the travel of bacteria, viruses, chemical, organic, and mineral matter through groundwater. It is necessary to consider all potential sources of pollution likely to contaminate an aquifer. Table 4-1 indicates some of the sources of groundwater pollution and the types of pollution that may result from these sources. Table 4-2 lists specific contaminants, the point of origin through which the contaminant entered the groundwater, and the distance the pollution is known to have traveled. Table 4-3 is similar to Table 4-2, except that additional information was available on the type of geological formation and the time it took the pollution to travel a specific distance. The distance a particular pollutant will travel depends on a number of factors which will be indicated in this section. It is important to determine the extent of the purification by the ground, a matter which is far from easy to determine. It clearly

TABLE 4-1. Sources and Types of Pollution of Groundwater Supply

Source of Pollution	Type of Pollution
Pit privy	Bacteria, viruses, chemicals
Leaking vault privy	Bacteria, viruses, chemicals
Flooded abandoned well	Bacteria, viruses
Open garbage dump	Garbage leachings
Sanitary landfill	Garbage leachings
Septic tank drainfield	Bacteria, viruses, detergent
Abandoned wells*	Bacteria, viruses, pesticides
Waste stabilization pond	Bacteria, viruses, detergents
Seepage pits	Various wastes, chemicals
Waste disposal wells	Bacteria, viruses, chemicals
Infiltration surface water	Bacteria, viruses, pesticides, chemicals
Ash dumps	Mineral leachings
Sewage spray irrigation	Bacteria, viruses

*Particularly if contaminated by surface runoff after rains.

TABLE 4-2. Distances Various Pollutants Have Been Known to Travel in Groundwater

Contaminant	Point of Origin	Distance Traveled	Reference
Detergent	Sewage pond	$\frac{3}{4}$ mi	1
Chloride	Sewage pond	$\frac{3}{4}$ mi	1
Fluoride	Industrial lagoon	2 mi	2
Minerals	Garbage dump	4.75 mi	3
ABS	Sewage lagoon	800 ft	4
Weed killer	...	105,600 ft	4
Typhoid bacteria	River entering Abandoned well	800 ft	5
Coliform bacteria	Polluted trench*	232 ft	6
Uranium	Polluted trench*	450 ft	6
Garbage leachings	Old garbage dump	1,475 ft	7
Garbage leachings	Sand pit dump	2,000 ft	8
Industrial detergent	Spraying on gravel bed	1,800 ft	9
Bacteria	Washed into karst formation	10 km	10
Sugar refinery waste	Soaked into ground	4.5 km	10

*Intersecting ground water.

depends upon a number of factors including the following (10):

1. The nature of the pollution. The ground has no capacity for removing polluting materials which are not susceptible to filtration, biological attack, or absorption.
2. The nature and structure of the geological strata, particularly the number, size, and interconnection of the pores or other openings.
3. The depth of the water table. It would be expected that the farther a liquid has to descend before reaching the standing water level, the greater the opportunity for purification.

TABLE 4-3. Distances and Times of Travel of Some Contaminants Through Various Geological Formations

Contaminant	Geological Formation	Distance Traveled	Time of Travel	Reference
Picric Acid	?	15,800 ft	4–6 years	87
Gasoline	Fractured limestone	10,000 ft	5 years	88
Phenol	Sand and gravel	1,500 ft.	4–5 years	89
Gasoline	Sand and gravel	2,300 Ft	7 years	90
Microorganisms	Sand, particle size 0.17 mm, dry soil, uniformity coefficient 1.65	120 in	?	91
Coliform bacteria	Sand, effective size 0.13 mm	65 ft	27 weeks	6
Chemicals	Sand, effective size 0.03 mm	115 ft	27 weeks	6
Bacillus prodigiosus	Aquifer, porosity 32.8%	69 ft	9 days	92
Coliform bacteria	Fine sand	400 ft	?	93
Iron bacteria	Fine sand	1,500 ft	?	93
Sludge effluent	Fine sand	1,500 ft	?	93
Chromate wastes	Glacial drift	1,000 ft	3	94

4. The rate at which polluted water enters the ground.
5. The condition of the surface of the ground. A plowed field or a well-managed forest floor having large amounts of organic material will absorb water rapidly. It also depends on whether or not the land is kept flooded.
6. The rate of flow of water through the ground. This determines the time available for purification before a given point is reached.
7. The rate at which the water is pumped out. This is related to the volume of saturated ground from which the water is drawn.

There are many processes constantly at work to prevent the accumulation of underground contamination. Some of the factors mentioned by Stiles and Crohurst (6) as having influence upon the accumulation are:

1. The hydrogen-ion concentration of the soil.
2. The effect of drying because of changes in the level of the groundwater.
3. Competition between bacteria.
4. Protozoan competition or the devouring of bacteria by other organisms.
5. Temperature.
6. The presence or absence of food supply.
7. The natural death rate of bacteria.
8. The effect of capillarity, which may carry bacteria into the capillary fringe where they may die by drying.
9. The natural action of bacteria to seek an oxygen supply near the water table, where they may be subsequently unwatered and killed by drying.

Stead (11) has indicated that few sewage bacteria are able to survive in groundwater environments; more than 99% die within two or three weeks within this environment.

A study of the literature indicates that knowledge of the travel of pollutants in groundwater is not as complete as desirable for engineering design purposes. Investigators seem to agree that pollution travels farthest in the direction of groundwater flow, and that chemicals travel much farther than bacteria in a water-bearing stratum (12). Pollution will, under some circumstances, travel at varying angles from the source of pollution, including a low adverse hydraulic gradient (3, 12, 13, 14), depending on how the pollutant is injected into the groundwater.

4-2. Permeability. The size, shape, and interrelations of the interstices in the groundwater strata largely determine the rapidity with which water may pass through a formation and the ability of a formation to filter out pollution. Formations that exert great filtering action on water passing through them, such as most formations of sand or sandstone, furnish the greatest protection against the transmission of pollution (15).

As a rule, permeability should be determined by test, and not from other properties such as grain size (16). The general soil type and grain size are useful, however, in indicating the approximate range of permeability to be expected. The permeability and drainage characteristics of soil and calculations pertinent to permeability are available in the literature (16, 17, 18, 19, 20, 21).

There is fairly general agreement among investigators that formations containing fissures and joints, solution channels, as in creviced limestones, and materials with large interstices, such as clean gravel, provide little filtering action. Water may move rapidly, transmitting contamination over a long distance from the initial source. Volcanic rocks have a wide range of hydrologic characteristics. Some recent basalt aquifers have close to the highest transmissivities known (17).

4-3. Hydraulic Gradient. Under a given hydraulic gradient, water will flow in the most permeable materials at a rate about 450,000,000 times that of the least permeable (22). The lowest rate at which water has been observed to move through a natural material in laboratory tests was about 1 foot in 10 years (23, 24). Under laboratory conditions, gravel with a hydraulic gradient of 10 ft/mi has been observed to carry water at the rate of 60 ft/day (22). Recognized water-bearing formations, from which wells obtain their water supplies, have a groundwater rate of movement that is generally not greater than 5 ft/day and not less than 5 ft/yr (22). The hydraulic properties of geological formations vary extensively (25) and influence the rate of flow of groundwater within a formation of a given permeability.

4-4. Sorption. Sorption is the ability to take up and hold by either adsorption or absorption. Many contaminants are retained on earth material by chemical and physical sorption, and the extent to which sorption occurs is not easily determined (14). Clays tend to have a greater sorptive capacity than sands; thus there is an inverse relation between sorption and permeability (14).

The ground sediments of the salt lakes are capable of adsorbing bacteria (26). The various types of sediments differ in their adsorption capacities for the same bacterial species. Adsorption is not limited to salt lakes. It was shown by Kruger (27) that the number of bacteria present in water may be reduced by the addition of coke, sand, clay, brick dust, magnesium oxide, and other powder-like substances. Adsorption is greater for substances with a low specific gravity than for those where the specific gravity is high. The degree of adsorption depends upon the size of the soil particles (28, 29). The smaller their size, the higher is the percentage of adsorbed bacteria. There is conflict in the literature about the limit, if any, beyond where there is a sharp decline in the degree of adsorption (29, 30, 31, 32).

4-5. Vertical Percolation of Water Containing Bacteria. Butler et al. (33) indicate that observations of coliform organism counts in the soil itself produced evidence that the bacteria are drastically reduced in numbers through the surface mat in the first 0.5 centimeters of soil, but that a subsequent build-up of *E. coli* occurs at lower levels. They state that this can be accounted for only if organisms are filtered out of the percolate at a rate greater than their death rate within the soil.

Krome et al. (34) noted that the problem of bacteria removal was resolved into two distinct areas of investigation. One area is concerned with the behavior of bacteria in the subsurface regions in the absence of particulate matter. Near the surface of the soil an abrupt drop in coliform bacteria occurred (12, 34); but farther down the column, below the first few centimeters of the clogged surface stratum, the decrease appeared to be a logarithmic function of the distance traversed by the contaminated percolate (34).

Tables 4-4 and 4-5 have been constructed from part of the information contained in Tables II and IV of Roebeck (35). The variations of bacterial travel at two different flow rates in a fine sand and a single flow rate in a coarser sand are noted. Roebeck's article contains substantial detail on the two sands and the results of experiments with and without ABS (alkyl benzene sulfonate).

4-6. Pollution From Pit Privies. A study made by Kligler (36) of 50 pit privies in different soils showed no evidence of pollution in wells 25 and 40 ft from the pits. The study covered both rainy and dry seasons. Soil tests indicate pollution may extend only approximately 5 ft. A study by Mom and Schaafsma (37) found coliform organisms in soil 20 years after contamination, in decreasing numbers down to a depth of 9 to 13 ft, but increasing again as the groundwater level was approached. In inhabited areas, sewage bacteria were found as deep as 33 ft, leading the investigators to conclude that purification in the soil is not as great as has been claimed.

Beard (38) reported that typhoid bacteria live from 10 to 84 days in excreta. He also found that the typhoid bacillus survived for 1 year in feces at freezing temperatures. Kligler (36) found that *E. typhosa* and *B. dysenteriae*

TABLE 4-4.* Arrival Time in Days for Coliform at Various Levels in Sand in an Upflow Column

Distance (feet)	Type of Sand	Flow Rate (feet/day)	Arrival Time‡ (days)
1	Newton†	0.4	15
2	Newton	0.4	70
4	Newton	0.4	114
6	Newton	0.4	231
8	Newton	0.4	417
0.5	Newton	0.8	9
1	Newton	0.8	41
2	Newton	0.8	60
4	Newton	0.8	82
6	Newton	0.8	119

*Adapted from Roebeck (35).
†Newton sand—effective size 0.18 mm, uniformity coefficient 1.8, porosity as placed 43%.
‡Arrival time—based on when at least 2 coliform per ml showed up in a liquid sample, and feed water contained 10,000 to 20,000 coliform per ml.

TABLE 4-5.* Arrival Time in Days for Coliform at Various Levels in Sand in an Upflow Column

Distance (feet)	Type of Sand	Flow Rate (feet/day)	Arrival Time‡ (days)
1	Chillicothe*	0.3	11
2	Chillicothe	0.3	62
4	Chillicothe	0.3	69

*Adapted from Roebeck (35).
†Chillicothe sand—effective size 0.38 mm, uniformity coefficient 1.37, porosity as placed 51%.
‡Arrival time—based on when at least 2 coliform per ml showed up in a liquid sample, and feed water contained 10,000 to 20,000 coliform per ml.

remain alive in feces up to 10 and 8 days, respectively. A somewhat longer survival was found in solid stools than in liquid stools.

Caldwell (40, 41, 42) conducted an extensive series of experiments with pit privies. The first experiment (43) investigated the pollution flow from a pit privy where permeable soils of considerable depth exist below the pit. The soils were fairly well assorted sands, typical of terrace plain deposits. It was found that "a pollution stream describes a path, the resultant of the interacting forces of linear flow with the groundwater toward the discharge stream and variations in densities of the influent flow in comparison to the groundwater." It further emphasized the importance of initial and changing conditions with time. Chemical pollution was traced 325 to 350 ft by odors and pH variations. The *B. coli* stream formed an inner core of less width and marked less extent than the chemical (43). The *B. coli* stream "expanded in width from 1.5 ft in the first 3 months to a maximum of 4 ft for a brief period at 5 ft after 6 months, and then rapidly contracted to less than a foot." Anaerobes of the *Cl. welchii* type probably reached 50 ft in significant concentration.

In the second experiment (40), an impervious stratum closely underlaid the flow. The soils were coarser than in the first experiments. The first 5 ft was 14.0% fine gravel, 34.7% coarse sand, 23.3% medium sand, 19.8% fine sand, and small amounts of very fine sand, silt, and clay. From 5 to 8 ft the proportion of gravel and coarse sand increased. Beyond 8 ft, fine and very fine sand predominated. This privy pit reached the groundwater. It was concluded that there exist no essential differences in the principles of pollution flow from latrines of different types reaching into groundwater motions. The differences are not of kind but of degree, due to variations of groundwater flow and latrine conditions. *B. coli* in significant numbers were carried 80 ft. After 16 months, the *B. coli* stream had regressed to between 40 and 60 ft.

In a third paper on pit privies, Caldwell (41) concluded:

1. Intestinal organisms are confined to the depository except as they are mechanically transported by other agencies, primarily fluid flow. Transportation is essentially downward with gravity flow, if soils below are pervious.
2. Transfer by lateral capillarity is normally insignificant, but from pits subject to seepage from rains or to which considerable volumes of water are added regularly, lateral seepage will occur as percolation is retarded and liquids impounded, the area of saturation being dome-shaped.

The limitation of pollution is due to rapid "defense"—clogging of pore spaces of the soil with fine fecal particles and aggregates of colloids and organisms—which, in addition to the accumulation of sludge, retards flow and hence increases the time for the death rate of fecal organisms to operate (41).

The annual per capita production of fecal solids is about 1.5 cubic feet for adults and about half this amount for children (44). The annual production of urine is about twelve times as great.

4-7. Pollution from Privy Vaults. The principles set forth in the pit privy and septic tank drain fields also apply to pollution from privy vaults. Levy and Kayser (45) reported the longevity of typhoid organisms in feces or a private vault as over five months, including a two-week period in which the feces were used as manure in a garden.

4-8. Movement of Water From Polluted Trenches. Stiles and Crohurst (46) conducted experiments with sewage-polluted trenches intersecting groundwater. In a sand of effective size 0.13 mm, coliform bacteria traveled 65 ft in 27 weeks, while chemicals traveled 115 ft in the same period. In earlier experiments the same authors (6) observed the movement of coliform organisms and uranium from polluted trenches intersecting the groundwater. They found bacteria 232 ft and uranium 450 ft from the trench, with both types of pollution persisting $2\frac{1}{2}$ years. In both their experiments, they found movement in the direction of groundwater flow only, and more extensive travel in wet weather than in dry. They also found a tendency for pollution to stay in the capillary fringe of groundwater when the groundwater table lowered. See Fig. 4-1 for a visual presentation.

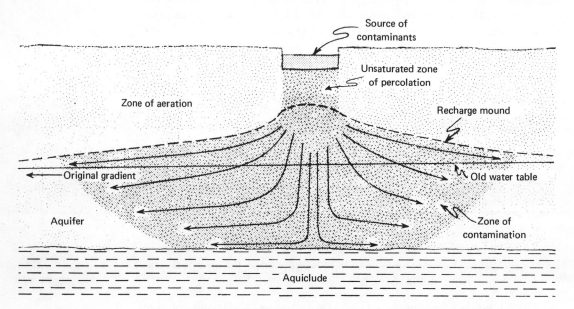

Figure 4-1. Schematic Diagram Showing Percolation of Contaminants Through Zone of Aeration and into Isotropic Aquifers. (Courtesy M. Deutsch, U.S. Geological Survey)

4-9. Bored-Hole Latrine. The bored-hole latrine usually consists of a cylindrical boring 10 to 25 in. in diameter, reaching a depth of 15 to 26 ft in order to penetrate the groundwater (47). This is for the purpose of obtaining septic decomposition of the fecal matter. The deliberate penetration into groundwater in this type of construction is in direct contravention of the accepted hygienic viewpoint that the safe disposal of excreta should preclude contamination of water.

Caldwell (47) accumulated a great mass of experimental data indicating that fecal organisms do not travel significant distances in all directions from a concentrated source, as commonly believed, but are only carried with the groundwater flow. The latrine was essentially a septic tank in immediate contact with the filtration bed (47). The effluent (initially a 5-ft column) containing fecal organisms and particles in suspension and chemical products in solution, issued into all strata subtended and was carried with the groundwater flow toward the drainage channel, forming a pollution stream indistinguishable from the normal groundwater (47). *E. coli* reached 5 ft in 3 days.

As the test continued, an occasional *E. coli* organism would be recovered at a distance of 35 ft. The chemical stream could be traced 85 ft but not 100 ft. The test soil ranged from 34.2 to 51 % coarse sand and 25.8 to 16.4 % medium sand.

4-10. Envelope-Pit Privy. The envelope-pit privy consists of a fine barrier envelope of fine-textured soil, around and under the pit privy, to intercept the outflow from such a latrine (48). The top of the envelope in this experiment consisted of clayey sand, and more than half of the lower section consisted of fine, silty sand. The envelope was approximately 1 ft thick. Caldwell (48) concluded that this type of construction suggests a practical application in safeguarding the groundwater from dangerous contamination in areas of suspected or observed high rates of flow. *B. coli* in this brief experiment were detected at a distance of 2.0 ft, but not at 2.6 ft.

4-11. Septic-Tank Drainage Fields. Septic tanks and cesspools have returned effluents to the soil for centuries. However, the steady growth of population has increased the number of septic-tank installations into the multimillions, and there are many reports of groundwater contamination where septic tanks serve a high population load per square mile. The principles discussed for the pit privy and the bored-hole latrine are applicable to this problem.

Calaway, Carrol, and Long (49) investigated the removal of bacteria in an intermittent sand filter whose effective sand size was 0.31 mm. Substantial reduction was found in coliform counts after settled sewage was passed through 30 inches of sand. The University of California (50) conducted a study in which one of the objectives was to determine the extent and degree to which mineral and organic matter (including bacteria) in the water would penetrate different soils. The most probable number (MPN) was substantially reduced with both final effluent and settled sewage by the time the liquid had percolated 4 ft through the four spreading basins used in the experiment. A paper by Butler, Orlob, and McGaulhey (51) discusses the same study (50) and indicates that the results seem to substantiate the 5-ft distance considered reasonably safe by Ligler (52) and Caldwell (41), as well as the general conclusions that the addition of wastewaters to groundwater may be bacterially safe.

Based on the above facts and unpublished additional work by the author, it is recommended that not less than 5 ft of sand, or finer material, be placed between a septic-tank drain and the point where an effluent can enter underground media of potentially higher transmissivities.

4-12. Infiltration From Waste-Stabilization Ponds. An unlined waste-stabilization pond serves as a large seepage pit with several feet of hydraulic head. It is this hydraulic head that differentiates the pond from other pollution sources discussed. Many ponds are operating with no adverse effect on groundwater supplies; however, serious problems have developed in a number of instances (53). Most reported instances of groundwater contamination have taken place in relatively humid environments east of the Mississippi River, where the depth to groundwater is not great, frequently less than 50 feet, and where unconsolidated materials such as sands, gravels, and clays are the principal porous media through which the fluids must move (54, 55). Water infiltrated through the bottom of a pond will have received the same purification treatment as that discharged from a surface overflow outlet (56). The bacterial count of the effluent is commonly less than one percent of the influent value (53). Most of the bacterial treatment of the effluent takes place in the several inches of material immediately beneath the pond bottom (34), and this is the zone in which the sealing effects are greatest.

4-13. Sewage Irrigation. Sewage effluents are being used for irrigation of parks, golf courses, baseball diamonds, and other recreational facilities. Table 4-6 shows a partial list of locations using sewage effluents. A California survey (57) contains additional locations and provides complete details on a number of sites.

TABLE 4-6. Typical Recreational Use of Sewage Effluents

Activity Using Effluent	*Location*	*Source of Effluent*	*Reference*
Golf course	El Toro Marine Air Base, Calif.	Trickling filter	57
Golf course	Camp Pendleton Marine Base, Calif.	Trickling filter	57
Park lawns	San Francisco Golden Gate Park	Activated sludge	58
Baseball diamonds	Twenty Nine Palms Marine Base	Oxidation pond	57
Wild fowl hunting lake	San Antonio, Texas	Activated sludge	59
Wildlife refuge	Maddock, North Dakota	Raw sewage lagoon	60
Golf course	Santa Fe, New Mexico	Trickling filter	57
Hotel grounds	Flamingo Hotel, Las Vegas	Trickling filter	57
Campus lawns	University of Florida	Trickling filter	61
Lake–boating, fishing	Santee Project, San Diego, Calif.	Activated sludge, additional treatment	62
Golf course	Naval Ordnance Test Station, China Lake, Calif.	Oxidation pond and storage reservoir	63
Golf course	Palo Alto, Calif.	Trickling filter	63

Sewage spray irrigation is a perfectly feasible means of wastewater disposal if certain precautions are taken. The effluent should be well treated, such as the effluent from an oxidation ditch. It should be well chlorinated or otherwise disinfected. Figure 4-2 shows a spray system where treatment plant effluent is being sprayed for irrigation of roadside vegetation along a busy interstate highway. Figure 4-3 indicates one of the sprinkler heads being used in the project shown in Fig. 4-2. Figure 4-3 shows sewage plant effluent being used to irrigate agricultural crops.

The subject of spray irrigation is treated from a design standpoint in

Figure 4-2. Sewage Effluent Used to Irrigate Roadside Vegetation.

Figure 4-3
Close-Up of Sprinkler Head Used
in Fig. 4-2.

Figure 4-4
Agricultural Use of Sewage Effluent
by Spray Irrigation.

Chapter 11. It is necessary to employ good engineering judgment in the layout of the spray field. Climatic conditions must be taken into consideration. The effluent should not run off into drainage ditches, but should be distributed over an adequate area so that it can soak into the ground.

4-14. Wastewater Wells. Unfortunately, a commonly held opinion that microorganisms rarely, if ever, penetrate more than 100 ft through continuous underground formations has assumed almost sacrosanct proportions (64). Obviously, the nature of the underground media through which groundwater and sewage are free to move will determine the extent of bacterial penetrations.

Whether or not a wastewater well is practical depends simply upon what you intend to put into it, the condition of the effluent, and the geological and hydrological conditions that prevail where the well is located. Domestic sewage or stormwater runoff that has been well treated and filtered through either a 30-in. sand filter or a suitable diatomite filter can be safely injected into aquifers that are several hundred feet from streams, limestone formations, or heavy gravel deposits. The latter formations can be used if the physical distance to water supplies is sufficiently great. The engineer who plans to use a wastewater well for chemical or radioactive contaminants, or salt water, must make a very rigorous investigation, which is beyond this presentation. In general, wastewater wells are not recommended and should be used only as a last resort where the alternative solutions are even less desirable.

The practical recharge rate has been found to be about half the safe yield of the recharge well (65). Periodic injection of chlorine, followed by short redevelopment of the recharge well, maintained its ability to receive injected water. According to the authors, the results show that direct recharge of groundwaters is a safe and feasible method of wastewater reclamation and groundwater replenishment.

4-15. *Leaching of Ash Dumps.* An ash dump is not of concern in sanitary engineering as a source of pathogenic bacterial pollution of groundwater. The percolation of natural precipitation or the movement of groundwater through an incinerator ash dump will leach soluble salts and alkalies (66). The rate will be very slow, even when water volumes far in excess of those derived from maximum precipitation or normal groundwater flow are applied to the ash surface.

Haupt (67) reported considerable increase in hardness of the water in wells 1,180 ft from an ash dump in a worked-out sand pit. He concluded that ash heaps should not be allowed in districts used for the collection of water unless it is certain that no infiltration from the surface of the ground can take place.

4-16. *Leaching of Sanitary Landfills.* Anderson and Dornbush (68) stated that a review of the literature indicated that pH, specific conductance, total hardness, alkalinity, and chloride and nitrate measurements would adequately describe variations in the chemical quality of the groundwater. Their preliminary conclusions from measurements conducted on a landfill were that groundwater in the immediate vicinity of, or in direct contact with, a refuse landfill can exhibit a significant increase in the concentration of dissolved minerals as determined by specific conductance measurements. They also indicated that the three most significant parameters are (of those used and in order of importance) chloride, sodium, and specific conductance.

The American Public Works Association (69) indicated that, for pollution of groundwater by refuse leaching, three basic conditions must exist:

1. The site must be over or adjacent to an aquifer.
2. There must be supersaturation within the fill caused by the flow of groundwater into the fill from percolation of precipitation and surface water runoff, by water decomposition, or by an artificial source.
3. Leached fluids must be produced, and the leachate must be capable of entering an aquifer.

In the Riverside Study (70), it was found that a sanitary landfill in intermittent or continuous contact with groundwater causes the groundwater in the immediate vicinity to become grossly polluted and unfit for domestic or irrigation use: "Concentration of mineral elements varying from 20 times those commonly found in unpolluted groundwater—up to 10,000 times in the case of ammonia nitrogen—are possible." The quality of the aquifer in this study was affected after 7 years.

One of the variables in a sanitary landfill is the water retention capacity per foot of refuse bed. One source (72) estimated 4.5 inches of moisture could be absorbed by 600 pounds per cubic yard of compacted dry refuse. A California study (73) indicated a retention of 2.5 inches of water per foot of wet refuse, whereas a British study (74) showed a retention of 1.5 inches of water per foot of wet refuse. Cook (75) summarized the water retention problem by stating that the type and amount of leachate depend on the type of refuse deposited, its compaction, the site, and other environmental conditions, especially rainfall and the ability of water to percolate through the earth covering layer.

4-17. Pollution by Open Dumps. Open dumps are the source of a number of public health and safety problems—disease, air and water pollution, fires, mosquitos, rodents, and insects (69). Whenever refuse is deposited on land, the potential impact on surface waters or aquifers may be significant. This can better be appreciated when one considers that ordinary community refuse may have a five-day BOD of 14,000 to 180,000 parts per million (76).

Open dumps are not recommended in any case. Figure 4-5 shows what can result from dumping chemical waste on the surface of the earth. Figure 4-6 shows how products of air pollution falling back on the surface of the earth act like an open dump of chemicals.

SOURCES AND TRANSMISSION OF POLLUTION

4-18. Farm Wastes. Farm drainage may contain high levels of soluble plant nutrients, such as nitrogen, phosphorus, and potassium, which under favorable

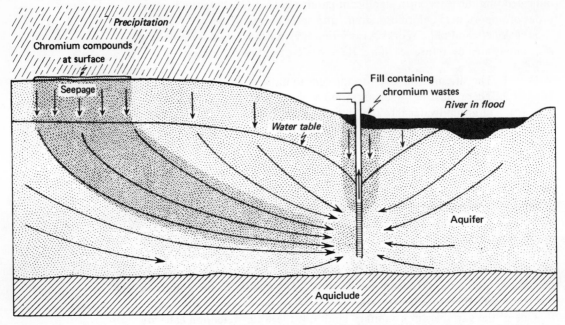

Figure 4-5. Schematic Diagram Showing Leaching of Soluble Contaminants into Aquifer from Precipitation and Flood Waters. (Courtesy M. Deutsch, U.S. Geological Survey)

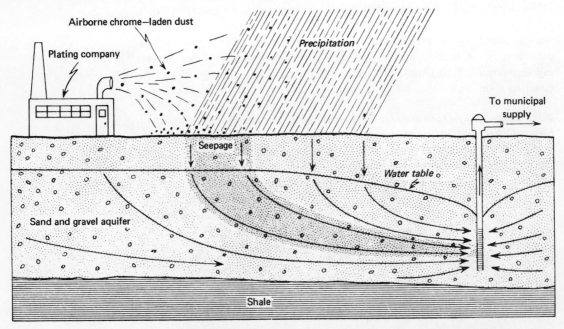

Figure 4-6. Diagram Showing Possible Mode of Entry of Airborne Wastes to Aquifer. (Courtesy M. Deutsch, U.S. Geological Survey)

weather conditions often result in rapid and prolific growth of phytoplankton and zooplankton beyond that necessary for good biological balance (77). Substances serving as a source of plant nutrients tend to promote eutrophication which results in objectionable algal growth. Nitrogen and phosphorus are considered potential water pollutants because of their nutrient or fertilizing capacity.

The ultimate disposal of farm animal manure is almost exclusively on land surfaces. Salter and Schollenberger (78) estimated that one half of the excrement from farm stock in the U.S. is dropped on pastures and uncultivated ground. The remaining half from pens, corrals, and loafing areas may be spread on the land, disposed of by liquid manure systems, or anaerobically composted. Henderson (79) indicates that the Zoonotic disease organisms, for which animals commonly serve as a reservoir of passive or clinical infection, include fifteen human disease entities that can be contracted via the oral route and hence, in theory, can be waterborne. It therefore appears that rural land drainage from watersheds with farm animal population can be important in terms of total oxygen demand on stream resources.

The Holstein cow produces approximately 1.25 cubic feet (10 gallons) of feces and urine per day (80). Fuller (81) found that the average dairy cow produces 86 pounds of waste daily per 1000 pounds of live weight and that the waste production of steers is 41.2 pounds per day per 1000 pounds of live weight. Table 4-7 indicates the percentage of BOD of cattle waste settleable

TABLE 4-7.* Percent BOD of Cattle Waste

	Settleable Solids	Suspended Solids
Dairy bull	12.0	57.8
Dairy cow	27.5	52.8
Beef	41.1	73.2

*Adapted from Witzel et al. (77).

TABLE 4-8.* Percent COD of Cattle Waste

	Settleable Solids	Suspended Solids
Dairy bull.	35.1	71.5
Dairy cow	49.8	79.5
Beef	45.2	76.0

*Adapted from Witzel et al. (77).

solids and suspended solids for three different classifications of cattle. Table 4-8 indicates the percentage of COD of cattle waste settleable solids and suspended solids for the same classifications of cattle.

Public health authorities object to runoff entering a stream if it exceeds 20 parts per million BOD. Wastes carried in runoff from barnyards and feedlots may vary in BOD from 100 to 10,000 parts per million, depending on the dilution and degree of deterioration of wastes (82).

Stegeman (83) found a value of 0.31 lb/day per hen of wet manure production. He selected a total solids content of 0.078 lb/day per hen and the total solids content of 0.078 lb/day per hen; the total settleable solids was found to be 0.046 lb/day per hen. The volatile solids production figure was reported as 0.059 lb/day per hen. He found 0.0144 lb/day per hen for the five-day BOD.

Taiganides (84) has pointed out that, for wet hog manure, the average value obtained by most investigators appeared to be 7.0 lb/day per 100-lb hog. Studies show that 68% of the total solids are present in the settleable solids (83). Raw hog waste contains approximately 0.92 lb/day per 100-lb hog of volatile solids. Also, 0.67 lb/day per 100-lb hog of volatile solids is present in the settleable solids. Table 4-9 shows the relative fecal output in grams per capita per day, the relative BOD per unit of waste, and comparative population equivalent of several domestic farm animals compared with those of man.

TABLE 4-9.* Population Equivalent of the Fecal Production by Animals, in Terms of Biochemical Oxygen Demand (BOD)

Biotype	Fecal (g/cap/day)	Relative BOD Per Unit of Waste (lb)	Population Equivalent
Man	150	1.0	1.0
Horse	16,100	0.105	11.3
Cow	23,600	0.105	16.4
Sheep	1,130	0.325	2.45
Hog	2,700	0.105	1.90
Hen	182	0.115	.14

*Adapted from Wadleigh (82).

Another source (85) indicates on a BOD basis that one dairy cow is equivalent to 20–25 people, one beef animal to 18–20 people, one hog to 2–3 people, and 10–15 chickens to 1 person. Reptiles can also be a source of contamination. Darrasse (86) found a water supply reservoir was contaminated by lizard droppings. Five serotypes of *Salmonella* were found in the tap water.

4-19. Pesticides and Herbicides. Pesticides and herbicides are sources of pollution when they find their way into surface or ground waters. The lethal level of these chemicals to any given species depends upon the size of the creature, the time of concentration (the length of exposure), the temperature, and often other environmental considerations. At least some species have the capability of building a degree of tolerance to toxic chemicals under certain conditions. The subject requires more space than can be given to it in this book. The U.S. Fish and Wildlife Service has annotated lists of investigations that have been conducted on the subject. Additional bibliographies are available from the U.S. Public Health Service.

4-20. Birds as Carriers of Pathogens. Wastes from sea birds and ducks have been implicated in the transmission of disease to man.

4-21. Fish as Carriers of Pathogens. Fish can serve as carriers of pathogens.

4-22. Pollution from Motorboats. Pollution due to motorboats may occur from hydrocarbon fuel constituents, either unaltered or partly oxidized, lead from antiknock fluid, and from discharge of fecal matter from boat toilets.

USEFUL EQUATIONS

4-23. Velocity of Flow in an Aquifer.

$$V = \frac{K_p h}{7.48 pL}$$

where V = velocity of flow in feet/day,
K_p = coefficient of permeability in Meinzer units,
h = difference in elevations between which flow is being measured in feet,
p = porosity expressed as a decimal,
L = distance in ft.

4-24. Soil Permeability at Different Temperatures.

$$\frac{K_p}{K_f} = \frac{u_{60}}{u_f}$$

where K_p = coefficient of permeability in Meinzer units,
K_f = coefficient of porosity at field temperature,
u_{60} = kinematic viscosity of water at 60°F in sq ft/s,
u_f = kinematic viscosity of water at field temperature (°F) in sq ft/s.

4-25. Field Coefficient of Permeability.

$$K_f = \frac{1 \text{ gal of water at field temperature per day}}{(1 \text{ ft} \times 1 \text{ mi})(1 \text{ ft/mi})}$$

4-26. Darcy.

$$1 \text{ darcy} = \frac{(1 \text{ centipoise} \times 1 \text{ cm/s})/1 \text{ cm}}{1 \text{ atm/cm}}$$

REFERENCES

1. NEEL, JOE K., and HOPKINS, G. J., "Experimental Lagooning of Raw Sewage." *Sew. & Ind. Wastes*, 38, 11, 1236 (1959).

2. DAVIES, S., and SANDERSON, W. W., "Fluoride Pollution of Ground Water by Industrial Wastes." *Proc. 10th Ind. Waste Conf.*, Purdue Univ., Ext. Ser. 89, 449 (1956).

3. ROESSLER, B., "The Influencing of Ground Water by Garbage and Refuse Dumps." *Vom Wasser*, 18, 43 (1950–51).

4. NEWELL, I. L., and ALMQUIST, F. O. A., "Contamination of Water Supplies." *Jour. AWWA*, 52, 786 (June, 1960).

5. WARRICK, L. F., and TULLEY, E. J., "Pollution of Abandoned Well Causes Fond DuLac Typhoid Epidemic." *Eng. News Record*, 104, 410 (March 6, 1930).

6. STILES, C. W., et al., *Experimental Bacteria and Chemical Pollution of Wells Via Ground Water, with a Report on the Geology and Ground-Water Hydrology of the Experimental Area at Fort Caswell, N.C.* U. S. Public Health Service, Hyg. Lab., Bull. No. 147, June, 1927.

7. LANG, A., "Pollution of Water Supplies, Especially of Underground Streams, by Chemical Wastes and by Garbage." *Z. Gesundheitstech. u. Stadtehv.* (Ger.), 24, 5, 174 (1932).

8. LANG, A., and BRUNS, H., "On Pollution of Ground Water by Chemicals." *Gas u. Wasser.* (Ger.), 83, 6 (Jan. 6, 1940).

9. "Survey of Ground Water Contamination and Waste Disposal Practices." *Jour. AWWA*, 52, 9, 1211 (1960).

10. BUCHAN, S., and KEY, A., "Pollution of Ground Water in Europe." *World Health Organiz.*, 14, 949 (1956).

11. STEAD, F. M., *A Discussion of Factors Limiting the Bacterial Pollution of Underground Waters by Sewage.* Report of Interior Fact-Finding Committee on Water Pollution to the Assembly of the State of California (1949), p. 138.

12. McCAUHEY, P. H., and KRONE, R. B., *Report on the Investigation of Travel of Pollution.* California State Water Pollution Control Board, Pub. No. 11, 1954.

13. DENNIS, J. M., "Infectious Hepatitis Epidemic in Delhi, India." *Jour. AWWA*, 51, 1288 (1959).

14. LeGRAND, HARRY E., "System for Evaluation of Contamination Potential of Some Waste Disposal Sites." *Jour. AWWA*, 56, 8, 959 (July, 1964).

15. PARKER, HOMER W., "Sanitary Engineering of Wild Land Recreation Areas." Dissertation, West Virginia University. Copyrighted. Published by University Microfilms, Ann Arbor, Michigan.

16. CEDERGREN, HARRY R. *Seepage, Drainage, and Flow Nets.* John Wiley & Sons, Inc., New York, 1967.

17. DAVIS, STANLEY N., and DeWIEST, ROGER J. M., *Hydro Geology.* John Wiley & Sons, Inc., New York, 1966.

18. DeWIEST, ROGER J. M. *Geohydrology.* John Wiley & Sons, Inc., New York, 1967.

19. JUMIKIS, A. R., *Introduction to Soil Mechanics.* D. Van Nostrand Co., Inc., Princeton, N.J., 1967.

20. TERZAGHI, KARL, and PECK, RALPH B., *Soil Mechanics in Engineering Practice*, 2nd ed. John Wiley & Sons, Inc., New York, 1967.

21. TODD, DAVID KEITH, *Ground Water Hydrology.* John Wiley & Sons, Inc., New York, 1967.

22. MEINZER, O. E., and WENZEL, LELAND K., "Movement of Ground Water and its Relation to Head, Permeability, and Storage, Physics of the Earth." Chapter IX, *Hydrology*. McGraw-Hill Book Co., Inc., New York, 1942.

23. MEINZER, O. E. "Movements of Ground Water." *Amer. Assn. Petroleum Geologists Bull.*, 20, 704 (1936).

24. STEARNS, N. D. "Laboratory Tests on Physical Properties of Water-Bearing Materials." *U. S. Geol. Survey Water Supply Paper*, 596, 121 (1927).

25. WEIDENBACH, FRITZ, "Geological Basis for Fixing the Area of the Protective Zones for Water Supplies." *Gas und Wasser.*, 92, 18, 229 (1951).

26. RUBENTSHIK, L., et al., "Adsorption of Bacteria in Salt Lakes." *Jour. of Bact.*, 32, 11 (1936).

27. KRUGER, W., *Zeitsch. f. Hyg.*, 7, 135 (1889). Cited in Reference 26.

28. BURRIAUZ, R., and MOSSELL, D. A., "The Significance of Various Organisms of Fecal Origin in Foods and Drinking Water." *Jour. Appl. Bact.*, 24, 353 (1961).

29. DIANOW, E. W., and WOROSHILOWA, A. A., *Nautschno-agron. Jour.*, 2, 520 (1925).

30. BECHHOLD, H., *Kolloid-Zeitschr.*, 23, 35 (1918).

31. CHUDIAKOW, N. N., *Nautschno-agron. Jour.*, 2, 742 (1925); *Centrbl. f. Bakt., Abt.* 2, 68, 345; *Pedology*, 2, 98 (1926).

32. KARPINSKA, N. S. *Nautschno-agron. Jour.*, 3, 587 (1925).

33. BUTLER, R. G., et al., "Underground Movement of Bacterial and Chemical Pollutants." *Jour. AWWA*, 46, 2, 97 (1954).

34. KRONE, R. B., et al., "Movement of Coliform Bacteria Through Porous Media." *Sew. & Ind. Wastes*, 30, 1, 1 (1958).

35. ROEBECK, G. G. *Ground Water Contamination Studies at the Sanitary Engineering Center*. Proceedings of 1961 Symposium, Ground Water Contamination, Robert A. Taft San. Eng. Center, Technical Report W61-5 (1961).

36. KLIGLER, I. J., *Investigation of Soil Pollution and Relation of Various Types of Privies to Spread of Intestinal Infections*. Monograph 15, Rockefeller Inst. of Medical Research, New York, Oct. 10, 1921.

37. MOM, C. P., and SCHAAFSMA, N. D., "Disposal of Fecal Mater and Pollution of Soil in the Tropics." *Mededeel. Diensl Volksgezondheid Nederland*, Indie, 22, 161 (1933).

38. BEARD, PAUL J. "The Survival of Typhoid in Nature." *Jour. AWWA*, 30, 124 (1938).

39. ALLEN, L. A., at al., "Effect of Treatment at the Sewage Works on the Numbers and Types of Bacteria in Sewage." *Jour. Hygiene*, 47, 303 (1949).

40. CALDWELL, ELFREDA L., "Pollution Flow From Pit Latrines When an Impervious Stratum Closely Underlies the Flow." *Jour. of Infectious Diseases*, 62, 3, 225 (1938).

41. CALDWELL, ELFREDA L., "Studies of Subsoil Pollution in Relation to Possible Contamination of the Ground Water from Human Excreta Deposited in Experimental Latrines." *Jour. of Infectious Diseases*, 62, 3, 272 (1938).

42. CALDWELL, ELFREDA L., "Study of an Envelope Pit Privy." *Jour. of Infectious Diseases*, 62, 3, 264 (1937).

43. CALDWELL, ELFREDA L., "Pollution Flow From a Pit Latrine When Permeable Soils of Considerable Depth Exist Below the Pit." *Jour. of Infectious Diseases*, 62, 3, 225 (1938).

44. "Drinking Water Standards—1946." *Jour. AWWA*, 38, 361 (March, 1946).

45. LEVY and KAYSER, *Cent. f. Bakt.*, 33, 489 (1903).

46. STILES, C. W., and CROHURST, HARRY R., "The Principles Underlying the Movement of Bacillus Coli in Ground-Water, with Resulting Pollution of Wells." *Public Health Reports*, 38, 1350 (June, 1923).

47. CALDWELL, ELFREDA L., and PARR, LELAND W., "Ground Water Pollution and the Bored Hole Latrine." *Jour. of Infectious Disease*, 61, 148 (1937).

48. CALDWELL, ELFREDA L., "Study of an Envelope Pit Privy." *Jour. of Infectious Diseases*, 62, 3, 264 (1937).

49. CALAWAY, WILSON T., et al., "Heterotrophic Bacteria Encountered in Intermittent Sand Filtration of Sewage." *Sew. & Ind. Wastes*, 47, 303 (1949).

50. GREENBERG, A. E., *Field Investigation of Waste Water Reclamation in Relation to Ground Water Pollution*. California State Water Pollution Control Board, Publ. No. 6, 1953.

51. BUTLER, R. G., et al., "Underground Movement of Bacterial and Chemical Pollutants." *Jour. AWWA*, 46, 2, 97 (Feb., 1954).

52. KLIGLER, I. J., *Investigation of Soil Pollution and Relation of Various Types of Privies to Spread of Intestinal Infections*. Monograph 15, Rockefeller Inst. of Medical Research, New York, Oct. 10, 1921.

53. FITZGERALD, G. P., and ROHLICH, GERARD A., "An Evaluation of Stabilization Pond Literature." *Sew. & Ind. Wastes*, 30, 10, 1213 (Oct., 1958).

54. KAY, L. A., "The Construction and Operation of Open-Air Swimming Pools and Bathing Places." *Canadian Jour. Public Health*, 51, 411 (1960).

55. PHELPS, EARLE B., *Public Health Engineering*, John Wiley & Sons, Inc., New York, 1948.

56. MEYER, GERALD., "Geologic and Hydrologic Aspects of Stabilization Ponds." *Jour. WPCF*, 32, 8, 820 (Aug., 1960).

57. *Survey of Direct Utilization of Waste Waters*. California Water Pollution Control Board, Publ. 12, 1955.

58. MARTIN, B., "Sewage Reclamation at Golden Gate Park." *Sew. & Ind. Wastes*, 23, 3, 319 (March, 1951).

59. WELLS, W. N., "Irrigation as a Sewage Re-Use Application." *Public Works*, 92, 8, 146 (Aug., 1961).

60. VAN HEUVELEN, WILLIS, and SVORE, J. H., "Sewage Lagoons in North Dakota." *Sew. & Ind. Wastes*, 26, 6, 771 (June, 1954).

61. KIKER, JOHN E., JR., *Reclamation of Water from Sewage*. Eighth Florida Municipal and Public Health Engineering Conference, March, 1955.

62. Askew, J. B., et al., "Microbiology of Reclaimed Water from Sewage for Recreational Use." *Amer. Jour. Public Health*, 55, 3, 453 (March, 1965).

63. *Report on Continued Study of Waste Water Reclamation and Utilization.* California Water Pollution Control Board, Pub. 15 (1956).

64. Bogan, R. H., *Problems Arising from Ground Water Contamination by Sewage Lagoons at Tieton, Washington.* Proceedings of 1961 Symposium, Ground Water Contamination. Robert A. Taft Sanitary Engineering Center Tech. Report W61-5 (1961).

65. Krone, R. B., et al., "Direct Recharge of Ground Water with Sewage Effluents." Jour. San. Eng. Div., *Proc. ASCE*, 83, No. 4, 1335 (Aug., 1957).

66. *Investigation of Leaching of Ash Dumps.* California State Water Pollution Control Board, 2 (1952).

67. Haupt, H., "Harmful Effect of Ash Deposits on Groundwater." *Gas-u. Wasserfach*, 78, 526 (1935).

68. Anderson, John R., and Dornbush, James N., "Influence of Sanitary Landfill on Ground Water Quality." *Jour. AWWA*, 59, 4, 457 (April, 1967).

69. *Municipal Refuse Disposal.* Prepared by the Amer. Public Works Assn., Public Administration Service, Chicago, 1966.

70. *Report on the Investigation of Leaching of a Sanitary Landfill.* State of California Water Pollution Control Board, Pub. 10, 1954.

71. *Report on Continuation of an Investigation of Leaching from Dumps.* University of Southern California Engineering Center Report 72–3, June, 1960.

72. *County Sanitation Districts of Los Angeles County, Planned Refuse Disposal,* 1955.

73. Greenberg, A. E., *Field Investigation of Waste Water Reclamation in Relation to Ground Water Pollution.* California State Water Pollution Control Board, Publ. No. 6, 1953.

74. *Pollution of Water by Tipped Refuse.* Report of the Technical Committee on the Experimental Disposal of House Refuse in Wet and Dry Pits. Ministry of Housing & Local Gov., H. M. Stationary Office.

75. Cook, H. A., *Microbiological and Chemical Investigation of Seepage from a Sanitary Landfill.* Unpublished Master of Science Thesis, West Virginia University, Morgantown, 1966.

76. Weaver, L., *Refuse Disposal, Its Significance.* Proceedings of 1961 Symposium, Ground Water Contamination, Robert A. Taft San. Eng. Center, Tech. Report W61-5, 1961.

77. Witzel, S. A., et al., *Physical, Chemical and Bacteriological Properties of Farm Wastes (Bovine Animals).* National Symposium on Farm Animal Wastes, Michigan State University, ASAE Publ. No. SP-0366, May 5–7, 1966.

78. Salter, R. M., and Schollenberger, C. J., "Farm Manure." *Yearbook of Agriculture*, U. S. Dept. of Agriculture, Washington, D.C. (1938).

79. Henderson, John M., "Agricultural Land Drainage and Stream Pollution." *Proc. ASCE*, Paper No. 3329, SA6, 61 (Nov., 1962).

80. HART, S. A., et al., "Dairy Manure Sanitation Study." *California Vector News*, 7, 7 (July, 1960).

81. FULLER, JAMES E., *Some Physical and Physiological Activities of Dairy Cows.* New Hampshire Agr. Exp. Sta. Tech. Bull. 35 (June, 1928).

82. WADLEIGH, CECIL H., *Wastes in Relation to Agriculture and Forestry.* U. S. D. A. Miscellaneous Publ. No. 1050, March, 1968.

83. STEGEMAN, GARY L., "Poultry and Hog Waste Characteristics." Unpublished Master's Thesis (Civil Engineering), University of Wisconsin, Madison, 1966.

84. TAIGANIDES, E. P., *Disposal of Animal Wastes.* 19th Ind. Waste Conf., Purdue Univ., 1964.

85. LOEHR, RAYMOND C., "Effluent Quality from Anaerobic Lagoons Treating Feedlot Wastes." *Sew. & Ind. Wastes*, 39, 3, 384 (March, 1967).

86. DARRASSE, H., et al., "Isolation of Several Salmonellas from a Water Distribution System. Origin of the Contamination." *Bull. Soc. Pathol. Exotique* (France), 52, 1, 53 (1959).

87. "Underground Waste Disposal and Control." Report of AWWA Task Group 2450R, *Jour. AWWA*, 49, 1334 (1957).

88. MULLER, J., "Bedeutsame Feststellungen bei Grundwasserverunreinigungen durch Benzin." 93, 205 (1952). Abstract *Jour. AWWA*, 45, 3, 66 P & R (1953).

89. "Survey of Ground Water Disposal Practices." Report of AWWA Task Group 2450R, *Jour. AWWA*, 52, 1211 (1960).

90. FRICKE, K., and KRAUSS-WICHMANN, "Starkere. Grundwasserverunreinigungen durch Benzin bei Wesel," Gesundheitsing. 74, 394 (1953); *Jour. AWWA*, 48, 4, 64, P & R (1956).

91. BAARS, J. K., "Travel of Pollution and Purification Enroute in Sandy Soils." *Bull. World Health Organ.*, 16, 4, 727 (1957).

92. DITTHORN, F., and LAUERSSEN, A. "Experiments on the Passage of Bacteria Through Soil." *Eng. Record*, 60, 642 (Dec., 1909).

93. DAPPERT, A. F., "Tracing Travel and Changes in Composition of Underground Pollution." *W. W. & S.*, 79, 8, 265 (1932).

94. "Well Pollution by Chromates in Douglas, Michigan." *Mich. Water Works News*, 7, 3, 16 (July, 1947).

Sewers 5

This chapter explains the use of lasers for sewer alignment, considers new sewer materials, and notes several other new developments involving sewer design. The subject of sewers, however, is long established and much of the material contained in this chapter has attained the status of unquestioned tradition. There are deficiencies in our knowledge, for example, on routing flow through sewers where stormwater is involved and on flow at sewer junctions. Thus, the policy in this chapter is to present at least one method that has been tried and tested insofar as basic sewer design is concerned, but to supplement this where new technologies have appeared.

Portions of this chapter have been extracted from *Clay Pipe Engineering Manual* by authorization of the National Clay Pipe Institute. If the reader is going to be involved in the design of sewers, it is recommended that he secure a copy of ASCE *Manual of Practice No. 37* (*WPCF Manual of Practice No. 9*), the *Clay Pipe Engineering Manual*, Handbook of *Concrete Culvert Pipe Hydraulics* by Portland Cement Association, and *Design Data* by American Concrete Pipe Association. Although the last three publications are written for the purpose of promoting particular materials, the reader may find the information useful.

GENERAL

5-1. Sanitary Sewers. A sanitary sewer carries only wastewater from the homes, commercial establishments, and industries served by the sewer system.

Ideally, no water from a rainstorm would find its way directly into the sanitary sewer.

5-2. Storm Sewers. A storm sewer is presumed to carry only waters from precipitation on the watershed served by the sewer. A classical assumption has been that this type of water would not be contaminated in a form such as sewage. Actually, any flow from urban communities as a result of a storm produces a runoff that may be as heavily contaminated in terms of BOD, COD, and suspended solids as the municipal wastewater in a sanitary sewer. The only counterbalancing factor is that during a storm the total runoff may be adequate to offset this pollution by dilution. This can be determined only by appropriate investigation.

5-3. Combined Sewers. Many sewer systems, particularly in the older cities, are combined sewer systems through which both sanitary and storm flows pass. Thus, each period of precipitation delivers large flows of water to the wastewater treatment facility. The treatment plant must either be of adequate capacity to handle this combined sewer flow, or part of the flow must be bypassed directly into the receiving stream or a storage reservoir. If the latter is used, the runoff can later be processed through the wastewater treatment plant.

ENGINEERING REPORT

5-4. General. The following comments on engineering reports have been taken directly from *Recommended Standards for Sewage Works* (1) and will be applicable to sewers and treatment plants.

The engineering report assembles basic information; presents design criteria and assumptions; examines alternative projects with preliminary layouts and cost estimates; describes financing methods giving anticipated charges for users; reviews organizational and staffing requirements; offers a conclusion with a proposed project for client consideration; and outlines official actions and procedures to implement the project.

The concept, factual data, and controlling assumptions and considerations for the functional planning of sewerage facilities are presented for each process unit and for the whole system. These data form the continuing technical basis for detail design and preparation of construction plans and specifications.

Architectural, structural, mechanical, and electrical designs are usually excluded. Sketches may be desirable to aid in presentation of a project. Outline specifications of process units, special equipment, etc., are occasionally included.

The content of the report is detailed and is explained in the *Recommended Standards for Sewage Works*. The report involves a detailed engineering investigation. It will indicate such things as the existing area, expansion, annexation, intermunicipal service, ultimate area, drainage basin, portion of drainage basin covered, population growth and anticipated increase during the life of the facility, residential, commercial, and industrial land use, topography, general geology, meteorology, precipitation, runoff, flooding, economic aspects, regulations applicable, existing collection system, existing site, existing process facilities, existing wastewater characteristics, assimilative capacity of receiving stream, and similar information.

DESIGN OF SEWERS

5-5. *Types of Sewers.* In general, the reviewing authority today will approve plans for new systems, extensions, or replacement sewers only as separate sanitary and storm sewers. For sanitary sewers, rainwater from roofs or streets and groundwater from foundation drains are excluded. Where combined sewers are proposed, they must facilitate adequate interception of sewage for treatment (1).

5-6. *Design Period.* In general, sewer systems should be designed for the estimated ultimate tributary population, except when considering parts of the systems that can be readily increased in capacity. Similarly, consideration should be given to the maximum anticipated capacity of institutions, industrial parks, and similar developments.

5-7. *Design Factors.* In determining the required capacities of sanitary sewers, the following factors should be considered: maximum hourly sewage flow, additional maximum sewage or waste flow from industrial plants, groundwater infiltration, topography of the area, location of waste treatment plant, depth of excavation, and pumping requirements.

5-8. *Design Basis.* New sewer systems should be designed on the basis of an average daily per capita flow of sewage of not less than 100 gal/day. This figure is assumed to cover normal infiltration, but an additional allowance should be made where conditions are unfavorable (1).

Generally, the sewers should be designed to carry, when running full, not less than the following daily per capita contributions of sewage, exclusive of sewage or other waste flow from industrial plants: laterals and sub-main sewers—400 gallons; main, trunk, and outfall sewers—250 gallons; and intercepting sewers, in the case of combined sewer systems—the above requirements of trunk sewers with sufficient additional capacity for the necessary increment of stormwater. Normally no interceptor should be designed for less than 350% of the gauged or estimated average dry-weather flow.

When deviations from these per capita rates appear justified, a description of the procedure used for sewer design may be required by the reviewing authorities.

5-9. *Minimum Size.* No sewer should be less than 8 inches in diameter (1).

5-10. Depth. In general, sewers should be sufficiently deep to receive sewage from basements (if used) and to prevent freezing.

5-11. Slope. All sewers should be so designed and constructed to give mean velocities, when flowing full, of not less than 2.0 ft/s, based on Kutter's formula using an *n* value of 0.013. Use of other practical *n* values may be permitted by the plan reviewing agency if deemed justifiable on the basis of research or field data presented (1). Table 5-1 indicates the minimum slopes which should be provided; however, slopes greater than these are desirable.

TABLE 5-1. Minimum Sewer Slope (1)

Sewer Size	Minimum Slope in Feet per 100 Feet
8-inch	0.40
10-inch	0.28
12-inch	0.22
14-inch	0.17
15-inch	0.15
16-inch	0.14
18-inch	0.12
21-inch	0.10
24-inch	0.08
27-inch	0.067
30-inch	0.058
36-inch	0.046

Under special conditions, if detailed justifiable reasons are given, slopes slightly less than those required for the 2.0 ft/s velocity when flowing full may be permitted. Such decreased slopes will only be considered where the depth of flow will be 0.3 of the diameter or greater for design average flow (1). Whenever such decreased slopes are elected, the design engineer must furnish with his report computations of the depths of flow in such pipes at minimum, average, and daily or hourly rates of flow. It must be recognized that decreased slopes may cause additional sewer maintenance expense. Sewers should be laid with uniform slope between manholes, and those with a 20% slope or greater should be anchored securely.

5-12. Alignments. Sewers 24 inches or less should be laid with straight alignment between manholes (1).

5-13. Increasing Size. When a smaller sewer joins a larger one, the invert of the larger sewer should be lowered sufficiently to maintain the same energy gradient. An approximate method for obtaining these results is to place the 0.8 depth point of both sewers at the same elevation.

5-14. High-Velocity Protection. Where velocities greater than 15 ft/s are attained, special provision should be made to protect against displacement by erosion and shock (1).

5-15. *Joints and Infiltration.* The method of making joints and the materials used should be included in the specifications. Sewer joints should be designed to minimize infiltration and to prevent the entrance of roots. Leakage tests should also be specified (1). This may include appropriate water or low-pressure air testing. The leakage outward or inward (exfiltration or infiltration) should not exceed 500 gallons per inch of pipe diameter per mile per day for any section of the system (1). The use of television cameras or other visual methods for inspection prior to placing the sewer in service is recommended.

Figure 5-1 indicates one method of testing by exfiltration of a vitrified clay sewer line. The drop of water in the vertical riser is measured accurately. References (2) through (11) contain additional information on infiltration and exfiltration.

Manhole test

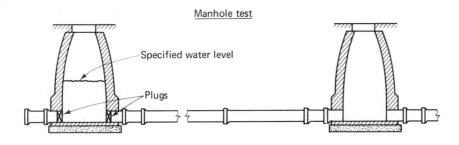

Line test

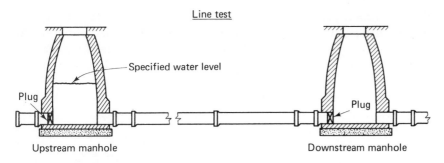

Figure 5-1. Exfiltration Test on Vitrified Clay Sewer Line. (Courtesy National Clay Pipe Institute)

5-16. *Inverted Siphons.* Inverted siphons should have not less than two barrels, with a minimum pipe size of 6 inches, and shall be provided with necessary appurtenances for convenient flushing and maintenance. The manholes should have adequate clearances for rodding, and in general, sufficient head should be provided and pipe sizes selected to secure velocities of at least 3.0 ft/s for average flows. The inlet and outlet details should be arranged so that the normal flow is diverted to one barrel, and so that either barrel may be cut out of service for cleaning.

5-17. Sewer Extensions. In general, sewer extensions should be allowed only if the receiving sewage treatment plant is either capable of adequately processing the added hydraulic and organic load or provision of adequate treatment facilities on a time schedule acceptable to the approving agency is assured.

5-18. Water Supply Interconnections. There should never be any physical connection between a public or private potable water supply system and a sewer which would permit the passage of any sewage or polluted water into the potable supply (1).

5-19. Relation to Waterworks Structures. While no general statement can be made to cover all conditions, it is generally recognized that sewers should meet the requirements of the approving agency with respect to minimum distances from public water supply wells or other water supply sources and structures.

5-20. Relation to Water Mains. Whenever possible, sewers should be laid at least 10 feet, horizontally, from any existing or proposed water main. Should local conditions prevent a lateral separation of 10 feet, a sewer may be laid closer than 10 feet to a water main if it is laid in a separate trench or is laid in the same trench with the water mains located at one side on a bench of undistrubed earth. In either case, the elevation of the crown of the sewer should be at least 18 inches below the invert of the water main.

5-21. Vertical Separation. Whenever sewers must cross under water mains, the top of the sewer should be at least 18 inches below the bottom of the water main. When the elevation of the sewer cannot be buried to meet this requirement, the water main should be relocated or reconstructed with slip-on or mechanical-joint cast-iron pipe, asbestos-cement pressure pipe or prestressed concrete cylinder pipe for a distance of 10 feet on each side of the sewer. One full length of water main should be centered over the sewer so that both joints will be as far from the sewer as possible.

HYDRAULICS OF SEWERS AND DRAINS*

5-22. Flow Characteristics Chart. Figure 5-2 shows graphically the theory and terminology applied to flow in open channels. It is assumed at point D a constant supply of water or sewage is being supplied. Between D and E the slope of the conduit is greater than required to carry the water at its initial velocity and is also greater than the retarding effect of friction; therefore acceleration occurs. At any point between E and F the potential energy of the water equals the loss of head due

*With some modifications the material in this subdivision has been taken from the *Clay Pipe Engineering Manual of* the National Clay Pipe Institute.

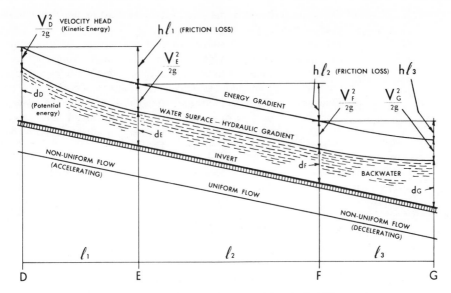

Figure 5-2. Flow Characteristics Chart. (Courtesy National Clay Pipe Institute)

to friction, and the velocity remains constant. Between F and G conditions have occurred which decrease the velocity with an invert slope less than critical, and as such this deceleration causes a backwater curve.

5-23. *The Hydraulic Profile.* In hydraulic design of sewers, three distinct slope lines are commonly referred to:

1. The slope of the invert* of the sewer or drain. This is fixed in location and elevation by construction.
2. The slope of the hydraulic gradient (H.G.). This is sometimes referred to as the hydraulic surface. This is the same, in gravity flow sewers, as the slope of the top surface of the liquid flowing in the sewer.
3. The energy gradient (E.G.). Its elevation is variable, depending upon the elevation of the water surface and the velocity of flow. For uniform flow, the slope of the energy gradient and the slope of the hydraulic surface are parallel to one another and to the slope of the invert, but at different elevations.

The vertical height of the energy gradient above the hydraulic gradient is always equal to the square of the flow velocity in feet per second, divided by 64.4. This quotient is called the velocity head ($v^2/2g$).

5-24. *Important Design Requirements.* In sewer system design it is imperative that the following requirements be met:

*The invert is the bottom point of the wetted cross-section of the sewer.

1. The wetted cross-sectional area of the pipe must be sufficiently large to pass the flow at the designed velocity.
2. The velocity must be sufficiently high to prevent the deposition of entrained solids in the pipe, but not high enough to induce excessive turbulence or excessive thrust at changes of direction. Where slopes are very steep, the computed velocity must be less than critical.
3. Where changes are made in the horizontal direction of the sewer line, in the size of the pipe, or in the rate of discharge, invert elevations must be matched in such a manner that the energy gradient does not change elevation abruptly, or slope upward.

This requirement is important because, wherever the energy gradient slope is opposite to the direction of flow, there will be backwater with consequent impedance of flow.

5-25. Accuracy in Determining Pipe Sizes.

a. Sanitary Sewers. Since the design of sanitary sewers and drains is usually based on estimates of future flow, it follows that computations to determine the required pipe size cannot be made with extreme accuracy.

b. Storm Sewers. On the other hand, with the aid of rainfall formulas for the area, it is possible to forecast future stormwater flow with much greater accuracy than future sanitary flow. For this reason the sizing of storm drains can be more precise than the sizing of sanitary sewers.

Since stormwater drains operate at peak flow for only a very few moments during any year, it is safe practice to design them to run full, provided the flow estimate is sound.

5-26. Velocity and Discharge Formulas.
After flow estimates have been prepared, including all allowances for future increases, and the topography of the system has been determined, the next step is to size each line and fix invert elevations throughout the system. To accomplish this, the designer uses a reliable discharge formula.

5-27. Kutter Flow Diagram for Full Flow.
For convenience, problems of full flow are solved by the use of a flow diagram based upon Kutter's formula. A diagram for this purpose is given in Fig. 5-3. It applies to pipes flowing full with $n = 0.013$. It also provides conversion factors for other values of n.

The Kutter flow diagram can be used directly to solve for V and Q, where D and s are given, or inversely, to solve for D and s when V and Q are given.

5-28. Determination for Part-Full Flow.
After determination of the factors on the Kutter flow diagram for pipes flowing full, v and q are determined for the particular case of part-full flow by reference to the diagram of hydraulic properties of circular sewers in Fig. 5-4. This is a graph of the variation in ratio of part-full flow to full flow (q/Q) and velocity part-full to velocity full (v/V) with the variation of depth of flow to pipe diameter (d/D).

5-29. Manning Formula.
Many designers find it convenient to use the Manning approximation of Kutter's formula. It is more convenient for direct arithmetical solution and for the preparation of diagrams. Its accuracy has been justified in a

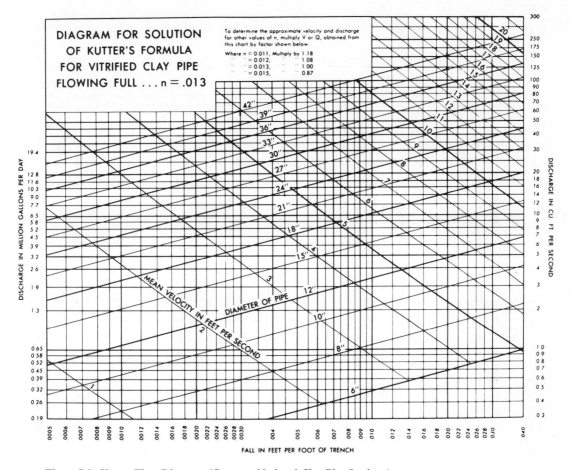

DIAGRAM FOR SOLUTION
OF KUTTER'S FORMULA
FOR VITRIFIED CLAY PIPE
FLOWING FULL ... n = .013

To determine the approximate velocity and discharge
for other values of n, multiply V or Q, obtained from
this chart by factor shown below:

Where n = 0.011, Multiply by 1.18
      " = 0.012,   "   1.08
      " = 0.013,   "   1.00
      " = 0.015,   "   0.87

Figure 5-3. Kutter Flow Diagram. (Courtesy National Clay Pipe Institute)

wide range of pipe-flow problems. The Manning approximation is

$$v = \frac{1.486}{n} r^{2/3} s^{1/2} \qquad (5\text{-}1)$$

The formula

$$q = (a)(v)(0.646) \qquad (5\text{-}2)$$

gives the flow in millions of gallons per day.

The graphical solution of Manning's equation is given in Fig. 5-5 for $n = 0.011$ and in Fig. 5-6 for $n = 0.013$.

In this subdivision of the text, letters have the following meanings.

  s represents the slope of the hydraulic gradient. Though numerically equal, in most cases, to the slope of the invert and the slope of the hydraulic surface, the designer should not confuse these slopes. s is measured by an abstract fraction which is the quotient obtained when the length of any straight portion of the hydraulic gradient is divided

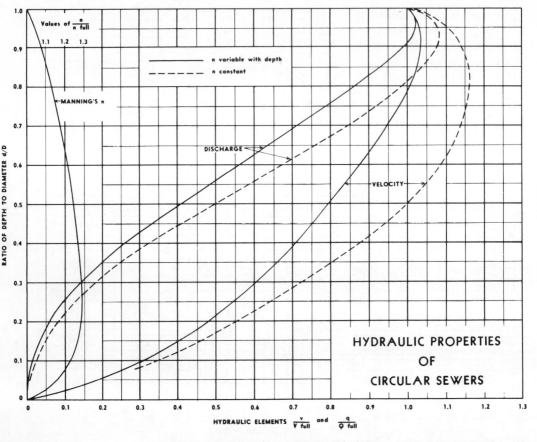

Figure 5-4. Hydraulic Properties of Circular Sewers. (Courtesy National Clay Pipe Institute)

into the difference of elevation between the ends of that section. The length and difference of elevation must be measured in the same units, usually feet.

D represents inside pipe diameter measured in feet.

d represents the depth of flow in feet.

a represents the area of the wetted cross section of the pipe measured in square feet.

A is a special case of a representing that the pipe runs full.

r represents the hydraulic radius of the wetted cross section of the pipe. It is obtained by dividing into a (or A if the pipe runs full) the length of the line of contact between the wetted cross section and the pipe. The length of this line of contact (called wetted perimeter) is also measured in feet.

R is a special case of r indicating that the pipe runs full or half full. R for pipes is always one-fourth the inside diameter and is measured in feet.

n is an empirically derived coefficient which is used as a measure of the interior surface characteristics of a pipe designed for the transmission

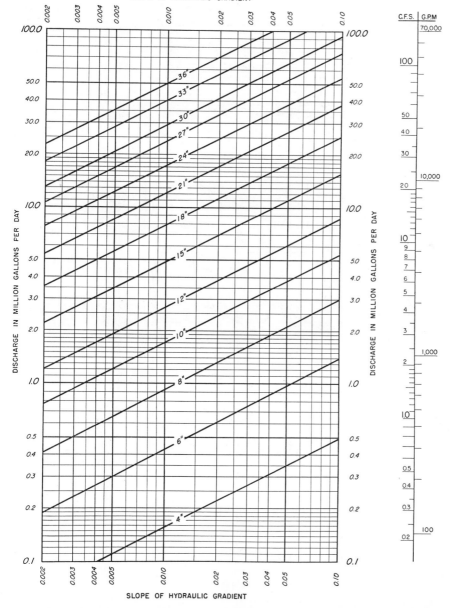

Figure 5-5. Nomograph Solution of Manning's Equation ($n = 0.011$).
(Courtesy National Clay Pipe Institute)

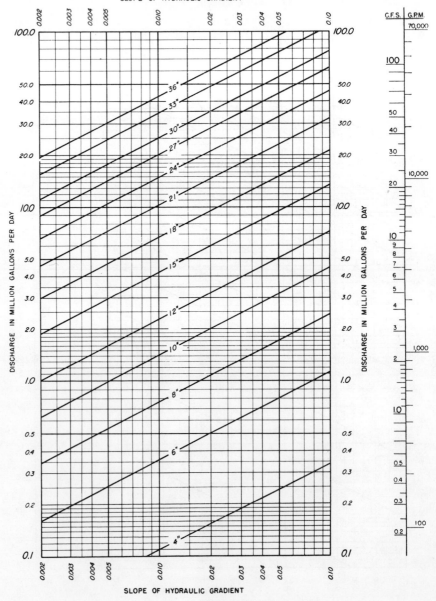

Figure 5-6. Nomograph Solution of Manning's Equation ($n = 0.013$).
(Courtesy National Clay Pipe Institute)

of liquids. This coefficient comes into use in determining the frictional losses in the conduit when transporting a liquid flow. The value of n is affected by pipe size, depth of flow, velocity or slope, and quality of construction. The actual value of n in an installed line may be increased appreciably by debris, grit deposits, and branch connections into the pipe. In controlled experiments, using clean water, values of n under 0.009 have consistently been obtained for vitrified clay pipe. However, because of the variations in n due to variable flow conditions, it is recommended that a conservative value of n be selected. A value of 0.013 is commonly used by experienced sanitary engineers. Recognized authorities point out that numerous tests have definitely established the n factor as the same for all materials commonly used in gravity flow sanitary sewer lines.

v is the velocity of flow (averaged over the wetted cross section) measured in feet per second. For sewers and drains flowing half full or full, v should, if possible, exceed 2 ft/s to prevent settlement of solids in the pipe. On the other hand, v should not exceed 10 ft/s unless special conditions are met.

V is a special case of v indicating that the pipe runs full or half full.

q represents the rate of discharge measured in convenient units. Unless otherwise stated, q should be measured in millions of gallons per day.

Q is a special case of q indicating that the pipe runs full.

Reductions for cases involving flow at less than full depth are made as previously described by use of the hydraulic properties diagram in Fig. 5-4.

5-30. Determination of Pipe Sizes. Before pipe sizes can be determined, it is necessary to know the grade and expected flow of the line. To obtain this information, the following steps are usually taken.

After a layout map of the area has been prepared (Fig. 5-9), working profile sheets are drawn. These show the variation of surface elevation along each trunk and sub-trunk line. Subsurface structures are also shown. A typical profile for sewer and drain design is shown in Fig. 5-7. Beginning at the downstream end of the profile for the main trunk line, the elevation of the energy gradient is noted and marked on the profile at that point.

Where the sewer discharges into a stream, this elevation will be the open water surface. Where discharge is into another sewer, it will be the energy gradient of that sewer.

A line to represent a tentative location for the energy gradient for the first section of sewer being designed is then drawn upstream to the next critical point on the profile. This may be a street intersection, an abrupt change of surface slope, or a sudden change in the estimated flow. In most cases this line should be drawn parallel to the average surface slope. Care must be taken to see that the final design of the sewer provides adequate but not too much cover and that the sewer clears all subsurface obstructions. The tentative energy gradient elevation at the upstream end of this first section of line is then marked on the profile.

A computation sheet is then prepared to include data in the form shown in Fig. 5-8. All available information is entered on this sheet at once. The stations, length of the section, and energy gradients and slope are computed from the profile for the section. Sanitary flow, industrial flow, and infiltration are estimated.

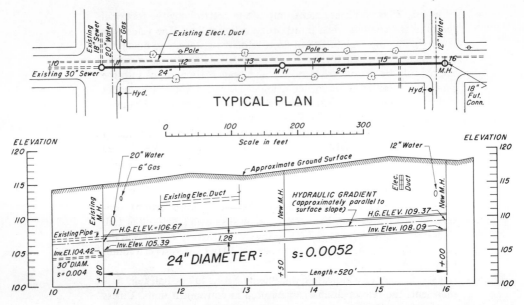

Figure 5-7. Typical Profile. (Courtesy National Clay Pipe Institute)

Street **ANY STREET**	Upstream E.G. Elev.	**109.37 FT.**
Upstream from point **14**	Downstream E.G. Elev.	**106.67 FT.**
From Sta. **10+80** to Sta. **16+00** Length **520 FT.**	Loss	**2.70 FT.**
"n" = **0.013**	E.G. Slope	**0.0052**
Sanitary Sewage **180,000 GPD**	Pipe Size	**24 IN.**
Industrial Wastes **1,500,000 GPD**	% Depth of Flow	**45**
Infiltration **1,200 GPD**	Depth of Flow	**0.90 FT.**
	Velocity	**4.94 FPS**
Total **1,681,200. GPD**	Upstream Invert. Elev.	**108.09**
Future Increases **500,000 GPD**	Downstream Invert. Elev.	**105.39**

Total Fut. Av. Daily Flow **2,181,200 GPD**

Total Fut. Max. Daily Flow **4.36 MGD**

Figure 5-8. Typical Computation Sheet. (Courtesy National Clay Pipe Institute)

a. Method of Computation. Pipes of various sizes are tried successively using the Kutter or Manning flow diagrams until the correct conditions for depth of flow and velocity are fulfilled, based on a proper coefficient *n*. Usually two trials will suffice. Seldom are more than three trials necessary.

In making these computations, the following factors should be watched: (1) v should not be less than 2 or more than 10 ft/s at half or full flow. (2) The depth of flow for sanitary sewers should not be over half the diameter of pipes 24 inches or less in size. Where larger pipes are used, the depth of flow should not be over seventenths of the pipe diameter. Stormwater drains should be designed to run full.

When the pipe size is found which answers all these factors, the size, depth of flow, and velocity are entered on the computation sheet. Velocity head is computed using the formula:

$$h_v = \frac{v^2}{64.4}$$

and is added to flow depth. The total of these two ($d + v^2/64.4$) is subtracted from upstream and downstream elevations of the energy gradient to give invert elevations at each end of the sewer section.

5-31. Typical Computations. As an aid to the reader, the following typical example is worked out to show determination of pipe size by use of the charts and the data shown in Table 5-2 and Figs. 5-3 and 5-8.

TABLE 5-2. Determination of Quantity of Flow

Sanitary sewage:	
Area = 30 acres	
Density = 20 persons per acre	
Per capita flow = 300 GPD	
30 acres @ 20 persons @ 300 GPD	= 180,000 GPD avg
Industrial wastes	1,500,000 GPD
Infiltration allowance	= 1,200 GPD
Total	1,681,200 GPD avg
Future increase (estimated)	500,000 GPD
Total future average daily flow	2,181,200 GPD
Maximum daily future flow	
@ 200%	4,362,400 GPD
	= 4.36 MGD
	(= 4.36 × 1.55 = 6.8 CFS)*
Slope of E.G. (from preliminary profile) = 0.0052	

*1 MGD = 1.55 CFS   0.646 MGD = 1 CFS

a. First Trial to Determine Pipe Size. Assume that the pipe will be 24 inches or less in diameter, in which case it is to be designed to run not more than half full. If it runs exactly half full at 6.8 CFS, then full flow would be 2 × 6.8 = 13.6 CFS.

On Kutter's formula diagram, locate point of intersection of the vertical line to show $s = 0.0052$ and horizontal line to show $Q = 13.6$. The corresponding pipe size is 22.5 inches and the velocity of flow is 5 ft/s. The requirement for velocity between 2 and 10 ft/s is satisfied.

There is no 22.5-inch pipe manufactured. The next larger available pipe size is 24 inches. On a slope of 0.0052 a 24-inch pipe running full will carry 16.1 CFS at a velocity of 5.2 ft/s.

To determine the actual depth of flow and actual velocity for $q = 6.8$ CFS, we find the ratio of q/Q thus: 6.8/16.1 = 0.42. Turning to Fig. 5-4, hydraulic properties of circular sewers, we find for the q/Q ratio a value of 0.42, the d/D ratio is 0.45, and the corresponding v/V ratio is 0.95. Thus the actual depth of flow for 6.8 CFS through a 24-inch pipe on a slope of 0.0052 will be $0.45 \times 24 = 10.8$ inches, which is 1.2 inches less than half full. The actual velocity will be $0.95 \times 5.2 = 4.94$ ft/s.

b. Second Trial (Using Next Smaller Pipe Size). The next available size of pipe smaller than 24-inch is 21-inch. On the specified slope of 0.0052, a 21-inch pipe running full will carry 11.3 CFS (Fig. 5-3). For this size, the q/Q ratio is $6.8/11.3 = 0.60$, and the corresponding d/D ratio is 0.55. The flow depth is $0.55 \times 21 = 11.55$ inches, which is 1 inch more than half full.

By this trial, it is shown that no size smaller than 24-inch satisfies design requirements. For the 24-inch size, the flow depth has been shown to be 10.8 inches or 0.90 feet. The actual velocity is 4.94 ft/s. The velocity head is then $(4.94)^2/64.4 = 0.38$ feet.

Adding depth of flow and velocity head ($0.90 + 0.38 = 1.28$), we find that the invert elevation of the pipe at every point should be 1.28 feet lower than the energy gradient.

This example has been used for completing the typical profile in Fig. 5-7 and the computation form in Fig. 5-8. The use of charts for the Manning approximation of Kutter's formula is identical with the procedure described above.

c. Energy Gradient for Adjacent Sections. When two sewer sections connect without a change of horizontal direction of the flow, the energy gradient elevation at the downstream end of the second section is the upstream energy gradient elevation of the first (or downstream) section.

There are exceptions to this rule in the cases of angle manholes, angle sewer intersections, and drop manholes. If the flow changes in horizontal direction when discharging into a downstream section of sewer, allowance should be made for a drop in E. G. elevation across the turn to compensate for the additional friction.

For a 90° turn, raise the energy gradient elevation of the entering sewer by an amount equal to the velocity head $v^2/2g$ for the stream section. For lesser turns, raise the elevation a proportionate amount.

d. Work One Section at a Time. In this way each section of sewer is successively sized on a computation sheet in the same manner as outlined here for the first section. Each successive section is completed before beginning the next.

Invert elevations do not necessarily match at connecting manholes. There is no requirement that they should be at the same elevation. It is only required that the slope of the invert in the manhole be downhill in the direction of flow.

The profile can be completed most conveniently after a computation sheet showing computed invert elevations has been prepared for each section. Clearance between existing subsurface structures and the bottoms or tops of pipes in the new line should be shown. The energy gradient should also be plotted as a working line. It should be examined to see that it slopes downhill continuously, and checked to be sure that the energy gradient elevations of all sewers coming into the line being designed are not lower than its energy gradient at the point of connection.

5-32. Layout of System. The design of a sanitary sewer includes the layout of a total system or part of a sewer system. Many factors should be included in the development of the layout:

    1. Of principal consideration is the determination of the total drainage district that will contribute flows into the system.

2. The location of an outlet is usually determined by conditions that will allow all flows from the drainage area to convene at a common outlet for treatment or discharge into an intercepting sewer.
3. Location of trunk lines.
4. Location of lateral lines.
5. Tributary areas outlined.
6. Acreage of tributary area.
7. System of designation.

5-33. Include Entire Drainage System.   The initial step in the layout is preparing a map which includes the entire drainage system in which the proposed project lies. If the present project does not include total drainage area, the general layout map must include plans for future extensions which would include total drainage area. The scale of this map should not be smaller than 300 feet to one inch, preferably larger. The entire drainage area referred to here designates the total territory within which it is possible to find a continuously downhill surface route from any point to the established outlet or discharge point of the proposed line. This is illustrated by the typical layout map shown in Fig. 5-9. On this layout map are drawn:

1. Contour lines at intervals small enough to permit estimation of surface elevations to the nearest foot.
2. Street locations.
3. All future sewer lines which may be needed within the drainage area.

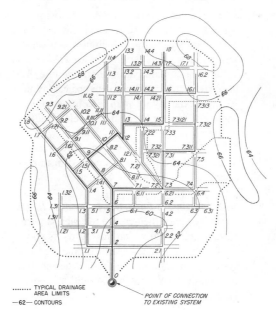

Figure 5-9
Typical Layout Map for Sewer and Drain Design. (Courtesy National Clay Pipe Institute)

.... TYPICAL DRAINAGE AREA LIMITS
—62— CONTOURS

POINT OF CONNECTION TO EXISTING SYSTEM

5-34. Location of Trunk Lines. Trunk and main sewer lines are located in the valleys. The location of each tentative line is confined to a street or available right-of-way. Beginning at the outlet (usually a point of lowest elevation) for the entire drainage area under consideration, tentative sewer routes are tried on the layout map. Each line is evaluated by progressing uphill along the route of minimum surface slope, and without reference to possible branch line locations.

To satisfy all design criteria, the line must reach the edge of the drainage area (ridge line) where future extension is impossible without working downhill. The first line which meets these criteria will indicate the tentative location of the trunk line for the entire drainage area. This tentative route is then examined to see if it is not possible to shorten its length by relocating some portions of the line. This may necessitate moving the line out of the valley location to an adjacent position. In relocating the line, avoid excessive "grade bucking," (opposition of sewer slope to ground slope), which will result in more costly excavations.

5-35. Location of Lateral Lines. Lateral sewers are located in the same manner as described for trunk lines. Beginning at the point of connection to previously established trunk or sub-trunk line, a new line, representing the lateral sewer, is drawn uphill along the route of minimum slope. It is confined to streets or rights-of-way and is extended until it reaches the end of the upgrade or the limit of the drainage area. This process is continued until tentative lateral lines are located on all streets or rights-of-way within the entire drainage area. These lateral lines will resemble somewhat the arrangement of a tree with its trunk and branches.

The final test of the layout is made by checking each route, working downstream, to see that the flow follows the steepest, immediately available downhill surface grade from every point to the outlet of the entire drainage area. This condition will follow automatically if the lines in the original layout were located in an upstream direction along the routes of minimum available surface grades.

5-36. Tributary Areas Outlines. Division lines are drawn on the map outlining the surface area tributary to each sewer location. Each division line will completely enclose an area within which flow is toward the point where its boundary line intersects the sewer line serving the area. No such area will contain more than one sewer. If the area contains more than one line, it is an indication that additional division lines are required to complete project layout map.

When the layout map is completed, every section of the drainage area will be enclosed in one of the subdistricts outlining local tributary areas.

5-37. Review for Design Economy. A critical review of each line should now be made to insure that each line is of a minimum length to cover the entire drainage area.

5-38. Acreage of Each Area is Measured. After the map is subdivided into individual tributary areas, the acreage for each is measured and noted. This provides the total surface area from which flow finds its way to the outlet point in each sub-area.

For convenience, it is recommended that in layout work the following decimal system of numbering be used to designate these outlet points. This system is used for layout work only and should not be used as a numbering system for the final plans.

5-39. A Convenient System for Designation. Starting at the outlet of the main trunk line, number that point "0" (see Fig. 5-9). Then working up the main trunk line only, designate each point where a branch line connects with the main line by a whole number—1, 2, 3, etc., consecutively. When this is completed, each branch line is treated in the same way, using a decimal after the whole number to designate connecting lines. For example, take point 6 on the map, Fig. 5-9. Point 6.1 represents the first connection on the number 6 branch; point 6.2 is the second connection; etc. Branches to branches are numbered by a second numeral behind the decimal, such as 6.11, 6.21 and so forth. It is standard practice to use a number to designate the end of each line.

In this way, the flow of the system is easily followed. On the typical layout map (Fig. 5-9), the main trunk line route is 18, 17, 16, 15, and so on to point 0, the point of outlet. The longest simple branch line is identified as 1.8, 1.7, 1.6, 1.5, and so on to point 1 where it connects to the trunk line. The most complex branch line illustrated is 7.3121, 7.312, 7.311, 7.31, 7.3, 7.2, 7.1, to point 7 where it connects to the trunk line.

Any section of the system may be identified quickly without the use of street names, stations, or reference to landmarks. The one-block long sewer line 9.2 to 9.1 shows by its identifying numbers that it is along the line of a principal branch, its downstream end one section removed from the location of the main line.

Next, a table is prepared listing all such reference points and corresponding total tributary areas. If the reference numbers are arranged in descending numerical order, accumulation of the acreage of subtributary areas will be convenient, and the completed sheet will be of great help in estimating future flow for both sanitary and storm design. A partial sheet of this type would be as shown in Table 5-3. In this table each column represents a continuous sewer length. The points designated by

TABLE 5-3. Total Tributary Area in Acres

Point	Area	Point	Area	Point	Area	Point	Area
17	27.3	7.4	38.2	7.3121	16.0	7.21	7.0
16	57.9	7.3	44.9	7.312	36.5	7.20	13.9
15	60.1	7.2	44.9	7.311	60.0		
14	64.3	7.1	190.2	7.31	81.4		
13	99.0	7.0	193.6	7.30	112.1		
12	114.1						
11	115.3						
10	175.4						
9	201.5						
8	235.7						
7	257.9						
6	432.3						

a whole number represent where the flow ultimately reaches the trunk line, and the points designated by a decimal represent where the flow reaches a branch line. This arrangement also indicates at a glance the area for which each sewer section should be designed. For instance, the sewer upstream from 7.1 and 7.2 should be large enough to serve 190.2 acres. This acreage is the total of the acreages which drain to each point in the system upstream from point 7.1. The sewer between 7.3 and 7.31 should serve 112.1 acres.

5-40. Rational Method of Design for Municipal Stormwater System. Methods for estimating peak rates of rainfall and stormwater runoff have been improved greatly during the past quarter century. A complete description of these advances would fill a good-sized textbook. The *rational method* of design described here is the foundation for more advanced concepts and methods required in the solution of critical stormwater flow problems. It is good engineering practice and is widely used in stormwater drainage design. It gives safe results well within the accepted limits of engineering precision and economy.

5-41. Rainfall Rates Are Approximate. Rainfall rates are usually expressed in terms of inches per hour and are available from the U. S. Weather Bureau. The rate of rainfall is affected by many factors which are not subject to numerical measurement. All "laws" of rainfall are, therefore, empirical approximations based on previous records statistically interrelated.

Rainfall behavior varies with location and topography. The maximum intensity of fall during individual storms is greater for short storms than for long. This is shown in the general relation

$$i = \frac{X}{t + Y}$$

where i is rate of fall in inches per hour, t is total duration of the storm in minutes, and X and Y are constants determined from previous records. As stated before, X and Y vary with location and topography, and they also vary according to the relative frequency of storms. For example, many Atlantic seaboard cities have stormwater drainage systems based on $X = 120$ and $Y = 20$.

A well-designed drainage system provides maximum protection against flood damage at minimum cost. In areas where some protection must be provided but where the effect of flooding is not too damaging, the drainage system may be designed to take care of ordinary storms but be insufficient to handle infrequent but more intense ones. On the other hand, in valuable city areas the system must be adequate to handle even the more severe storms.

It is known that the more severe a storm is, the less frequently it occurs. If a sufficiently long record of rainfall measurements in a particular location is analyzed (by grouping storms of equal duration according to frequency of occurrence), it will be found that maximum rates of rainfall have a definite relation to the average time between storms of equal intensity. Specific values of X and Y apply to storms of a definite frequency.

5-42. Selection of Rainfall Frequency. As far as can be determined, there is no absolute limit to the intensity of rainfall which may occur, but obviously it is uneconomical to design a drainage system to handle extreme storms which may occur on an average of only once in a hundred years. Drainage systems are designed to handle storms having frequencies of from one-per-year average to one-per-five (or ten) years average, depending on the value of the properties in the drainage area and on the nuisance value of flooding. Design for two-year frequency is common in most cities for all areas but the most expensively developed ones. Designers rarely use rainfall frequencies greater than once in ten years.

For purposes of design, the rainfall frequency is adopted first, and then previous records of rainfall are analyzed to determine correct values for X and Y. In the absence of adequate rainfall records, the results of similar analyses made by others for localities having similar rainfall conditions may be used. (See Fig. 5-10.)

Generally, it will be found that X will be nearly 120 and Y 20, for two-year frequencies and that for less frequent storms X and Y will be such that i increases on the order of 50%.

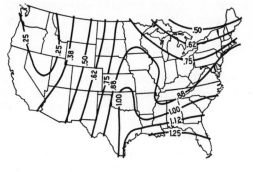 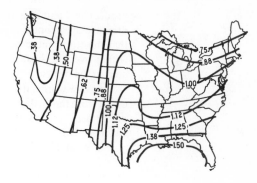

Fifteen-minute rainfall, in inches, to be expected once in 2 years.   *Fifteen-minute rainfall, in inches, to be expected once in 5 years.*

From *Rainfall Intensity-Frequency Data*, Miscellaneous Publications No. 204,
U. S. Department of Agriculture.

Figure 5-10. Rainfall Intensity-Frequency.

The relation $i = X/(t + Y)$ is hyperbolic, and where X is about 120 and Y about 20 (as is usually the case) and when t is very small, it gives an infinitely high value for i. In practice, there is an upper limit to the value of i which usually corresponds to a value of t ranging from 5 to 15 minutes, depending on the location of the installation. In the design of stormwater sewers, t may be safely limited to 15 minutes minimum. Storms of lesser duration and with correspondingly higher fall-intensities are less frequent than storms of longer duration and show the same values for X and Y in the above relation. In other words, storms of less than 15 minutes' duration are not considered in standard practice.

5-43. Determining Surface Runoff. When rain falls upon any surface, part of it flows over the surface to an outlet, part of it evaporates, part may be absorbed by the surface covering, part may penetrate through the surface covering, and part remains on the surface in pools and ponds. Unless the surface was previously covered with snow or ice, the runoff is less than the total rainfall, and the rate of runoff is correspondingly less than the rate of fall.

The percent of the rainfall which reaches the outlet varies widely with different surfaces. Flat turfed areas rarely deliver more than 30% while paved surfaces and roofs deliver about 90%. This percentage is called the runoff coefficient C. The amount of rainfall on one acre of surface at a rate of one inch per hour, if delivered to the outlet at $C = 100\%$ (with no loss for ponding, absorption, evaporation, etc.), results in a flow at the outlet close to 1 cubic foot per second.

$$\frac{43560 \times 1}{60 \times 60 \times 12} = 1.008$$

Hence the flow at the outlet of any drainage area, when delivering surface drainage water so that no loss occurs, may be written

$$Q = CiA$$

where Q is runoff in cubic feet per second, C is the runoff coefficient, i (as before) is the intensity of rainfall in inches per hour, and A is acres in the drainage area from which runoff takes place.

5-44. Time of Concentration. In the above equation, Q is maximum when the product of i and A is maximum. Except in rare cases only, the total acreage in a drainage area contributes runoff simultaneously when Q is maximum. This means that the storm must be of sufficient duration to allow time for the travel of water (either over the surface or in artificial channels) from every point in the drainage area to the outlet.

In other words, the storm must be of sufficient duration for water to flow to the outlet from the farthermost point in the area. This time depends on the size and shape of the area and on the hydraulic characteristics of the surface and the artificial channels. It is called the *time of concentration, t*, and is measured in minutes.

Since rainfall intensity i is higher for lower values of t, it follows that *the maximum runoff will occur when t in the relation $i = X/(t + Y)$ is exactly equal to the time of concentration*. Storms of longer duration will give lower intensities, and shorter storms will not permit the total drainage area to contribute runoff to the outlet. This reasoning is the essence of the *rational method*.

5-45. Application of Rational Method. Application of the rational method of design for storm sewers comprises 13 steps which must be taken in proper order:

1. Appraisal of the location of the drainage area to determine the proper storm frequency.
2. Determination of X and Y from previous rainfall records.
3. Preparation of tentative layout of the proposed drainage system.*
4. Numbering the control point of the system.
5. Determination of the total drainage area A in acres tributary to each control point.
6. Determination of the average runoff coefficient C for the total area tributary to each point.
7. Determination of the time of concentration. Starting at the upstream end of the proposed drainage system (the point farthest from the outlet of the whole drainage area), determine the time of travel of surface water from the farthest point in the tributary drainage area to the upstream end of the drain. If it is less than 15 minutes, use $t = 15$ in the following steps 8 and 9, but retain actual value of t for use in step 12.
8. Determine i from $i = X/(t + Y)$.
9. Determine Q from $Q = CiA$.
10. Determine the necessary size of the drain to the next downstream controlling point.
11. Determine the time of travel through the drain.
12. Add this increment of time of travel to the *actual* time determined in step 7.
13. Repeat all these steps successively for each control point in the drainage area.

In the above process of determining t by successive addition, record totals for each control point accurately to the nearest tenth of a minute. Notice that setting a minimum of 15 minutes for t in step 7 is applicable only when determining i. When

---

*Steps 3, 4 and 5 are the same as those described in detail for the design of sanitary sewers.

determining runoff, be sure that *A* includes *all* the area tributary to the control point and that *C* is the *average* runoff coefficient for the entire tributary area. In the design of a comprehensive storm drain system, drains must have adequate capacity to transmit peak flow.

APPURTENANCES

5-46. Manholes. Manholes should be installed at the end of each line; at all changes in grade, size, or alignment; at all intersections; and at distances not greater than 400 ft for sewers 15 in. or less in diameter, and 500 ft for sewers 18 in. to 30 in., except that distances up to 600 ft may be approved in cases where adequate modern cleaning equipment for such spacing is provided. Greater spacing may be permitted in larger sewers and in those carrying a settled effluent (1). Lampholes may be used only for special conditions and shall not be substituted for manholes nor installed at the end of laterals greater than 150 ft in length.

A drop pipe should be provided for a sewer entering a manhole at an elevation of 24 in. or more above the manhole invert. Where the difference in elevation between the incoming sewer and the manhole invert is less than 24 in., the invert should be filleted to prevent deposition of solids.

The minimum diameter of manholes should be 42 in.; larger diameters are preferable (1). Flap gates are desirable in manholes at the upstream end of laterals which are at minimum grades and which are not to be extended at an early date. The flow channel through manholes should be made to conform in shape and slope to that of the sewers.

Figure 5-11 illustrates a typical line manhole, Fig. 5-12 shows typical drop manhole details, and Fig. 5-13 provides typical manhole details.

TYPICAL LINE MANHOLE

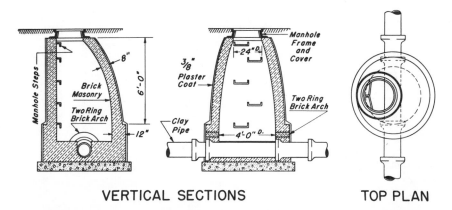

VERTICAL SECTIONS   **TOP PLAN**

Figure 5-11. Typical Line Manhole. (Courtesy National Clay Pipe Institute)

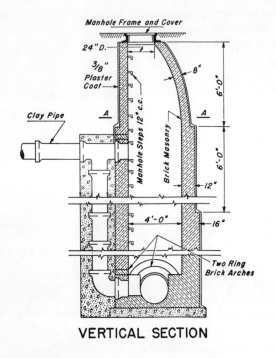

VERTICAL SECTION

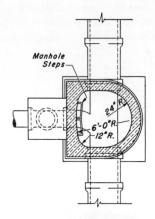

HORIZONTAL SECTION A-A

Figure 5-12
Typical Drop Manhole Details.
(Courtesy National Clay Pipe Institute)

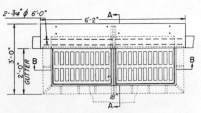

PLAN

NOTE;
   PIPE MAY ENTER INLET AT ANY ANGLE OR GRADE TO SUIT LOCAL CONDITIONS.
   IF INLETS ARE MADE OF BRICK, MAKE WALL THICKNESS LENGTH OF ONE BRICK OR 8 INCHES CONCRETE FOR WALLS, CLASS A-1-2-3-4 MIX MORTAR 1 TO 1½

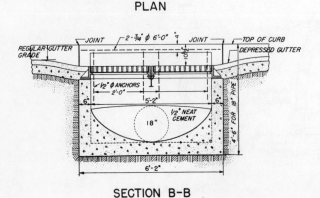

SECTION B-B

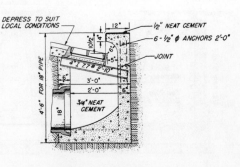

SECTION A-A

Figure 5-14. Typical Drop Inlet. (Courtesy National Clay Pipe Institute)

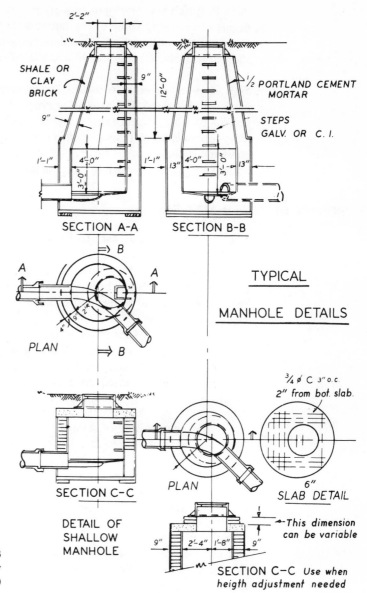

Figure 5-13
Typical Manhole Details. (Courtesy
National Clay Pipe Institute)

5-47. Manhole Covers. Watertight manhole covers should be used wherever
the manhole tops may be flooded by street runoff or high water. Manholes
of brick or segmented block should be waterproofed on the exterior with
plaster coatings, supplemented by a bituminous waterproof coating where
groundwater conditions are unfavorable.

5-48. Drop Inlet. Figure 5-14 illustrates the construction of a typical drop inlet. Research and development work is in progress seeking better inlets and means of removing some of the undesirable features of the existing grates. But until this information is available in a usable form, the configuration shown probably will continue to find use on residential streets at least.

5-49. Appurtenances in General. Appurtenances in general are treated in more detail in the ASCE *Manual of Practice No.* 37 (*WPCF Manual of Practice No.* 9). This reference shows typical manholes for small sewers fabricated from concrete, several types of curb inlets, how to calculate siphons, how to construct sanitary sewage diversion and stormwater overflow devices, flap or backwater gates, and various other details not contained in this book.

CORROSION OF SEWER MATERIALS

5-50. Hydrogen Sulfide and Acid Formation. Some of the materials used in sewer construction are subject to corrosion from both liquid and gaseous materials present in the sewers. The attack by sewage materials or their by-products on wetwells will be discussed in Chapter 12. The discussion here will be confined solely to sewers and is based on a presentation in the *Clay Pipe Engineering Manual*.

The protection of the sewer system from the ravages of sewer gas attack is of fundamental importance in designing and providing permanent trouble-free lines. Failure to fully and properly evaluate any of the contributing factors may lead to subsequent failure of the sewer line. Some factors contributing to sulfide generation and evolution are:

1. Temperature of sewage.
2. Strength of sewage.
3. Velocity of flow.
4. Age of sewage.
5. pH of sewage.
6. Sulfate concentration.

Sulfides are generated in the slime layer which forms between the sewer pipe and the flowing sewage. This action takes place by the bacterial conversion of sulfates to sulfides. The sulfides form hydrogen sulfide gas which first diffuses into the sewage and then escapes into the sewer atmosphere.

Once the gas is created within the line and released to the atmosphere above the sewage, it comes in contact with the moist surfaces in the upper part of the pipe and is oxidized very rapidly, by the action of bacteria, into dilute sulfuric acid. The sulfuric acid collects on the exposed arch of the pipe and sets up a chemical attack unless the pipe material is chemically inert and hence invulnerable to corrosive acid action. When the arch becomes too thin to support the earth load above, it caves in, causing the sewer to become inoperative. The process is shown in Fig. 5-15.

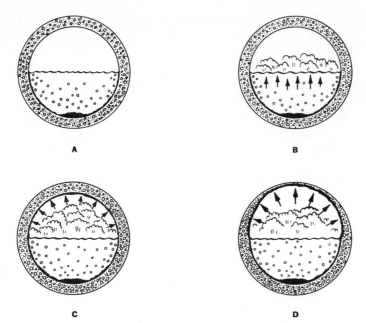

A. Bacteria in the slime under flowing sewage converts sulfates to sulfides.

C. Hydrogen sulfide gas in atmosphere makes contact with slime in arch of pipe which contains more bacteria. Bacterial action converts gas to sulfuric acid.

B. Sulfides in the liquid make their way to surface and are released into the sewer atmosphere as hydrogen sulfide gas.

D. If pipe is of corrodible material, sulfuric acid attacks it causing ultimate failure. Vitrified clay pipe is chemically inert and hence invulnerable to acid.

Figure 5-15. Stages of Sewer Corrosion. (Courtesy National Clay Pipe Institute)

Most of the harmful consequences of sulfides in sewage are due to hydrogen sulfide gas. In solution this is in equilibrium with its partly ionized form HS. The two together comprise what is called *dissolved sulfide* in analytical reports. The proportion of dissolved sulfide existing as hydrogen sulfide varies with pH. At pH = 6, the hydrogen sulfide concentration is 83% of the dissolved sulfide; at pH = 7, the proportion is 33%; and at pH = 8, the proportion is 5%.

Actual field investigations of hydrogen sulfide and acid formation in sewers reveal the crown moisture to have a pH of 2, even though the pH of the sewage was close to neutral (pH = 7).

The factor which determines whether sulfide buildup occurs in a stream of sewage is whether or not oxygen is absorbed at the surface of the stream fast enough to oxidize the hydrogen sulfide diffusing out of the slimes. The oxygen need varies from one sewage to another. Oxygen absorption depends principally upon velocity of flow. Thus a good velocity of flow may help prevent sulfide buildup.

On the other hand, velocity may be damaging because, if any hydrogen sulfide is present in a stream of sewage, its rate of release increases with increased rate of flow. Turbulence due to junctions, changes of pipe size, drops, etc., will cause a relatively rapid release of hydrogen sulfide gas if any is present in the liquid.

Force mains are a major source of sulfide problems in sewers, particularly if the sewage is retained for any appreciable length of time. High sulfide concentrations will not damage the interior of the filled pipe, but downstream sewers may have odor nuisances and damage to structures, particularly if the sewage empties out of the force main with much turbulence.

Sewage temperature is another contributing factor in the rate of development of sewer gas. The tremendous increase in the use of household appliances such as dishwashers and washing machines has resulted in large quantities of hot water being discharged into the sewer system. Average sewer temperatures today are considerably higher than they were only a few years ago. Consideration should be given to the fact that, for every 10°C of increase in sewer temperature over 20°C, there is a 100% increase in the effective BOD. This indicates why it is difficult to control hydrogen sulfide generation in sewers.

STRUCTURAL

5-51. Marston Formula. One of the more important structural considerations of sewer design is that of trench width which is illustrated in Fig. 5-16. The material that follows was extracted from the *Clay Pipe Engineering Manual*. The author has substantially condensed the original presentation.

The Marston formula applies to the calculation of loads on pipes. Actual tests have been made with many kinds of soils to determine their weights and frictional characteristics and to determine the relative settlement of each. These measurable quantities have been combined into a single expression to produce for each case an exact computation of the total weight supported by the pipe.

The factors taken into consideration in the following Marston formula are:

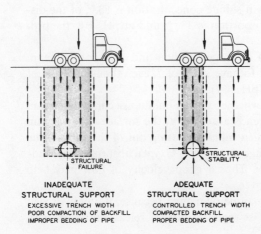

INADEQUATE
STRUCTURAL SUPPORT
EXCESSIVE TRENCH WIDTH
POOR COMPACTION OF BACKFILL
IMPROPER BEDDING OF PIPE

ADEQUATE
STRUCTURAL SUPPORT
CONTROLLED TRENCH WIDTH
COMPACTED BACKFILL
PROPER BEDDING OF PIPE

Figure 5-16
Trench Width. (Courtesy National Clay Pipe Institute)

1. Depth of backfill cover over the top of the pipe.
2. Width of trench measured at the level of the top of the pipe.
3. Unit weight of backfill.
4. Values for frictional characteristics of the backfill material.

By substitution of available data in the Marston formula for pipe in trenches

$$W_c = C_d w B_d^2$$

where W_c = the vertical external load on a closed conduit due to fill materials, in pounds per foot of length,

C_d = load calculation coefficient for conduits completely buried in ditches, abstract number. (See computation diagram in Fig. 5-17.),

w = the unit weight of fill materials, in pounds per cubic foot, and

B_d = horizontal breadth of ditch at top of conduit, in feet,

a direct result is obtained for the load on the pipe in terms of pounds per linear foot. The computation of loads is made very easy by this formula and the computation diagram in Fig. 5-17.

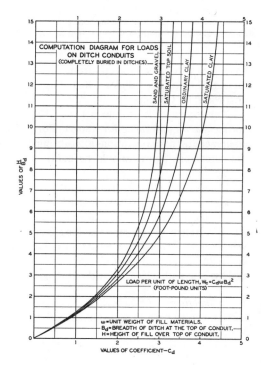

Figure 5-17
Computation Diagram for Loads
on Ditch Conduits (Completely
Buried in Ditches). (Courtesy
National Clay Pipe Institute)

5-52. Live Loads. When live loads are encountered, they are added to the predetermined dead load. However, only a fraction of any surface load is transmitted to the pipe.

The percentages of surface wheel loads which reach the pipe, according to outside diameters of pipe and trench depths, are given in Table 5-4. These percentages

TABLE 5-4. Percentage of Wheel Loads Transmitted to Underground Pipes*

Tabulated figures show percentage of wheel load applied to one linear foot of pipe.

Depth of Backfill Over Top of Pipe in feet	Pipe Size Inches	6	8	10	12	15	18	21	24	27	30	33	36	39	42
	Outside Diam. of Pipe in Feet (Approx.)	0.64	0.81	1.0	1.2	1.5	1.8	2.1	2.4	2.7	3.0	3.3	3.5	3.9	4.2
1		12.8	15.0	17.3	20.0	22.6	24.8	26.4	27.2	28.0	28.6	29.0	29.4	29.8	29.9
2		5.7	7.0	8.3	9.6	11.5	13.2	15.0	15.6	16.8	17.8	18.7	19.5	20.0	20.5
3		2.9	3.6	4.3	5.2	6.4	7.5	8.6	9.3	10.2	11.1	11.8	12.5	12.9	13.5
4		1.7	2.1	2.5	3.1	3.9	4.6	5.3	5.8	6.5	7.2	7.9	8.5	8.8	9.2
5		1.2	1.4	1.7	2.1	2.6	3.1	3.6	3.9	4.4	4.9	5.3	5.8	6.1	6.4
6		0.8	1.0	1.1	1.4	1.8	2.1	2.5	2.8	3.1	3.5	3.8	4.2	4.3	4.4
7		0.5	0.7	0.8	1.0	1.3	1.6	1.9	2.1	2.3	2.6	2.9	3.2	3.3	3.5
8		0.4	0.5	0.6	0.8	1.0	1.2	1.4	1.6	1.8	2.0	2.2	2.3	2.5	2.6

*These figures make no allowance for impact. For moving loads (particularly construction equipment) on unsurfaced areas, the percentage figures above should be multiplied by an impact factor of 1.5 Highway, 1.75 Railway, 1.00 Airfield Runways, 1.50 Airfield Taxiways.

have been determined directly from data contained in "Theory of External Loads on Closed Conduits," Bulletin 96, published by the Engineering Experiment Station at Iowa State College.

Attention is called to the fact that the percentage figures shown in the table *do not include* an allowance for impact. For moving loads (particularly construction equipment used for operations after the pipe has been installed) on unsurfaced areas, the percentage figures should be multiplied by an impact factor of 1.00 to 1.75 (Table 5-4).

Unless other data are available, it is safe to estimate that truck wheel loads are the greatest live loads to be supported. H-20 wheel loadings are standard for highway and bridge design and are equally applicable for estimating live loads on sewers.

H-20 refers to loadings resulting from the passage of trucks having a gross weight of 20 tons, 80% of which is on the rear axle, with axle spacing of 14 ft 0 in. center to center, and a wheel gauge of 6 ft 0 in., each rear wheel carrying one half this load or 16,000 lbs each without impact.

5-53. Superimposed Loads. Sewers located in yards of industrial plants or under earth fills may have to withstand superimposed loads in addition to trench backfill loads.

When superimposed loads (Fig. 5-18) are encountered, they are added to the predetermined backfill load. To compute the superimposed load on a pipe, the following Marston formula is used.

$$W_{us} = C_{us} B_d U_s$$

W_{us} = the average total vertical load on a section of a conduit due to U_s, in pounds per linear foot,

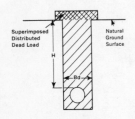

Figure 5-18
Superimposed Dead Load.

C_{us} = a coefficient for superimposed loads on ditch conduits, abstract number (see Table 5-5),

B_d = horizontal breadth of ditch at top of conduit, in feet.

U_s = uniformly distributed, superimposed load, in pounds per square foot;

H = vertical height from top of conduit to upper surface of fill, in feet.

Only some of the basics have been treated in this text. The reader should consult the *Clay Pipe Engineering Manual* or the ASCE *Manual of Engineering Practice No. 37* (*WPCF Manual of Practice No. 9*) for more complete information.

TABLE 5-5. Values of C_{us}

H/B_d	Sand or Damp Top Soil	Saturated Top Soil	Dry Clay	Wet Clay
0.0	1.00	1.00	1.00	1.00
0.5	0.85	0.86	0.88	0.89
1.0	0.72	0.75	0.77	0.80
1.5	0.61	0.64	0.67	0.72
2.0	0.52	0.55	0.59	0.64
2.5	0.44	0.48	0.52	0.57
3.0	0.37	0.41	0.45	0.51
4.0	0.27	0.31	0.35	0.41
5.0	0.19	0.23	0.27	0.33
6.0	0.14	0.17	0.20	0.26
8.0	0.07	0.09	0.12	0.17
10.0	0.04	0.05	0.07	0.11

SEWER PIPE MATERIALS

5-54. General. Clay, concrete, asbestos cement, cast iron, steel, and a number of plastic compounds account for most of the sewer pipe in use. There is no single material that is best for all applications nor any material that does not have some undesirable property. However, on a relative merit basis for a particular project, there may be one material that clearly stands out as the most desirable.

5-55. Clay Pipe. Vitrified clay pipe is extremely resistant to corrosion by acids and alkalies. It is not subject to damage from hydrogen sulfide generating sulfuric acid, as would be the case with a concrete pipe if the sewer were running only partly full with septic sewage. Clay resists erosion and scour, but it is susceptible to breakage. If the trenches are of the proper width, the bedding under the pipe is properly constructed, and all the other precautions of installation are taken, clay forms a long-term sewer. The old-style joints are susceptible to root penetration.

Clay is normally limited to a size range of 4 through 36 in. It is heavy and more brittle than some of the other materials, and if the workmen are careless, the sewer may not be properly placed. New connections for joining clay pipe should lower construction cost for clay. Cost is one of the detrimental factors of clay. References 12 through 19 should be consulted.

Clay is a satisfactory material for gravity flow sewers placed below ground, and it has been used on support trestles, trusses, and suspension structures; however, the author prefers flanged cast-iron construction for above-ground sewers. The author has most frequently used clay in sizes of 24 in. and smaller, except in unusual circumstances. If heavy tree roots prevail, cast-iron construction with mechanical joints below ground is preferable.

If clay is used, indicate on the plans or in the specifications the pipe diameter (either full or nominal), the type of joint, the class or strength, and applicable specifications.

5-56. Concrete Pipe. Both nonreinforced concrete (4- to 24-in. diameter) and reinforced concrete (12- to 144-in. diameter) are readily available in lengths varying from 4 to 24 ft, depending on pipe type and the manufacturer. A dense concrete, which may employ limestone or dolomite aggregate, should be used. Concrete provides a structurally strong pipe and is available at a low cost compared to some of the other materials. Its primary disadvantage is that it may be subject to corrosion in the presence of acid or hydrogen sulfide. There are many localities where corrosion is not a problem.

When the author is involved in a project in an area where the behavior of any material is in doubt, a careful engineering investigation is made. If the investigation reveals problems, then the next step is to explore how various preventative measures have performed. The economics of the case must be evaluated, and the end product is the application of good engineering judgment.

Describe in the specifications, or on the plans, the pipe diameter, class or strength, method of manufacture when important to the installation, method of jointing, and all special lining and concrete requirements (including aggregate if limestone or dolomite is required) applicable to the particular job. Alternative specifications and supplemental information are contained in references (2) through (33). Concrete can be used for gravity sewers, pressure mains, inverted siphons, and similar applications if proper joints are used.

Figure 5-19 indicates a concrete sewer being constructed, Fig. 5-20 shows the construction of a change in grade on a project, Fig. 5-21 is a photograph of a major conduit, and Fig. 5-22 shows another view of a large concrete sewer under construction. Figure 5-23 illustrates pieces of concrete pipe used for prefabricated manhole construction, and Fig. 5-24 indicates a concrete manhole installed prior to completion of grading work. Concrete is probably the most widely used single material used in sewer construction at the present time.

5-57. Cast Iron. Cast-iron pipe ranges in size from 2 to 48 in. and can be joined by a variety of methods. The *Handbook of Cast Iron Pipe* by the Cast

Figure 5-19
Concrete Pipe Sewer Construction.
(Courtesy American Concrete Pipe
Association)

Figure 5-20
Change in Grade Construction.
(Courtesy American Concrete Pipe
Association)

Iron Pipe Research Association, Chicago, Ill., should be part of every designer's library. Cast iron is a high-strength pipe particularly suitable for certain types of difficult laying situations. It can withstand greater pressure than the other normally used sewer materials and is particularly suitable for pressure mains, river crossings, etc. A cement lining with a bituminous seal coating is often specified for the interior of the pipe.

In specifying cast iron, the engineer should indicate pipe diameter, thickness, strength or class, type of joint, type of lining, and type of exterior coating. Cast iron will work well in most soils, but there are locations where the

Figure 5-21
Concrete Aqueduct. (Courtesy
Wayne Smith)

Figure 5-22
Large Sewer in Process of
Construction. (Courtesy American
Concrete Pipe Association)

Figure 5-23. Unassembled Concrete Manhole Components. (Courtesy
American Concrete Pipe Association)

Figure 5-24
Concrete Pipe Manhole. (Courtesy
American Concrete Pipe
Association)

corrosion is severe enough to make it an undesirable material. Cast iron is particularly good where there are heavy root problems. It is not recommended for brackish seawater where electrolytic action might be set up.

5-58. Plastics. Plastics are in a constant state of development and improvement. Like any product in its early stage of development, there are trials and tribulations in the process. Some of the problems that have come to the author's attention are indicated below.

The human element is probably one of the most important. Plastic pipe is light and easy to handle, and workers usually like the light weight during installation. However, the solvent compound used to fuse some types of plastic pipe (i.e., weld sections together) has a tendency in warm weather to emit fumes that are disagreeable to a worker having to make the joint at the bottom of a narrow trench. This has resulted in poor workmanship and leaky joints. This characteristic may not be exhibited by all solvents; however, it is a point to check. Slip joint construction would avoid this problem.

Some of the materials have had a tendency to be brittle and develop cracks, particularly in large pipes. Materials are in a constant state of development, and it is possible that this problem has already been resolved. Another problem that may occur is structural deformation. The author obtains a list of jobs installed with a given product under similar conditions and investigates rather carefully to determine how a product is performing.

Under laboratory conditions there are plastics in existence that have shown exceptional resistance to corrosion and attack by substances in sewers. One of the solutions may be a wedding of materials. Up until now, a single pinhole in a coating, poor coating at joints, or poor adherence of plastic coatings due to poor application have been problems of concern to the author. It appears that it may be possible to make coatings for steel and concrete that will be superior to anything that has been available in the past.

5-59. Cost Comparison. Labor costs and material costs vary from one location to another; thus assignment of exact dollar-and-cent values is meaningless. Figure 5-25 indicates relative costs for some of the different materials under a given set of conditions. Despite the changes that may occur in exact costs, the figure may be useful in obtaining an overall concept of price and why certain materials may not be used more than they are at the present time.

LASER SYSTEM

5-60. Definition. The word *laser* is an abbreviation for "*l*ight *a*mplification by *s*timulated *e*mission of *r*adiation." The unit that will be described in this section is a helium-neon gas laser. This means that the light emitted from the laser is generated from a gas mixture of helium and neon, which gives a light of the characteristic red color (wavelength of 6328 angstroms). The unique characteristic of the light which makes it so useful for alignment purposes is that it is a very pure (single wavelength) light, which can be collimated to maintain a very small diameter spot for long distances.

5-61. Power Used. The unit described is classified as a low-power-output laser, and it is rated at less than 0.002 watts (2 mW). The beam size is expanded to be a little less than 1 cm² in area, so the beam intensity is approximately 2 mW/cm². By the same definition, normal sunlight is about 100 mW/cm²; a 100-W light bulb at 1 ft is 7.0 mW/cm²; a 250-W infrared heat lamp at 1 ft is 80 mW/cm²; and an automotive headlamp at 5 ft is 20 mW/cm².

5-62. Safety Precautions. As with any tool, the laser can be dangerous. The primary thing to know and remember is never to look directly into the light. In general, the intensity is such that incidental exposure to the eye will not do damage; however, play it safe, and do not look into the beam. General exposure to any part of the body, other than the eye, is in no way harmful. Always use the targets provided with the system for viewing the spot.

5-63. System Description. Use of the laser beam-aligner system can be compared to shooting a rifle at a target. In this case, however, the "rifle" is held stationary and the target is moved until you score a bulls-eye on a translucent target. The "rifle" used is a laser which projects a tight, intense beam of light along the centerline of the pipe. It provides a centerline and grade line for pipe

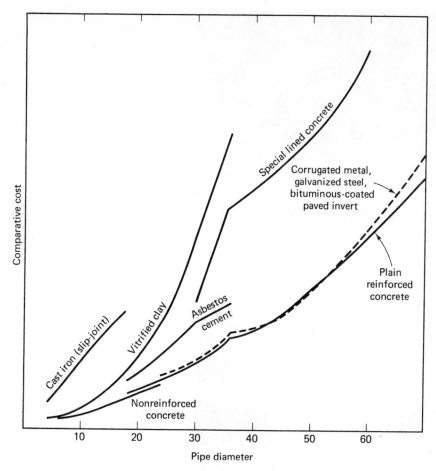

Figure 5-25. Comparative Cost of Sewer Installations Using Different Pipe Materials.

laying and also provides a reference in the ditch where the subgrade for the pipe can be checked. Figure 5-26 shows the laser unit in the manhole. It emits a beam along a line where the center of the pipe should be located. Figure 5-27 shows a workman reading a target at the end of the pipe. Figure 5-28 is a sketch of one of the possible setups that may be used, and Fig. 5-29 shows a phantom view of the laser system in a manhole.

The system is faster than any of the older methods. It is possible to maintain accuracies of line and grade better than $\frac{1}{100}$ of 1%. The published operating range of the setup indicated is 1000 ft; however, it has been used as far as 1100 ft in a similar underground arrangement and as far as 1300 ft on the surface in connection with pipe work. The system can be used to lay pipe uphill or downhill. The manufacturer claims that the system is adaptable to

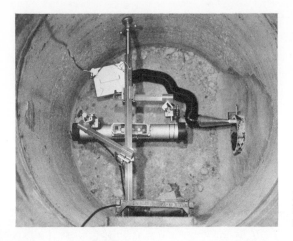

Figure 5-26
Laser Unit Mounted in Manhole.
(Courtesy Laser Alignment Inc.)

Figure 5-27
Workman Reading Laser Target at
End of Pipe. (Courtesy Laser
Alignment Inc.)

over 95% of all field conditions and that an increase in productivity of 25% can be achieved over older methods. Figure 5-30 shows a laser mounted in an underground pipe during construction.

PROBLEMS

Note: These problems require engineering assumptions and judgment. Rainfall and runoff data can be obtained from appropriate federal reports. Show sources for all data and bases for all assumptions.

1. Obtain a topographic map of at least 1 mile square of a town containing streets throughout the map area selected for this problem. Design a sanitary sewer system for the entire section using a tributary population of 6,500. Indicate the placement of manholes, pipe sizes used, grade, depths of sewer at all manholes,

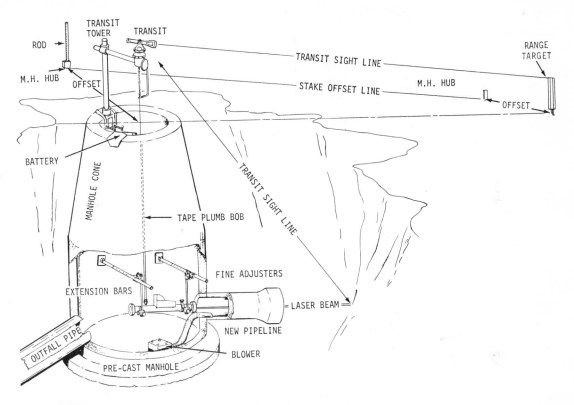

Figure 5-28. Steup Using Pressure Pads in Precast Manhole Base. (Courtesy Laser Alignment Inc.)

Figure 5-30. Laser Mounted in Underground Pipe. (Courtesy American Concrete Pipe Association)

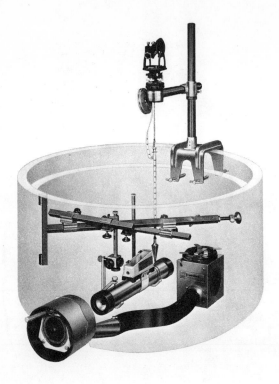

Figure 5-29
Phantom View of Laser System in
Manhole. (Courtesy Laser Aligment
Inc.)

and any other pertinent information necessary to effect a complete sewer system.

2. Repeat Problem 1, except on the side of highest elevation assume that a sanitary sewer line must enter the section and that the tributary population of this entering sewer is 8,300 persons. Calculate the size of the entire sewer system.

3. Design a storm sewer system for the same area selected for Problem 1. Identify the town and indicate the source of all data used in the problem, including design storm used.

4. Assume a combined sewer system and rework Problem 2 to determine the sewer pipe size, etc.

REFERENCES

1. Great Lakes-Upper Mississippi River Board of State Sanitary Engineers, *Recommended Standards for Sewage Works*, 1971 rev. ed. Health Education Service, P.O. Box 7283, Albany, N.Y. 12224.

2. BROOKS, J. N., "Infiltration of Ground Water into Sewers." *Trans. ASCE*, 76, 1909 (1913).

3. OLDER, C., and CONSOER, A. W., "Ground-Water Infiltration in Pipe Sewers." *Eng. News-Rec.*, 105, 695 (1930).

4. SEYMOUR, R. W., "Jointing Compounds for Sewer Pipe." *Pub. Works*, 82, 8, 74 (1951).

5. "Infiltration into Sewers." *Public Works*, 77, 7 (1946).

6. RAWN, A. M., "What Cost Leaking Manholes?" *Water Works & Sew.*, 84, 12, 459 (1937).

7. JEWETT, J. E., "Sewer Infiltration Control at Miami." *Eng. News-Rec.*, 106, 150 (1931).

8. HORNE, R. W., "Control of Infiltration and Storm Flow in Operation of Sewerage Systems," *Sewage Works J.*, 17, 2, 209 (1945).

9. NILES, A. H., "Sand and Silt Problems at Toledo, Ohio." *Sewage Works J.*, 24, 10, 1318 (1952).

10. HOUSE, H. W., and POMEROY, R., "Sewer Pipe Jointing Research—A Progress Report." *Sewage Works J.*, 19, 2, 191 (1947).

11. *Clay Pipe Engineering Manual.* Clay Sewer Pipe Association, Columbus, Ohio, 1962.

12. *Clay Pipe Engineering Manual.* National Clay Pipe Institute, 1968.

13. *Recommended Practice for Installing Vitrified Clay Sewer Pipe.* ASTM C12.

14. *Pipe, Clay, Sewer.* Federal Specification SS-P-361b.

15. *Methods of Testing Clay Pipe.* ASTM C301.

16. *Standard Strength Clay Sewer Pipe.* ASTM C13.

17. *Extra Strength Clay Pipe.* ASTM C200.

18. *Compression Joints for Vitrified Clay Bell and Spigot Pipe.* ASTM C425.

19. *Compression Couplings for Vitrified Clay Plain End Pipe.* ASTM C594.

20. *Concrete Sewers*, Portland Cement Assn., Chicago, Ill.

21. *Pipe, Concrete, (Nonreinforced, Sewer).* Federal Specification SS-P-371c.

22. *Pipe, Concrete, (Reinforced, Sewer).* Federal Specification SS-P-375b.

23. *Pipe, Pressure, Reinforced Concrete, Pretension Reinforcement (Steel Cylinder Type).* Federal Specification SS-P-381.

24. *Design Data.* American Concrete Pipe Association, Arlington, Va.

25. *Concrete Sewer, Storm Drain, and Culvert Pipe.* ASTM C14.

26. *Reinforced Concrete Culvert, Storm Drain, and Sewer Pipe.* ASTM C76.

27. *Reinforced Concrete Low-Head Pressure Pipe.* ASTM C361.

28. *Joints for Circular Concrete Sewert and Culvert Pipe, Using Flexible, Watertight, Rubber Gaskets.* ASTM C443.

29. *Determining Physical Properties of Concrete Pipe or Tile.* ASTM C497.

30. *Reinforced Concrete Arch Culvert, Storm Drain, and Sewer Pipe.* ASTM C506.

31. *Reinforced Concrete Elliptical Culvert, Storm Drain, and Sewer Pipe.* ASTM C507.

32. *Perforated Concrete Pipe.* ASTM C444.

33. *Precast Reinforced Concrete Manhole Sections.* ASTM 478.

Sewage
Treatment 6
Part I

A wastewater treatment system is merely a combination of unit processes put together in a manner to achieve a desired end result. There are many combinations that could be used to achieve the same degree of treatment. The designer's problem is to determine what combination of units will be the best engineering solution for a given system. Certain combinations are now widely used, but do not approach sewage treatment with the idea that only these systems should be considered.

Essentially, the designer's efforts are first concentrated toward separation of the solids from the liquid. Solids separation may be in a series of steps or only one or two steps. In a large municipal plant, we may separate the solids in a series of classes as follows: (a) screen out logs, blocks of wood, plastic bottles, chunks of metal, and refuse in general; (b) separate grit such as rocks, cinders, fly ash, and other heavy inorganic materials small enough to pass the screens; (c) use a grease flotation tank to remove grease and heavy oils; and (d) allocate sections of the plant to remove the organic solids. Once a given class of solids (or combination of classes) has been removed, the next problem is what to do with it. The era of primary treatment alone seems to be ending, and the trend is toward secondary treatment as a bare minimum.

Very few treatment plants are blessed with a uniform loading. It is unlikely that any domestic waste treatment plant would even approach uniformity. Figure 6-1 indicates one pattern of flow that might be experienced in a typical small municipality. Any measurements taken on a plant must con-

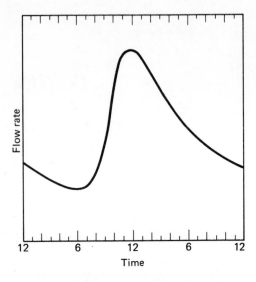

Figure 6-1
Typical Sewage-Loading Cycle.

sider daily flows, wet-weather flows, the variation of industrial waste flows, etc. The solids loading will normally follow the hydraulic loading as some function. The ratio of solids may be less at 3 a.m. if the flow is largely composed of infiltration water, for example, but it usually follows some pattern.

The information that follows will quote the Ten-State Standards in many places. In general, these standards are a regulatory guide, but you will find methods and equations that are not as yet generally accepted by the Ten-State Standards. The fact that they are not currently set forth in the Standards does not mean that they cannot be approved. The application may be supplemented by a detailed engineering explanation supported by calculations and documentation. If the particular equipment has never been used in the state concerned, obtain adequate operating data from other areas where it has been used. The more radical the deviation you propose, the more probability that you will have to appear in person to discuss the application and further defend the application against specific questions by the regulatory agency.

BASIC PLANT DESIGN

6-1. Hydraulic Flows. Investigate the problem and take adequate measurements if part of the system exists. Check the sources of flow rigorously, and consider wet-weather flows. Corrective action may be necessary to correct excessive wet-weather flows. The plant conduits must be of adequate size to handle the maximum expected loads including shock loads. Pockets and corners where solids can accumulate should be eliminated. Suitable gates should be placed in channels to seal-off unused sections which might accumulate solids (1). The use of shear gates or stop planks is permitted where they can be used in place of gate valves or sluice gates.

6-2. *Assimilation.* Analyze the receiving stream and the regulations pertaining to it. The ability of the receiving stream to assimilate the actual organic loads and shock loads is involved in the computation of the design organic loading. A consideration of the loading cycle of the plant is necessary.

6-3. *Arrangement of Units.* Component parts of the plant should be arranged for greatest operating convenience, flexibility, economy, and to facilitate installation of future units (1).

6-4. *Installation of Mechanical Equipment.* The specifications should be so written that the installation and initial operation of major items of mechanical equipment will be supervised by a representative of the manufacturer.

6-5. *Bypasses.* Except when duplicate units are available, properly located and arranged bypass structures must be provided so that each unit of the plant can be removed from service independently.

6-6. *Drains.* Means should be provided to dewater each unit (1). Due consideration should be given to the possible need for hydrostatic pressure-relief devices, particularly where the groundwater table is high.

6-7. *Construction Materials.* Consideration should also be given to the selection of materials which are to be used in sewage treatment works because of the possible presence of hydrogen sulfide and other corrosive gases, greases, oils, and other common constituents of sewage. This is particularly important in the selection of metals and paints. Dissimilar metals should be avoided to minimize galvanic action (1).

6-8. *Painting.* The use of paints containing lead should be avoided (1). To facilitate identification of piping, particularly in the large plants, it is suggested that the different lines have contrasting colors. The following color scheme is recommended for purposes of standardization (1):

> Sludge line—brown.
> Gas line—red.
> Potable water line—blue.
> Chlorine line—yellow.
> Sewage line—gray.
> Compressed air line—green.
> Water lines for heating digesters or buildings—blue, with 6-in. red
>   bands spaced 30 in. apart.

6-9. *Operating Equipment.* The specifications should include a complete outfit of tools and accessories for the plant operator's use, such as squeegees, wrenches, valve keys, rakes, and shovels. A portable pump is desirable. Readily accessible storage space and workbench facilities should be provided and consideration given to provision of a garage area which would also provide space for large equipment, maintenance, and repair.

6-10. Grading and Landscaping. Upon completion of the plant, the ground should be graded (1). Concrete or gravel walkways should be provided for access to all units. When possible, steep slopes should be avoided to prevent erosion. Surface water shall not be permitted to drain into any unit, and particular care should be taken to protect trickling filter beds, sludge beds, and intermittent sand filters from surface wash. Provision should be made for landscaping, particularly when a plant must be located near residential areas.

6-11. Outfalls. A provision should be made for effective dispersion of the effluent into the receiving body of water. The outfall sewer, where practicable, should be extended to the low water level of the receiving body of water in a way that insures the satisfactory dispersion of the effluent and should have its outlet submerged (1). Headwalls may be used when adequate dispersion is obtained without carrying the outfall into the stream. The outfall sewer should be protected against the effects of floodwater, tides, ice, and other hazards. A manhole should be provided at the shore end of all gravity sewers extending into the receiving waters (1). Hazards to navigation must be considered in designing outfall sewers.

6-12. Emergency Considerations. Evaluate what will happen under various emergency conditions that might occur. If public health is endangered, it may be necessary to provide a standby power source.

6-13. Water Supply. An adequate supply of potable water is normally necessary at any type of plant. Care must be taken to avoid cross connections or direct connections with nonpotable lines.

6-14. Safety. Adequate provision should be made to effectively protect the operator and visitors from hazards. The following should be provided to fulfill the particular needs of each plant (1):

1. Enclosure of the plant site with a fence designed to discourage the entrance of unauthorized persons and animals.
2. Installation of handrails and guards where necessary.
3. Provision of first-aid equipment.
4. Posting of "No Smoking" signs in hazardous conditions.
5. Provision of protective clothing and equipment such as gas masks, goggles, and gloves.
6. Provision of portable blower and sufficient hose.

Gasoline and other inflammable substances are sometimes dumped into sewers. Explosions have resulted in the sewers and fires have occurred at plants. Digesters have been blown up by ignition of dangerous gases resulting from bacterial action. Consider safety and potential vandalism in every segment of the plant design.

PRIMARY WASTE TREATMENT

6-15. Bar Screens. Protection for pumps and other equipment shall be provided by the installation of coarse bar racks or screens (1). All facilities should be readily accessible for maintenance. A bar rack having clear openings greater than $1\frac{3}{4}$ inches should precede mechanically cleaned grit-removal facilities. Manually cleaned screens located in deep pits shall be provided with stairway access, adequate lighting and ventilation, and convenient and adequate means of removing screenings (1). Screening devices installed in a building where other equipment or offices are located should be separated from the rest of the building, provided with separate outside entrances, and provided with adequate means of ventilation.

Manually cleaned screens should have clear openings from 1 to $1\frac{3}{4}$ in. between bars. Design and installation should be such that they can be conveniently cleaned. For manually raked bar screens, the screen chamber should be designed to provide a velocity through the screen of approximately 1 ft/s at average rate of flow. Manually cleaned screens, except those for emergency use, should be placed on a slope of 30° to 45° with the horizontal.

One of the most widely used equations for the calculations of screen hydraulics was suggested by Kirschmer (2):

$$h = \beta\left(\frac{w}{b}\right)^{4/3} h_v \sin\theta \qquad (6-1)$$

where h = head loss in ft,
β = a bar-shape factor (see Table 6-1),
w = maximum cross-sectional width of the bars facing the direction of the flow,
b = minimum clear spacing of the bars,
h_v = velocity head of the flow approaching the rack, in ft,
θ = angle of the rack with the horizontal.

TABLE 6-1. Kirschmer Factor*

Bar Type	β
Sharp-edged rectangular	2.42
Rectangular with semicircular upstream face	1.83
Circular bars	1.79
Rectangular with semicircular upstream and downstream face	1.67
Semiciruclar upstream face tapering in symmetrical curve to small, semicircular, downstream face (teardrop).	0.76

*Reference 2.

Another computation that has been used for this calculation is as follows (3):

$$h = \frac{V^2 - v^2}{2g} \times \frac{1}{0.7} \qquad (6\text{-}2)$$

where h = head loss, in ft,
  V = velocity through screen, in ft/s,
  v = velocity above screen, in ft/s,
  g = acceleration due to gravity (32.2 ft/s).

The minimum allowable acceptable loss through a manually cleaned bar screen is 6 in., and 1 ft is better design. Maximum head loss with clogged screen should not exceed 2.5 ft. Use a straight approach channel ahead of any screen in order to insure good velocity distribution across the screen. Velocities as high as 2 to 4 ft/s can be used, but 1 ft/s is preferred design. A manually cleaned screen should be used only in a small plant where mechanical screens cannot be justified.

6-16. Mechanical Screens. Clear openings for mechanically cleaned screens may be as small as $\frac{5}{8}$ in. For mechanically cleaned screens, maximum velocities during wet-weather periods should not exceed 2.5 ft/s (1). The velocity shall be calculated from a vertical projection of the screen openings on the cross-sectional area between the invert of the channel and the flow line.

The following design procedure is presented by courtesy of Rex Chain Belt Company (Envirex, Inc.):

Select bar size and spacing and determine the efficiency factor from Table 6-2. Determine the number of units desired. Then divide the total maximum daily flow or total maximum storm flow by the number of screens desired to obtain maximum flow per screen. The procedure is then as follows:

Max. daily flow in MGD $\times$ 1.547 = max. daily flow in ft^3/s
Max. storm flow in MGD $\times$ 1.547 = max. storm flow in ft^3/s

$\dfrac{\text{ft}^3/\text{s}}{2}$ = net area through bars for max. daily flow $\left.\vphantom{\dfrac{\text{ft}^3/\text{s}}{2}}\right\}$ use larger of two

$\dfrac{\text{ft}^3/\text{s}}{3}$ = net area through bars for max. storm flow $\quad$ values for final design

$\dfrac{\text{Net area in ft}^2}{\text{Efficiency coeff. for bars}}$ = gross area or channel cross-section wet area

Minimum width of bar rack 2 ft 0 in.   maximum width 14 ft 0 in.

$\dfrac{\text{Channel cross-section wet area}}{\text{Max. desired width or depth}}$ = corresponding depth or width

The above figures are based on recessing channel walls 6 in. each side for chain tracks and screen frame. The overall width of the screen frame is 12 in. greater than the width of the bar rack. If it is not possible to recess the walls, the channel should be made 1 ft 0 in. wider than calculated above.

The following is a set of typical computations:

Assume: Max. daily flow = 4 MGD = 4 $\times$ 1.547 = 6.188 ft^3/s

Assume: Max. storm flow $= 7$ MGD $= 7 \times 1.547 = 10.829$ ft³/s

Desired velocity through bar rack for max. daily flow $= 2$ ft/s.
Since $Q = Av$, $6.188 = A \times 2$ or net area through bars $= 3.094$ ft².
Desired velocity through bar rack for max. storm flow $= 3$ ft/s.
Net area through bars $= 10.829/3 = 3.61$ ft².

The gross area will be based upon the larger of the two net areas and the selected bar rack. For a rack consisting of $2 \times \frac{5}{16}$ in. bars spaced to provide clear openings of 1 in., the efficiency is 0.768 and the gross area is $3.61/0.768 = 4.72$ ft².

The channel width in this case might be established at 3 ft, in which case the water depth would be $4.72/3 = 1.57$ ft. This is a theoretical water depth which *may* be affected by subsequent plant units. For instance, a grit chamber may follow the screen chamber and be of such design that a head may be built up, the effect of which would be observed in the screen chamber. This is particularly true if the sewage flows by pipeline from the screen chamber to the grit chamber, and the flow in the grit chamber is subject to controlled velocity. Or a screen chamber may be followed immediately by a wet well in which case the sewage depth, especially at the low flows, would be less than computed. In fact, it is sometimes necessary to increase the sewage depth by installing stop planks across the channel, behind the screen. The planks serve as a weir, and the head built over this impromptu weir further serves to increase the sewage depth in the channel.

It is possible to compute the head loss through a bar rack from the formula.

$$h = \frac{V^2 - v^2}{2g} \times \frac{1}{0.7}$$

where $h =$ head loss in feet
$V =$ velocity through rack
$v =$ velocity above rack
$g =$ acceleration due to gravity (32.2 ft/s/s)

TABLE 6-2.* Efficiencies of Bar Spacings

Bar Size (inches)	Openings (inches)	Efficiency	Bar Size (inches)	Openings (inches)	Efficiency
$\frac{1}{4}$	$\frac{1}{2}$	0.667	$\frac{7}{16}$	$\frac{1}{2}$	0.534
$\frac{1}{4}$	$\frac{3}{4}$	0.750	$\frac{7}{16}$	$\frac{3}{4}$	0.632
$\frac{1}{4}$	1	0.800	$\frac{7}{16}$	1	0.696
$\frac{1}{4}$	$1\frac{1}{4}$	0.834	$\frac{7}{16}$	$1\frac{1}{4}$	0.741
$\frac{1}{4}$	$1\frac{1}{2}$	0.856	$\frac{7}{16}$	$1\frac{1}{2}$	0.775
$\frac{5}{16}$	$\frac{1}{2}$	0.615	$\frac{1}{2}$	$\frac{1}{2}$	0.500
$\frac{5}{16}$	$\frac{3}{4}$	0.705	$\frac{1}{2}$	$\frac{3}{4}$	0.600
$\frac{5}{16}$	1	0.768	$\frac{1}{2}$	1	0.667
$\frac{5}{16}$	$1\frac{1}{4}$	0.803	$\frac{1}{2}$	$1\frac{1}{4}$	0.715
$\frac{5}{16}$	$1\frac{1}{2}$	0.828	$\frac{1}{2}$	$1\frac{1}{2}$	0.750
$\frac{3}{8}$	$\frac{1}{2}$	0.572			
$\frac{3}{8}$	$\frac{3}{4}$	0.667			
$\frac{3}{8}$	1	0.728			
$\frac{3}{8}$	$1\frac{1}{4}$	0.770			
$\frac{3}{8}$	$1\frac{1}{2}$	0.800			

*Copyright 1957 by Chain Belt Company.

or

$$h = 0.0222(V^2 - v^2)$$

Referring to the foregoing example, and again making use of $Q = Av$, the velocity above the rack is $10.829/4.72 = 2.3$ ft/ps. Therefore:

$$h = 0.0222(3^2 - 2.3^2) = 0.0222 \times 3.7 = 0.082 \text{ ft}$$

or approximately 1 in.

Now suppose the screen became half plugged with screenings, leaves, and other debris. From $Q = Av$ it is seen that the area is directly proportional to the velocity. In other words, if the area is cut in half, the velocity must double. The head loss therefore is:

$$h = 0.0222(6^2 - 2.3^2) = 0.0222 \times 30.7 = 0.682 \text{ ft}$$

or approximately $8\frac{1}{4}$ in.

The increase in head loss is over $\frac{1}{2}$ ft as the screen becomes half plugged. The need for accurate control of the cleaning cycle and protection against surge loads are thus demonstrated.

6-17. *Data Required for Mechanical Screen Design.* In order to intelligently get a plant together and communicate with manufacturers on specific merchandise, it is necessary to establish certain information.

The volume of sewage to be handled by the proposed screen should be set forth as shown in Table 6-3. Figure 6-2 will assist in estimating the cubic

TABLE 6-3. Flow in MGD (If flow measured in other units, note units below.)

	Present	*Design*	*Future*
Maximum			
Average			
Minimum			

feet of screenings per million gallons of sewage. It is necessary to indicate the type of screenings to be handled, whether your sewer is separate or combined, the clear opening between bars that you prefer ($\frac{1}{2}$, $\frac{3}{4}$, 1, $1\frac{1}{4}$, or $1\frac{1}{2}$ in.), and whether the pumping operation is involved immediately before the screen chamber or immediately following the screen chamber (4).

What is the difference in elevation between the operating floor and the invert of the incoming sewer? Specify whether the proposed installation is to be made in a new channel or an existing channel: if in an existing channel, provide detail drawings and dimensions; if in a new channel, be sure to indicate all space limitations. Indicate what will control sewage depth in the channel. Provide the sewage depths for the items indicated in Table 6-3. Indicate whether duplicate units are required; electrical characteristics including voltage, phase, and frequency; and any other information considered pertinent (4).

Indicate whether a traversing or stationary rake superstructure will be required. Indicate whether the superstructure will have either an integrally

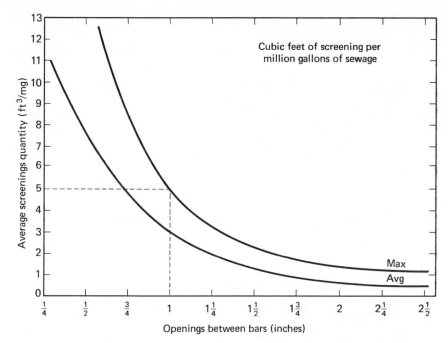

Figure 6-2. Mechanically Cleaned Bar Screen; Cubic Feet of Screenings per Million Gallons of Sewage. (Courtesy Rex Chainbelt Inc.)

mounted discharge chute/hopper or a removable discharge hopper, and show by whom it is furnished (5). If an integrally mounted hopper is desired, describe any unloading requirements; a sketch is desirable. If an integrally mounted chute is desired, describe the discharge requirement including a drawing or sketch. If a removable hopper is desired, describe the type and method of removal and provide a sketch if possible.

Indicate the rake-tooth length desired: 12, 18, and 24 in. are popular sizes. Also indicate the rake-tooth penetration into the face of the bar rack ($\frac{3}{4}$ in. tooth penetration is standard) (5). Next indicate the rake width desired, i.e., effective cleaning width (2-, 4-, 6-, 8-, and 10-ft widths are available as standard). Indicate the operating mechanism preferred (single-beam is standard with Rex Chainbelt; however, double-beam is available for unusual accumulations of debris buildup at the bottom of the bar rack). Indicate whether rake carriage shall be guided or nonguided. Nonguided rake carriage is recommended. If guided, indicate by whom the guides are to be furnished and attach drawings. If the bar rack already exists, indicate how it is presently cleaned and all pertinent details.

Figure 6-3 indicates how to set up a sketch and also pertinent dimensional data that must be established.

6-18. Screen Parameters. The screen channel invert should be 3 to 6 in. below the invert of the incoming sewers. To prevent jetting action, the length

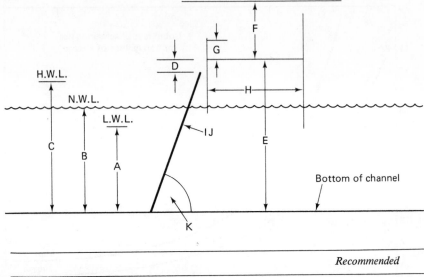

Figure 6-3. Mechanized Bar Design Data (5).

		Recommended
A. Low water depth		
B. Normal or average water depth		
C. High water depth		
D. Top of rack to operating floor		0
E. Operating floor to channel bottom		
F. Overhead clearance		Min: 14'0"*
G. Handrail height		Max: 3'6"*
H. Lateral clearance		Min: 8'0"
I. Size of bars		
J. Clear opening between bars		Normal 3"–6"
K. Bar rack angle of incline		75°–84°
L. Total width of rack to be cleaned		

*If handrail will not be installed, so indicate. This will allow for a reduction in headroom required for the hoist superstructure.

and/or construction of the screen channel should be adequate to reestablish the hydraulic flow pattern following the drop in elevation (1). The channel preceding and following the screen should be shaped to eliminate stranding and settling of solids. Fillets may be necessary. All mechanical units which are operated by timing devices should be provided with auxiliary controls which will set the cleaning mechanism in operation at predetermined high-water marks. Channels should be equipped with the necessary gates to divert flow from any one comminuting unit. Provisions must also be made for dewatering each unit. When mechanically operated screening or comminuting devices are used, auxiliary manually cleaned screens should be provided (1). The design should include provisions for automatic diversion of the entire sewage flow through the auxiliary screens should the regular units fail.

Amply sized facilities must be provided for removal, handling, storage, and disposal of screenings in a sanitary manner. Manually cleaned screening facilities should include an accessible platform from which the operator may rake screenings easily and safely (1). Suitable drainage facilities should be provided for both the platform and the storage areas. Grinding of screenings and return to the sewage flow is satisfactory; however, burial or incineration should also be considered. Open-area disposal is prohibited.

6-19. Fine Screens. The use of fine screens in lieu of sedimentation alone is not normally permitted under the Ten-State Standards. In special cases where it can be demonstrated that the features peculiar to this equipment may be used to advantage, the merits of such proposed installations will be determined individually by the reviewing authority.

6-20. Screen Types. A typical manually cleaned screen chamber is shown in Fig. 6-4. The principle of the cleaning mechanism of one type of mechanically cleaned screen is shown in Fig. 6-5. Another mechanically cleaned screen is illustrated in Fig. 6-6. Screen manufacturers frequently make a number of different types of units to meet various plant sizes and preferences of the designer.

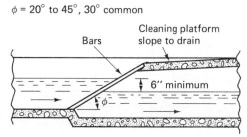

$\phi = 20°$ to $45°$, $30°$ common

Bars

Cleaning platform slope to drain

6" minimum

ϕ

Figure 6-4
Manually Cleaned Screen Chamber.

LOWERS RAISES DUMPS

Figure 6-5. Principle of Operation of One Mechanically Cleaned Screening Device. (Courtesy Rex Chainbelt Inc.)

Figure 6-6
Combination Mechanically Cleaned
Bar Screen and Comminutor
Device. (Courtesy Environmental
Equipment Division, FMC
Corporation)

6-21. Grit Removal Facilities. Grit consists of sand, gravel, silt, ashes, clinker, egg shells, coffee grounds, bone chips, seeds, and similar items. It tends to impose heavy wear on sewage pumps, sludge pumps, and other equipment in contact with it, as well as cause operating problems such as contributing to clogging of pipes. To relieve this problem, the trend in modern municipal plant design is to include a grit-removal chamber. The Ten-State Standards (1) indicate that grit-removal facilities should be provided for all sewage treatment plants, and they are required for plants receiving sewage from combined sewers or from sewer systems receiving substantial amounts of grit. If a plant, serving a separate sewer system, is designed without grit facilities, the design should include provisions for future installation. Grit-removal facilities, except in unusual circumstances, should be located ahead of pumps and comminuting devices. In such cases, coarse bar racks should be placed ahead of mechanically cleaned grit-removal facilities.

A study of municipal records indicates that the quantity of grit may range from $\frac{1}{3}$ ft³ to 24 ft³ per million gallons of sewage. In cities above the Mason-Dixon line with combined sewers, 7 ft³ can be used as a trial figure that can be increased or decreased according to detailed study of local conditions. South of the Mason-Dixon line, a starting figure of 3 ft³ may be used.

6-22. Grit Chamber Design. It is usually necessary to use several chambers to handle the flow range of the plant and provide spare capacity so that any one of the chambers can be repaired as needed. Analysis of the plant flow requirements is one of the first steps.

It has been found that velocities greater than 1.25 ft/s carry too much grit over into the treatment plant, whereas velocities under 0.75 ft/s result in too much organic matter settling out. It is common practice to use a velocity halfway between these extremes, i.e., 1 ft/s. Next we must decide what size particle we want to remove. It has been found that particles that can pass through a 60-mesh screen are less damaging than coarser materials. Thus it is common practice to endeavor to remove all particles coarser than 60-mesh, i.e., particles not less than 0.25 mm (0.01 in.). It is also necessary to select a specific gravity value; 2.65, which is the value for silica sand, is normally selected.

From a theoretical viewpoint, Stoke's law and Hazen's formula form the basis for the design. Equations associated with the phenomenon of settling are contained in the section on sedimentation later in the text. The particles settle according to these equations; however, the organics and grit are not concentrated at the surface. Thus, even though most of the organic material settles slower than the grit, it is found in practice that the grit and organics are mixed throughout the channel contents. Figure 6-7 illustrates the concepts. In the top sketch the assumption is that all grit and organics are at the surface, which of course is not true. If it were true, the lower sketch indicates how, at low flow, it would be important to cut the grit chamber off at a particular length. The middle sketch indicates more realistically what occurs in practice.

Obviously, some form of engineering assumption and compromise has to be employed. The simplest procedure is to assume all the grit and organics

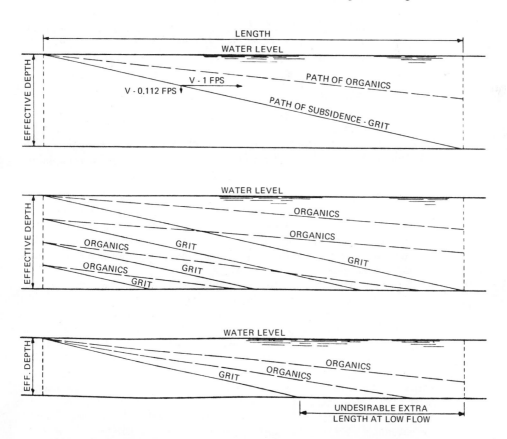

Figure 6-7. Gravimetric Separation of Grit and Organic Matter in a Grit Chamber. (Courtesy Rex Chainbelt Inc.)

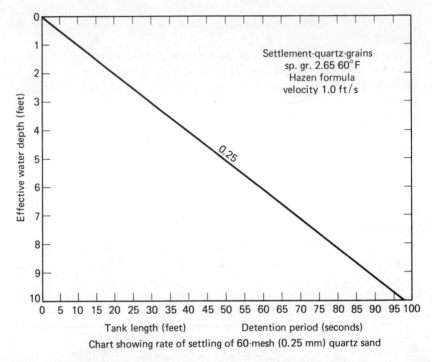

Chart showing rate of settling of 60-mesh (0.25 mm) quartz sand

Figure 6-8. Rate of Settling. (Courtesy Rex Chainbelt, Inc.)

are at the surface. Next, establish a practical wetted cross section for the plant flow rate to be put through a given channel. A quick and realistic way to solve the problem is to work with Fig. 6-8. The use of this graph is explained in the following explanation from Rex Chainbelt, Inc:

The diagonal line represents the path of subsidence of grit particles of 0.25-mm diameter. It is important to realize that this chart is based on the assumption that all grit enters the tank at the water surface. That is the only practical premise for the construction of a chart of this nature, in light of the use to which it is put. Stoke's law and the Hazen formula form the basis for the computations attendant to the plotting of the curve.

The length as determined from the graph respresents the optimum operating conditions. It is recommended that additional length be provided to compensate for irregularities in channel construction, varying specific gravities, varying temperatures, etc. This factor of safety should be 25 to 50%, depending on design conditions. This chart is a handy shortcut to the determination of the channel length, and its use is illustrated by the following example.

Assume a maximum flow of 10 MGD or 15.47 ft³/s, and assume it is desired to remove all grit of 60-mesh and larger. With the design velocity in the grit chamber being 1 ft/s, it is obvious that the wetted cross-sectional area of the proposed channel must be approximately 15.47 ft². If an arbitrary channel width of 5 ft is assumed, the maximum effective water depth would therefore be 15.5 divided by 5, or 3.1 ft. Now the chart may be used for determining the channel length, which must be sufficient to allow time for the grit particles to settle to the bottom before it reaches the outlet end of the channel.

In this case, notice that a quartz particle 0.25 mm in diameter settles at a rate of 0.112 ft/s. Since the water depth in the channel is 3.1 ft, it would take the particle 3.1 divided by 0.112, or 29 s to settle to the bottom. Since the sewage is flowing through the channel at a velocity of 1 ft/s, the channel would therefore have to be 29 ft long. This tank length can be found directly from Fig. 6-8 simply by following the 3.1-ft depth line horizontally to its intersection with the 0.25-mm diameter particles subsidence line, and noting the required length from the corresponding point along the bottom of the chart. This gives a channel length of 30 ft, which agrees substantially with the calculation. Allowing for the factor of safety, the recommended length would be approximately 44 ft.

A grit chamber is thus theoretically designed to remove particles above a certain size at a predetermined sewage depth and velocity. While a constant flow cannot be maintained, it is possible to maintain a velocity reasonably close to the desired value of 1 ft/s, and it is extremely important that this be done. There are two principal velocity-controlling devices in general use—the proportional weir and the Parshall flume.

A pure grit, free from organic material, cannot be obtained from a grit chamber, no matter how carefully it is designed. The questions are, how clean must the grit be to allow convenient disposal without nuisance and how shall the grit be conditioned to remove the objectionable organic material?

Practically the only acceptable test for ascertaining the nuisance value of grit is the Dazey churn test which is, in itself, a very inconclusive determination. Actually, the Dazey churn test measures the amount of *free* organic material in the grit and expresses the amount as a percentage of the entire mass. Large organic solids can be in the grit, however, and not be correctly reflected in the test results. For example, quantities of angle worms that might be present in the grit samples would not be siphoned off in the Dazey churn test, and therefore, theoretically, would not contribute to the putrescibility of the grit. Further, large quantities of nonputrescible organics might be left within the churn after washing and cause a false increase in the putrescibility reading.

It is generally accepted that a grit of less than 5% putrescibility is satisfactory for ordinary disposal, although some specifications call for a putrescibility of less than 1%. It has been our experience that any grit, no matter how carefully conditioned, is far from being beach sand. The difference between a grit of 1% and 3% putrescibility is not readily discernible by sight or smell, and the need for removing that extra 2 or 3% is very questionable, particularly from the economical viewpoint.

Figure 6-9 indicates a mechanized deep-channel grit chamber arrangement, and Fig. 6-10 illustrates one type of grit washer. Again from Rex Chainbelt:

Several types of grit washers are available and, with proper operation and maintenance, they can be made to deliver a grit containing less than 1% putrescible matter. But grit is very abrasive and difficult to handle, with the result that maintenance of such equipment is likely to be expensive. It is recommended that this equipment be reserved for the most exacting jobs and that grit conditioning by recirculation be substituted as a means of obtaining a satisfactory product for most requirements.

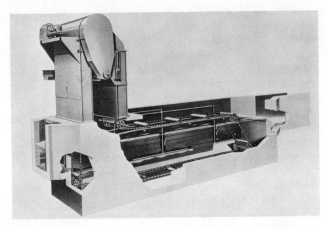

Figure 6-9
Deep Channel Chain and Bucket
Grit Collector. (Courtesy Rex
Chainbelt, Inc.)

Figure 6-10
One Type of Grit Washer.
(Courtesy Rex Chainbelt, Inc.)

As shown previously, organics are present in grit because some of the material enters the channel near the bottom and accumulates during low flow when the grit channel is out of proportion between length and water depth. Recirculation is merely a means for returning the grit to the channel when settling conditions are optimum, thus allowing the grit to resettle and the organics to flow out of the channel.

6-23. Control Weirs. Figure 6-11 illustrates how to design a control weir. A Sutro weir is merely one-half the proportional weir, i.e., from the vertical centerline to one side. Thus the Sutro weir has one vertical side. In the equations below, Q is the total discharge in ft^3/s past the weirs; Q_1 is the discharge in ft^3/s through the rectangular part of the weir; and a is the vertical dimension of the rectangular portion of the weir. The bottom of the weir is usually 4 to 12 in. above the bottom of the chamber, and the dimension of a should

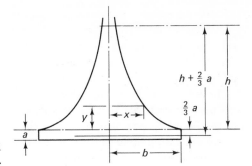

Figure 6-11
Proportional Weir.

be a minimum of 1 in. Free fall at the outlet of these weirs is essential (6, 7).

$$x = b\left(1 - \frac{2}{\pi} \arctan \sqrt{\frac{y}{a}}\right) \tag{6-3}$$

$$Q = b\sqrt{2ag}(h + \tfrac{2}{3}a) \tag{6-4}$$

$$Q_1 = \tfrac{2}{3}b\sqrt{2g}[(h + a)^{3/2} - h^{3/2}] \tag{6-5}$$

Soucek (8) offered a slightly different form of the equation:

$$Q = 4.97a^{1/2}b\left(h - \frac{a}{3}\right) \tag{6-6}$$

Parshall flumes are also used as grit chamber flow-control devices. These flumes are available as standard prefabricated catalog items for treatment plant use. Data on how to design these flumes is available from a number of sources (9, 10, 11, 12).

6-24. Grease and Oil Removal. Grease and oils in sewage may include waste automotive lubricants, free fatty acids, mineral oils, calcium and magnesium soaps, waxes, fats, and miscellaneous hydrocarbon or petrochemical products. The quantity of these materials is not adequate in most domestic sewages to warrant recovery as a by-product for sale. Every effort should be made to keep grease out of the sewers by means of devices such as the "Cascade" grease interceptors (Josam Mfg. Co.), the "Hydra-Filter" (Wade Mfg. Co.) or the "Greaseptor" (Zurn Industries) at restaurants or other establishments where significant contributions of grease can be expected. A simple device such as a white board has been used to track down suspected polluters. The board is suspended in appropriate manholes and checked periodically.

In industrial practice where significant quantities of oil are encountered, an API separator is usually used. The American Petroleum Institute (API) instructions for the design of these separators are contained in *Manual on Disposal of Refinery Wastes*, Vol. I. Domestic sewage seldom contains oil in adequate quantities to warrant such a separator at the plant. Following the API separator, the next unit is normally some type of air flotation unit. It is not suggested that the designer indiscriminately incorporate an air flotation unit; in fact, the majority of plants that the average engineer designs will not warrant inclusion of this equipment.

6-25. *Air Flotation.* These units are usually of the aeration type, pressure type, or the vacuum type. The aeration-type units can use an air diffuser on the bottom of the tank and permit air to bubble up through the tank. The agitation may cause increased grease and suspended-solids removal, but in the form described it is not considered a very effective system.

The Rex Float-Treat separator shown in Fig. 6-12 is one variation of the pressure-type unit. Air is fed into a pressure tank containing sewage liquid and the liquid is saturated with air. The liquid is reduced to atmospheric pressure, releasing dissolved gas which is in excess of saturation at atmospheric pressure. The fine bubbles that are released adhere to suspended matter and float it to the surface.

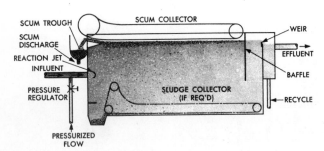

Figure 6-12
Air Flotation Unit. (Courtesy Rex Chainbelt Inc.)

The vacuum-type unit works on essentially the same principle, except that the liquid is saturated at atmospheric pressure and then a partial vacuum is pulled on the system. This causes bubbles to be released due to reduction of solubility of the gas in liquid principle.

Flotation units have some distinct advantages for resolving certain problems in industrial waste treatment. In domestic waste treatment, however, if the designer is not careful, it may result in additional equipment, higher operating cost, higher power requirements, plus the need for more skilled maintenance and operation personnel. This will have to be weighed against the advantages of removing grease, solids, grit, and other material at this stage of the plant process. Odor may be better controlled by such a unit, depending upon the plant design.

The amount of oil or grease that can be removed by such a unit depends to some extent on the concentration of the oil in the influent. The use of alum in the system can result in increased oil removal. If alum is used, it is usually necessary to regulate the pH of the system. Removal may be most efficient at a pH of 8.5 for the system, as an example, with the effectiveness of the alum addition dropping off if the pH is either increased or decreased.

Rother (13) found that oil removal will be reduced if the influent concentration is less than 150 ppm of oil. Oil removal was 90–95% when the influent had 150 ppm or more of oil. When the influent had 100 ppm oil, the oil removal decreased to 80–85%, and when the influent had only 25 ppm oil, the oil removal was down to 65–70%. Simonsen (14) reported 87% oil removal for an influent containing 32 ppm oil; thus there is not agreement on this subject.

To provide a flotation unit, a manufacturer will need to know the oil in ppm, grease in ppm, BOD, COD, suspended solids, temperature, and pH. It will also be necessary to provide design flow, future flow, and runoff period each for maximum, average, minimum, and peak flow. Additional things the manufacturer needs to know include: the treatment required and the waste characteristics, whether any chemicals are used ahead of the unit in the process, whether influent is pumped or flows by gravity, types of tanks (existing, to be built, or to be supplied by the manufacturer), head available, sketch of proposed location, compressed air capacity available, and its pressure as well as the available electrical power characteristics.

6-26. *Flocculation and Pre-aeration.* Flocculation of sewage by air or mechanical agitation, with or without coagulating aids, is worthy of consideration when it is desired to reduce the strength of sewage prior to treatment (1). Also, flocculation may be beneficial in pretreating sewage containing certain industrial wastes. The units should be designed so that removal from service will not interfere with the normal operation of the remainder of the plant. When air or mechanical agitation is used in conjunction with chemicals to coagulate or flocculate the sewage, the detention period should be about 30 minutes at the design flow (1). However, if polymers are used, this may be varied.

If the process is used with the intent of BOD reduction, then 45 minutes' detention time should be allowed at the design flow unless polymers are used. The paddles should have a peripheral speed of $1\frac{1}{2}$ to $2\frac{1}{2}$ ft/s to prevent deposition of solids. Aerators may consist of any type of equipment used for aerating activated sludge. The agitation should be controllable in order to obtain good mixing and maintain self-cleansing velocities across the tank floor (1). Inlet and outlet devices should be designed to insure proper distribution and to prevent short-circuiting. Convenient means should be provided for removing grit.

At plants where there are two or more flocculation basins utilizing chemicals, provision shall be made for a quick mix of the sewage with the chemical so that the sewage passing to the flocculation basins will be of uniform composition (1). The detention period provided in the quick-mix chamber should be very short, $\frac{1}{2}$ to 3 minutes.

Gentle agitation in the form of mechanical stirring or diffused aeration increases the floc formation of organic materials by increasing the number of collisions or contacts and by releasing entrapped gases. Mechanical units might include revolving and/or reciprocating paddles, draft tubes, or radial-flow turbine impellers. The spiral-roll aerators of the activated sludge system are a form of flocculator. One type of treatment process, sedimentation, may follow flocculation with a 1.5- to 2.0-hr detention time and an overflow rate of approximately 800 GPD/ft².

Two types of flocculators are shown in Fig. 6-13 and a third type is shown in Fig. 6-14. Argaman (15) concluded that:

    1. The physical parameters directly affecting the performance of a floc-

culation reactor are the residence time, its distribution, the total power input, and the properties of the turbulence field.

2. For a given stirring mechanism, the effect of the turbulence energy spectrum may be expressed by the rms velocity gradient, a paddle performance coefficient, and an energy spectrum coefficient.

3. For a given residence time, the performance increases almost linearly with the rms velocity gradient until a maximum value is reached beyond which any further increase results in a decrease in performance.

4. For any desired performance, there is a minimum residence time requirement. The rms velocity gradient associated with this residence time is an optimum value in the sense that any other value will result in lower performance.

5. Compartmentalization of a continuous-flow system exerts a considerable influence on the flocculation performance. Systems with equal overall residence times will perform better as the degree of compartmentalization increases.

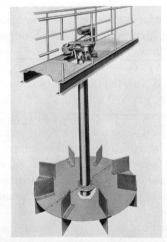

Figure 6-13. Flocculators. (Courtesy Rex Chainbelt, Inc.)

Figure 6-14
Vertical Paddle Flocculator.
(Courtesy Koppers Co., Inc.)

When chemicals are used, dolomite lime will probably be found to be the most satisfactory means of regulating the pH of the raw influent. Sodium carbonate is inclined in some cases not to give a satisfactory floc. Either alum or ferrous sulfate are likely to have a rather critical pH range of satisfactory operation. Below a certain range, floc formation may be poor or fail to form, and above a critical value the floc may settle either very slowly or not at all. Alum is more expensive than ferrous sulfate but usually requires considerably less chemical; hence, the cost differential for treatment is usually not significant. Ferrous sulfate usually requires more lime, i.e., alum will usually work best at a pH value lower than that of the ferrous sulfate.

Camp and Stein (16) generalized Smoluchowski's equation to include turbulent flow conditions by defining a root-mean-square (rms) velocity gradient, G, which they substituted for the velocity gradient existing in laminar flow. They then expressed the collision frequency as:

$$N_c = \tfrac{4}{3} n_i n_j R_{ij}^3 G \qquad (6\text{-}7)$$

where N_c = number of collisions per unit time and per unit time between particles of radii R_i and R_j,

n_i and n_j = number concentrations of the colliding particles,

R_{ij} = collision radius $R_i + R_j$,

G = rms velocity gradient.

Camp and Stein also indicated that the velocity gradient is given by:

$$G = \sqrt{\frac{\epsilon}{\nu}} \qquad (6\text{-}8)$$

in which ϵ is the total power dissipated per unit mass of fluid, and ν is the kinematic viscosity.

If paddle blades are made out of wood, it should be all heart redwood. A 3-to-1 speed adjustment range on paddles should be provided.

The equipment manufacturer will need the same flow data required for flotation units, along with information about suspended solids, COD, BOD, whether chemicals are to be used and if so what are they and at what point are they to enter the system, whether flash mixing equipment will be necessary, whether the tanks exist or are to be built or furnished, and any special features desired. If the unit is used later in the plant, particularly where coagulation is used as part of the advanced waste treatment phase, the manufacturer may need to have information such as carbonate hardness, noncarbonate hardness, maximum and minimum temperatures, turbidity, bicarbonate alkalinity, caustic alkalinity, total alkalinity, carbonate alkalinity, calcium as calcium carbonate, whether color or iron removal is required, and any other pertinent information.

6-27. Sedimentation. Sedimentation consists of allowing a liquid to flow through an appropriately designed tank at a sufficiently slow velocity that a significant amount of the solids can settle out. Unfortunately, there is substantial variation in the characteristics of sewage in different plants.

At the sedimentation tank itself, there are a number of factors that affect the design: (a) turbulence due to inlet design, (b) effluent weir design and associated flow paths, (c) flow variations in the influent, (d) wind-induced currents, (e) temperature variation between tank contents and the influent sewage, (f) density variations of the influent with respect to tank contents, and (g) sludge-scraper and removal-mechanism design. Granular material tends to act like a discrete particle, whereas organic material may tend to floc if agitated slightly. Particles are not spheres but a variety of shapes with different dimensions in different planes, and their settling may involve anything but the theoretical curved path assumed in some equations.

For a review of the different settling theories up to 1957, consult Behn's paper (17). Other suggested references will be set forth in this presentation. Unfortunately, despite all the theories and equations that have been suggested by a number of brilliant men, the state of the art of practical sedimentation-tank design still hinges heavily upon an accumulation of experience expressed as a series of guidelines.

If the solids removal efficiency of a larger number of sewage plant sedimentation basins were plotted on a graph, it would be found that substantial dispersion exists. If a median value were established for this dispersion, it would be found that about 73 % of the *solids* can be removed at an overflow rate of 400 GPD/ft²; 68 % at 500 GPD/ft²; and 58 % at 1,000 GPD/ft². It would also be found that the circular tanks hold somewhat of a performance edge on rectangular tanks. Where an extremely compact plant is required, it may be necessary to use rectangular tanks. Rectangular tanks can meet or exceed the performance of circular tanks when they are properly designed and operated.

Stoke's law may be written as follows:

$$v = \frac{1}{18} \frac{g}{\mu} (\rho_1 - \rho) d^2 \qquad (6.9)$$

where $v =$ the settling velocity in cm/s,
$\quad g =$ gravity in cm/s²,
$\quad \mu =$ absolute viscosity coefficient of the fluid in dyne-s/cm² or poises,
$\quad \rho_1 =$ the density of the particle in g/cm³,
$\quad \rho =$ the density of the fluid, g/cm³,
$\quad d =$ the particle diameter in cm.

Newton proposed a relationship for drag as follows:

$$D = C_D A \frac{\rho V^2}{2} \qquad (6\text{-}10)$$

where $D =$ the drag,
$\quad C_D =$ the drag coefficient,
$\quad A =$ the projected area of the body in the direction of motion,
$\rho V^2/2 =$ the dynamic pressure.

Equating the drag of Newton's equation to the weight of the sphere results in the general equation for the settling velocity of a sphere in terms of the drag coefficient.

$$v = \sqrt{\frac{4}{3} \frac{g}{C} \frac{\rho_1 - \rho}{\rho} d} \qquad (6\text{-}11)$$

which leads to the approximate expression

$$v = \sqrt{\frac{4}{3} \frac{g}{C} (S_s - 1) d} \qquad (6\text{-}12)$$

where S_s is the specific gravity of the particle.

It will be found that Stoke's law closely parallels experimental data for Reynold's numbers less than 1.0. Above 1, the drag coefficient has an increasing deviation with Stoke's law. This deviation is not overly serious until Reynold's numbers above 20,000 are encountered.

In Fig. 6-7 the settling of a grit particle as well as organic material is illustrated. The same figure illustrates the principle of material settling in the sedimentation basin. Under ideal conditions

$$r = \frac{v}{V_o} = \frac{Av}{Q} \qquad (6\text{-}13)$$

where $r =$ removal, as a ratio, of particles settling at any velocity v less than the overflow rate V_o,
$\quad A =$ the surface area of the tank corresponding to length l.

Removal in terms of the total suspension is

$$r \, dP = \frac{1}{V_o} v \, dP \qquad (6\text{-}14)$$

where P is the initial concentration of particles settling at velocity v or less, and the total removal of all particles is

$$R = 1 - P_o + \frac{1}{V_o} \int_0^{P_o} v \, dP \qquad (6\text{-}15)$$

The concentration at a point is

$$X = \int_0^{P_1} dP = P_1 \qquad (6\text{-}16)$$

where P_1 is the value of P for

$$v = \frac{h}{H} V_o \qquad (6\text{-}17)$$

where h is the distance the particle has settled at any time t, and H is the depth of the settling basin. These relations were first pointed out by Hazen (18) in 1904 and were later repeated and amplified by Camp (19).

When a suspension flows through a settling tank, the removal of particles may depend upon detention time, overflow rate, or both. Whether one or the other is more important depends primarily upon the settling properties of the suspension (20). The more shallow the basin, the more rapidly a particle of a given size and specific gravity will settle out. The use of trays and other schemes to make use of extremely shallow sedimentation has been investigated over a period of years (21, 22, 23, 24). Relatively small-diameter (1–4 in.) tubes 2–4 ft in length provide efficient sedimentation with detention times of 6 minutes and less (25). Tube length, diameter, and flow rate, the nature of incoming material, and the nature and quantity of chemicals added affect the performance of the tube settler. Raw-water turbidities of up to 1,000 Jackson units have been successfully treated in the tube units. The first significant application of these tubes will probably be in later stages of the treatment plant process, particularly in association with advanced wastewater treatment. These tubes have to be inclined in order to facilitate sludge removal.

In large sedimentation tanks, it is usually considered necessary to add some type of slow-moving mechanical scraper to aid in pulling the sludge to a low point where it can be carried off by a pump suction line or some other means of removal. Both rectangular and circular tanks have several manufacturers.

6-28. Rectangular Sediment-Tank Design. The tanks to be considered here are custom-designed by the design engineer.

a. Width. The chain-pulled rakes used to collect sludge on the bottom of the tanks come in widths up to 20 feet per unit. It is possible to use multiple pairs of chains and thus design tanks that are 100 or more feet wide insofar as the mechanization and construction are concerned. Increasing the tank width increases the possibility of disturbance by wind and short-circuiting due to temperature differential with the incoming sewage. It places greater stress on inlet and outlet design details. Tanks should be long and narrow to minimize the inlet and outlet disturbances, cross winds, density currents, and longitudinal mixing (22).

b. Tank Length. A maximum length of 300 ft is suggested. The minimum length of flow from inlet to outlet should be 10 ft unless special provisions are made to prevent short-circuiting (1). A floor slope of 1.0% is recommended. The author suggests a length-to-width ratio of 5/1.

c. Tank Depth. The liquid depth of mechanically cleaned settling tanks should be as shallow as practicable but not less than 7 ft (1). Final clarifiers for activated sludge should not be less than 8 ft.

d. Inlets. Inlets should be designed to dissipate the inlet's velocity, to distribute the flow equally, and to prevent short-circuiting. Channels should be designed to maintain a velocity of at least 1 ft/s at one-half design flow. Corner pockets and dead ends should be eliminated and corner fillets or channeling used where necessary (1). Provisions should be made for elimination or removal of floating materials in inlet structures having submerged ports.

The inlet problem is one of uniform distribution. Numerous devices have been used and the majority of them have one or more deficiencies. Some of the common schemes are: a series of inlet pipes (out of a manifold) across the end of the basin with upturned elbows, several inlet ports discharging against baffles, simple overflow weirs, and perforated baffles and an 180° return elbow combination discharging against the tank wall. If water enters under any velocity, such as a pump discharge, some form of energy-arresting may become necessary. The arrangement in Fig. 6-15 is not a time-tested scheme. It is an incomplete drawing combining portions of three different schemes which, when used individually, worked very well in the laboratory. The first part, the inlet energy dissipater, did not have tailwater as shown here. A hydraulic jump would be created by the tailwater, or backwater, at the entrance to the horizontal shelf, and the stream would jet out into the stilling basin instead of spreading uniformly over the 30° indicated. The addition of the sills S_1 and S_2 will help to spread the water and arrest the energy.

This scheme is offered here to stimulate thought and originality in looking for better inlets and is not offered as an optimum solution. In the scheme indicated, the wet-weather flow may be 2 or 3 times the maximum daily flow, and the baffles would have to be spaced with appropriate opening area to accommodate the flow, yet produce enough resistance to uniformly distribute the flow over the entire baffle area. The terms *slat* and *baffle* are used interchangeably here. The vertical slats (or baffles) provide future flexibility to the operator, but they are not a device one would want to have to change sizes upon very often.

If inlet ports are used, the gates should be provided to regulate flow. In general, an inlet scheme should be left open, or covered by removable gratings, to facilitate maintenance and inspection. Inlet devices that drop flows into sedimentation tanks should not be used.

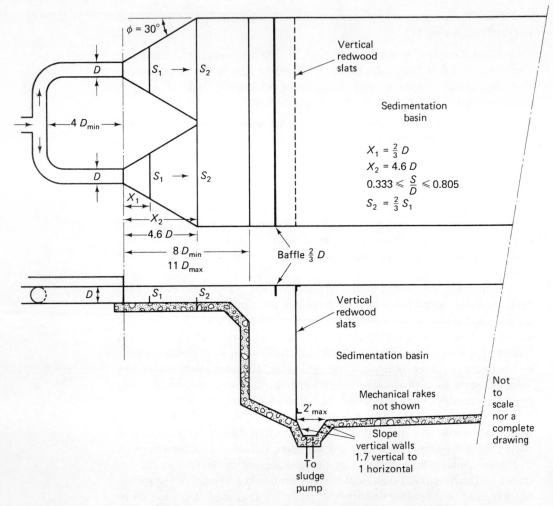

Figure 6-15. Energy Dissipater and Flow Distributor. (Energy Dissipater Design Based on Project 3-5-66-92, The Texas Highway Dept. in Cooperation with FHWA, by U. of Texas)

e. Outlets. A number of outlet schemes have been used. Adjustable horizontal weir plates with 90° V notches for low flows (or launders with multiple weirs) have been best in the author's experience. Overflow weir loadings should not exceed 10,000 GPD per linear foot for plants designed for average flows of 1.0 MGD or less, and a lower loading is preferable. Regulatory agencies often allow loadings up to 15,000 GPD per linear foot for plants over 1.0 MGD/day, but the 10,000 figure is preferable. If pumping is required, the pump capacity should be related to the tank design to avoid excessive weir loading (1).

f. Scum Removal. Effective scum collection and removal facilities including baffling should be provided ahead of the outlet weirs on all settling tanks. Provisions may be made for the discharge of scum with the sludge; other provisions, such as incineration, may be necessary to dispose of floating materials which may adversely affect sludge handling and disposal (1).

g. Submerged Surfaces. The tops of troughs, beams, and similar construction features which are submerged should have a minimum slope of 1.4 vertical to 1 horizontal. The underside of such features should have a slope of 1 to 1 to prevent the accumulation of scum and solids (1).

h. Multiple Tanks. Multiple units should be provided when removal from service of a single unit for a short period will result in objectionable conditions or material damage (1).

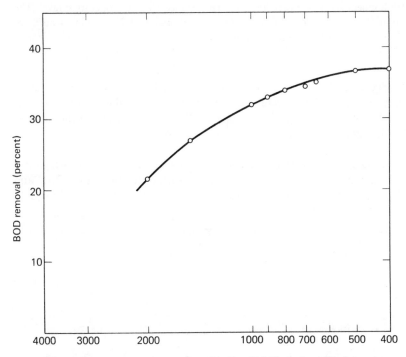

Figure 6-16. Settling Rate—Gallons Per Day/Ft² Tank Area (Design Flow) (1).

i. Surface Settling Rates. Surface settling rates for primary tanks (any type) not followed by secondary treatment should not exceed 600 GPD/ft² for plants having a design flow of 1.0 MGD or less (1). Higher surface-settling rates may be permitted for larger plants. Some indication of BOD removals may be obtained from Fig. 6-16; however, BOD removals for sewage containing appreciable quantities of industrial wastes should be determined by laboratory tests and consideration of the quantity and characteristics of the waste (1).

Surface settling rates for intermediate settling tanks (any type) should not exceed 1,000 GPD/ft² based on their design flows (1).

Surface settling rates for final settling tanks, based on significant flow periods, should not exceed 800 GPD/ft², except certain activated sludge units to be discussed later in the text (1).

j. Sludge Removal. Sludge removal is adequately expressed by the Ten States Standards: (1):

A sludge well should be provided or appropriate equipment installed for viewing and sampling the sludge. Provisions should be made to permit continuous sludge removal from final settling tanks when the sludge is returned to primary settling tanks. Each sludge hopper shall have an individual valved sludge withdrawal line at least 6 in. in diameter. Head available for withdrawal of sludge shall be at least 30 in. Sludge hoppers shall be accessible for maintenance from the operating level. The minimum slope of the side walls of sludge hoppers shall be 1.7 vertical to 1 horizontal. Clearance between the end of the sludge draw-off pipe and the hopper walls shall be sufficient to prevent "bridging" of solids. Hopper bottoms shall have a maximum dimension of 2 ft.

Sludge is collected by mechanical rake mechanisms operated at 2 to 3 ft/min. Figure 6-17 illustrates one type of heavy-duty sludge-collector raking mechanism. There are several types of these mechanisms on the market, including light-weight equipment for small tanks.

6-29. Rectangular Sedimentation Tank Design Aids. There are two easy ways to find the tank capacity: (a) If the required surface loading in GPD/ft² and the flow in MGD are known, use Table 6-4 to find the surface area in square

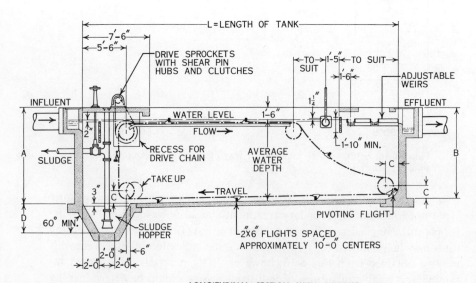

LONGITUDINAL SECTION WITH SKIMMER

Figure 6-17. Sludge-Raking Mechanism for Large Tanks. (Courtesy Environmental Equipment Division, FMC Corporation)

TABLE 6-4. Surface Area of Tank.*

Flow, MGD	Surface loading, GPD/ft^2							
	500	600	700	800	900	1000	1100	1200
	Required surface area, square feet							
.25	500	417	357	313	278	250	227	208
.33	660	550	472	413	367	330	300	275
.50	1,000	832	714	625	556	500	450	417
.66	1,320	1,100	943	825	734	660	600	550
.75	1,500	1,250	1,071	938	834	775	682	625
1.00	2,000	1,666	1,430	1,250	1,110	1,000	908	833
1.50	3,000	2,500	2,142	1,875	1,666	1,500	1,364	1,250
2.00	4,000	3,333	2,857	2,500	2,220	2,000	1,820	1,668
2.50	5,000	4,166	3,571	3,130	2,780	2,500	2,270	2,083
3.00	6,000	5,000	4,285	3,750	3,333	3,000	2,730	2,500
4.00	8,000	6,666	5,715	5,000	4,444	4,000	3,640	3,330
5.00	10,000	8,332	7,152	6,250	5,560	5,000	4,550	4,160
6.00	12,000	10,000	8,571	7,500	6,670	6,000	5,450	5,000
7.00	14,000	11,666	10,000	8,750	7,764	7,000	6,350	5,830
7.50	15,000	12,500	10,715	9,380	8,340	7,500	6,820	6,250
8.00	16,000	13,333	11,428	10,000	8,890	8,000	7,270	6,660
9.00	18,000	15,000	12,857	11,250	10,000	9,000	8,180	7,500
10.00	20,000	16,666	14,285	12,500	11,100	10,000	9,090	8,330
12.50	25,000	20,820	17,857	15,600	13,900	12,500	11,360	10,416
15.00	30,000	25,000	21,428	18,750	16,650	15,000	13,620	12,500
17.50	35,000	29,166	25,000	21,875	19,450	17,500	15,900	14,600
20.00	40,000	33,333	28,500	25,000	22,222	20,000	18,180	16,650
25.00	50,000	41,666	35,700	31,250	27,800	25,000	22,750	20,820
30.00	60,000	50,000	42,950	37,450	33,180	30,000	27,300	25,000
35.00	70,000	58,333	50,000	43,550	38,850	35,000	31,800	29,200
37.50	75,000	62,500	53,600	46,875	41,540	37,500	34,100	31,200
40.00	80,000	66,666	57,200	50,000	44,444	40,000	36,400	33,300
45.00	90,000	75,000	64,400	56,250	50,000	45,000	40,900	37,500
50.00	100,000	83,333	71,500	62,500	55,620	50,000	45,500	41,700

*Courtesy of Environmental Equipment Division, FMC Corporation.

feet and multiply this area by the water depth to determine the tank capacity. (b) Knowing the required period of detention in hours and the flow in MGD, find the tank capacity from Table 6-5. The maximum recommended tank lengths are given in Table 6-6. After determining the tank capacity, refer to Tables 6-7 and 6-8 and find the tank dimensions. In most cases, more than one selection can be made; but use the one best suited to local conditions. It is recommended that the length of the tank be three to five times the width. Selections meeting this recommendation are in the shaded areas in Tables 6-7 and 6-8.

6-30. Circular Settling Tank. Arguments exist for all the different tank schemes. Rectangular settling tanks have the advantage of lower equipment cost, low construction cost when several tanks are used with common walls, and acceptable solids removals. The disadvantages of rectangular tanks are

TABLE 6-5. Capacity of Tank*

Capacity, cubic feet

Flow, MGD	Period of detention, hours											
	.25	.50	.75	1.00	1.25	1.50	1.75	2.00	2.50	3.00	3.50	4.00
.25	348	696	1,044	1,392	1,740	2,088	2,437	2,785	3,481	4,177	4,874	5,570
.33	459	919	1,378	1,838	2,297	2,757	3,216	3,676	4,595	5,514	6,433	7,352
.50	696	1,392	2,088	2,785	3,481	4,177	4,874	5,570	6,963	8,355	9,748	11,140
.66	919	1,838	2,757	3,676	4,595	5,514	6,433	7,352	9,191	11,029	12,867	14,705
.75	1,044	2,088	3,133	4,177	5,222	6,266	7,311	8,355	10,444	12,533	14,622	16,711
1.00	1,392	2,785	4,177	5,570	6,963	8,355	9,748	11,140	13,926	16,711	19,496	22,281
1.50	2,088	4,177	6,266	8,355	10,444	12,533	14,622	16,711	20,889	25,066	29,244	33,422
2.00	2,785	5,570	8,355	11,140	13,926	16,711	19,496	22,281	27,852	33,422	38,992	44,563
2.50	3,481	6,963	10,444	13,926	17,407	20,889	24,370	27,852	34,815	41,778	48,741	55,704
3.00	4,177	8,355	12,533	16,711	20,889	25,066	29,244	33,422	41,778	50,133	58,489	66,844
4.00	5,570	11,140	16,711	22,281	27,852	33,422	38,992	44,563	55,704	66,844	77,985	89,126
5.00	6,963	13,926	20,889	27,852	34,815	41,778	48,741	55,704	69,630	83,556	97,482	111,408
6.00	8,355	16,711	25,066	33,422	41,778	50,133	58,489	66,844	83,556	100,267	116,978	133,689
7.00	9,748	19,496	29,244	38,992	48,741	58,489	68,237	77,985	97,482	116,978	136,474	155,971
7.50	10,444	20,889	31,333	41,778	52,222	62,667	73,111	83,556	104,445	125,334	146,223	167,112
8.00	11,140	22,281	33,422	44,563	55,704	66,844	77,985	89,126	111,408	133,689	155,971	178,252
9.00	12,533	25,066	37,600	50,133	62,667	75,200	87,733	100,267	125,334	150,400	175,467	200,534
10.00	13,926	27,852	41,778	55,704	69,630	83,556	97,482	111,408	139,260	167,112	194,964	222,816
12.50	17,407	34,815	52,222	69,630	87,037	104,445	121,852	139,260	174,075	208,890	243,705	278,520
15.00	20,889	41,778	62,667	83,556	104,445	125,334	146,223	167,112	208,890	250,668	292,446	334,224
17.50	24,370	48,741	73,111	97,482	121,852	146,223	170,593	194,964	243,705	292,446	341,187	389,928
20.00	27,852	55,704	83,556	111,408	139,260	167,112	194,964	222,816	278,520	334,224	389,928	445,632
25.00	34,815	69,630	104,445	139,260	174,075	208,890	243,705	278,520	348,150	417,780	487,410	557,040
30.00	41,778	83,556	125,334	167,112	208,890	250,668	292,446	334,224	417,780	501,336	584,892	668,448
35.00	48,741	97,482	146,223	194,964	243,705	292,446	341,187	389,928	487,410	584,892	682,374	779,856
37.50	52,222	104,445	156,667	208,890	261,112	313,335	365,557	417,780	522,225	626,670	731,115	835,560
40.00	55,704	111,408	167,112	222,816	278,520	334,224	389,928	445,632	557,040	668,448	779,856	891,264
45.00	62,667	125,334	188,001	250,668	313,335	376,002	438,669	501,336	626,670	752,004	877,338	1,002,672
50.00	69,630	139,260	208,890	278,520	348,150	417,780	487,410	557,040	696,300	835,560	974,820	1,114,080

*Courtesy of Environmental Equipment Division, FMC Corporation.

146

TABLE 6-6. Maximum Length of Tank

Detention, hours	Length of tank, feet
1.0	225
1.5	250
2.0	275
2.5	300

*Courtesy of Environmental Equipment Division, FMC Corporation.

TABLE 6-7. Single Tank and Single Logitudinal Collector Without Cross Collector.

8-foot water depth

Width, feet	Capacity, cubic feet								
	Length, feet								
	20	25	30	40	50	60	75	90	100
6	960	1,200	1,440	1,920	2,400	2,880	3,600	4,320	4,800
8	1,280	1,600	1,920	2,560	3,200	3,840	4,800	5,760	6,400
10	1,600	2,000	2,400	3,200	4,000	4,800	6,000	7,200	8,000

10-foot water depth

	25	30	40	50	60	75	90	100	125
8	2,000	2,400	3,200	4,000	4,800	6,000	7,200	8,000	10,000
10	2,500	3,000	4,000	5,000	6,000	7,500	9,000	10,000	12,500
12	3,000	3,600	4,800	6,000	7,200	9,000	10,800	12,000	15,000
14	3,500	4,200	5,600	7,000	8,400	10,500	12,600	14,000	17,500
16	4,000	4,800	6,400	8,000	9,600	12,000	14,400	16,000	20,000
18	4,500	5,400	7,200	9,000	10,800	13,500	16,200	18,000	22,500
20	5,000	6,000	8,000	10,000	12,000	15,000	18,000	20,000	30,000

12-foot water depth

	30	40	50	60	75	90	100	125	150
10	3,600	4,800	6,000	7,200	9,000	10,800	12,000	15,000	18,000
12	4,320	5,760	7,200	8,640	10,800	12,960	14,400	18,000	21,600
14	5,040	6,720	8,400	10,080	12,600	15,120	16,800	21,000	25,200
16	5,760	7,680	9,600	11,520	14,400	17,280	19,200	24,000	28,800
18	6,480	8,640	10,800	12,960	16,200	19,440	21,600	27,000	32,400
20	7,200	9,600	12,000	14,400	18,000	21,600	24,000	30,000	36,000

Shaded areas: recommended lengths.
Courtesy of Environmental Equipment Division, FMC Corporation.

high upkeep and maintenance costs of sprockets, chain, and flights; more complicated structural design; possible dead spaces; and limited control of floating material.

The circular tanks can be classified as center feed and peripheral feed types. The peripheral feed can further be subdivided into the types shown in Fig. 6-18(b) and 6-18(c). All of these types are called *clarifiers*. The center-feed types have low upkeep costs and are easy to design and to construct. They have the disadvantage of possible short-circuiting, lack of positive scum control, sludge creepage, and lower detention efficiency. The peripheral feed has a skirt around the inside of the tank, and the liquid must slow down under the skirt. At the bottom of the skirt, the flow enters a larger area where the velocity is further reduced.

There are various research studies that have been conducted on sedimentation, parts of which are referenced in this chapter; however, a study by

TABLE 6-8. Single Tank and Multiple Longitudinal Collectors With One Cross-Collector.

10-foot water depth

Width, feet	Designation ▲	Capacity, cubic feet — Length, feet									
		50	75	100	125	150	175	200	225	250	300
25	1-212-1	12,500	18,750	25,000	31,250	37,500	43,750	50,000	56,250	62,500	75,000
29	1-214-1	14,500	21,750	29,000	36,250	43,500	50,750	58,000	65,250	72,500	87,000
33	1-216-1	16,500	24,750	33,000	41,250	49,500	57,750	66,000	74,250	82,500	99,000
37	1-218-1	18,500	27,750	37,000	46,250	55,500	64,750	74,000	83,250	92,500	111,000
38	1-312-1	19,000	28,500	38,000	47,500	57,000	66,500	76,000	85,500	95,000	114,000
41	1-220-1	20,500	30,750	41,000	51,250	61,500	71,750	82,000	92,250	102,500	123,000
44	1-314-1	22,000	33,000	44,000	55,000	66,000	77,000	88,000	99,000	110,000	132,000
50	1-316-1	25,000	37,500	50,000	62,500	75,000	87,500	100,000	112,500	125,000	150,000
51	1-412-1	25,500	38,250	51,000	63,750	76,500	89,250	102,000	114,750	127,500	153,000
56	1-318-1	28,000	42,000	56,000	70,000	84,000	98,000	112,000	126,000	140,000	168,000
59	1-414-1	29,500	44,250	59,000	73,750	88,500	103,250	118,000	132,750	147,500	177,000
62	1-320-1	31,000	46,500	62,000	77,500	93,000	108,500	124,000	139,500	155,000	186,000
67	1-416-1	33,500	50,250	67,000	83,750	100,500	117,250	134,000	150,750	167,500	201,000
75	1-418-1	37,500	56,250	75,000	93,750	112,500	131,250	150,000	168,750	187,500	225,000
83	1-420-1	41,500	62,250	83,000	103,750	124,500	145,250	166,000	186,750	207,500	249,000

12-foot water depth

Width, feet	Designation	50	75	100	125	150	175	200	225	250	300
25	1-212-1	15,000	22,500	30,000	37,500	45,000	52,500	60,000	67,500	75,000	90,000
29	1-214-1	17,400	26,100	34,800	43,500	52,200	60,900	69,600	78,300	87,000	104,400
33	1-216-1	19,800	29,700	39,600	49,500	59,400	69,300	79,200	89,100	99,000	118,800
37	1-218-1	22,200	33,300	44,400	55,500	66,600	77,700	88,800	99,900	111,000	133,200
38	1-312-1	22,800	34,200	45,600	57,000	68,400	79,800	91,200	102,600	114,000	136,800
41	1-220-1	24,600	36,900	49,200	61,500	73,800	86,100	98,400	110,700	123,000	147,600
44	1-314-1	26,400	39,600	52,800	66,000	79,200	92,400	105,600	118,800	132,000	158,400
50	1-316-1	30,000	45,000	60,000	75,000	90,000	105,000	120,000	135,000	150,000	180,000
51	1-412-1	30,600	45,900	61,200	76,500	91,800	107,100	122,400	137,700	153,000	183,600
56	1-318-1	33,600	50,400	67,200	84,000	100,800	117,600	134,400	151,200	168,000	201,600
59	1-414-1	35,400	53,100	70,800	88,500	106,200	123,900	141,600	159,300	177,000	212,400
62	1-320-1	37,200	55,800	74,400	93,000	111,600	130,200	148,800	167,400	186,000	223,200
67	1-416-1	40,200	60,350	80,400	100,500	120,600	140,700	160,800	180,900	201,000	241,200
75	1-418-1	45,000	67,500	90,000	112,500	135,000	157,490	180,000	202,500	225,000	270,000
83	1-420-1	49,800	74,700	99,600	124,500	149,400	174,300	199,200	224,100	249,000	298,800

▲ Explanation of designations: **1-212-1**

Number of tanks
Number of longitudinal collectors
Width of each longitudinal collector
Number of cross collectors

Courtesy of Environmental Equipment Division, FMC Corporation.

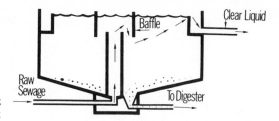

(a) Center Feed Circular Settling Tank

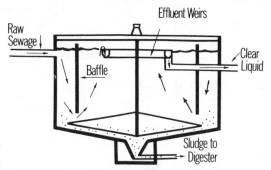

(b) Peripheral Feed Circular Settling Tank

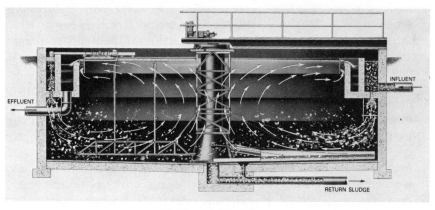

(c) Rim-Flo Clarifier

Figure 6-18. Circular Settling Tanks. (Courtesy Environmental Equipment Division, FMC Corporation)

Dague and Baumann concerned itself specifically with evaluating the two circular tank types indicated in 6-18(a) and 6-18(b). This indicated that the peripheral-feed clarifier is from two to four times more efficient, hydraulically, than the center-feed clarifier. The comparisons of model to prototype indicate that models operating in accordance with the Froud law can be used with confidence in settling tank studies.

A comparison of the flow curves obtained in the above work is shown in Fig. 6-19. The author has used both types with satisfactory results but has used the peripheral-feed type most often for units under 60 ft in diameter.

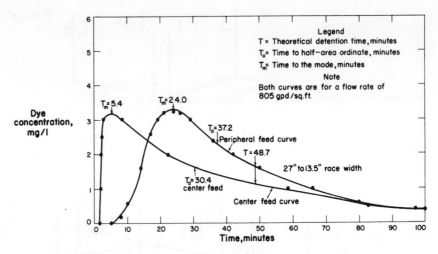

Figure 6-19. Comparison of Flow Curves. (Courtesy Richard R. Dague, E. Robert Baumann, and Lakeside Equipment Corp.)

The Rim-flo clarifier, shown in Fig. 6-18(c) seems to operate quite satisfactorily in larger sizes. There is an argument that, as the peripheral-feed type shown in Fig. 6-18(b) increases in diameter, a point is reached beyond which the inlet distribution into the clarfier is not uniform. The Rim-flo concept is supposed to retain the advantages of the peripheral feed in the larger-size units.

The engineer has the responsibility of selecting for his client the type of unit that in his judgment will be the best engineering solution considering all factors including economics. Changes in corporate management, management policy, or controlling parent corporation, may result in a change in quality of equipment construction in either direction. It is therefore advisable to check all structural design details, how the sludge-scraper drive mechanism is mounted, all rotating parts, cantilever devices, drive mechanisms, size of drive motors, speeds, etc. It is also advisable to check access platforms, ladders, safety provisions, hydraulic details of the unit selected, maintenance details, etc. Check the thickness of materials used, corrosion protection, and other pertinent details.

Table 6-4 can be used to determine the surface area of a circular tank required for a given flow in MGD. With the surface area known, Table 6-9 can be used to determine the inside diameter of a tank, tank volume in cubic feet, and tank weir length at periphery for the center-feed type of clarifier.

The peripheral-feed clarifier in Fig. 6-18(c) is treated in detail in Chapter 9.

SECONDARY WASTE TREATMENT

6-31. Activated Sludge. The activated-sludge process is the most versatile of the biological treatment systems, and it has the greatest number of variations

TABLE 6-9. Determination of Tank Diameter, Weir Length at Periphery and Tank Volume, When Surface Area is Known.

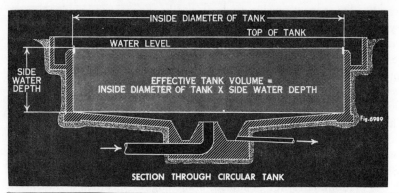

Surface area, sq. ft.	Inside diameter of tank, feet	Weir length at periphery, feet	Side water depth, feet								
			7	8	9	10	11	12	13	14	15
			Tank volume, cubic feet								
314	20	62.8	2,198	2,512	2,826	3,140	3,454	3,768	4,082	4,396	4,710
491	25	78.2	3,437	3,928	4,419	4,910	5,401	5,892	6,383	6,874	7,365
707	30	94.2	4,949	5,656	6,363	7,070	7,777	8,484	9,191	9,898	10,605
962	35	109.9	6,734	7,696	8,658	9,620	10,582	11,544	12,506	13,468	14,430
1,257	40	125.6	8,799	10,056	11,313	12,570	13,827	15,084	16,341	17,598	18,855
1,590	45	141.3	11,130	12,720	14,310	15,900	17,490	19,080	20,670	22,260	23,850
1,964	50	157.0	13,748	15,712	17,676	19,640	21,604	23,568	25,532	27,496	29,460
2,376	55	172.7	16,632	19,008	21,384	23,760	26,136	28,512	30,888	33,264	35,640
2,827	60	188.5	19,789	22,616	25,443	28,270	31,097	33,924	36,751	39,578	42,405
3,318	65	204.2	23,226	26,544	29,862	33,180	36,498	39,816	43,134	46,452	49,770
3,849	70	219.9	26,943	30,792	34,641	38,490	42,339	46,188	50,037	53,886	57,735
4,418	75	235.6	30,926	35,344	39,762	44,180	48,598	53,016	57,434	61,852	66,270
5,027	80	251.3	35,189	40,216	45,243	50,270	55,297	60,324	65,351	70,378	75,405
5,675	85	267.0	39,725	45,400	51,075	56,750	62,425	68,100	73,775	79,450	85,125
6,362	90	282.7	44,534	50,896	57,258	63,620	69,982	76,344	82,706	89,068	95,430
7,088	95	298.4	49,616	56,704	63,792	70,880	77,968	85,056	92,144	99,232	106,320
7,854	100	314.1	54,978	62,832	70,686	78,540	86,394	94,248	102,102	109,956	117,810
8,659	105	329.8	60,613	69,272	77,931	86,590	95,249	103,908	112,567	121,226	129,885
9,503	110	345.5	66,521	76,024	85,527	95,030	104,533	114,036	123,539	133,042	142,545
10,387	115	361.2	72,709	83,096	93,483	103,870	114,257	124,644	135,031	145,418	155,805
11,310	120	376.9	79,170	90,480	101,790	113,100	124,410	135,720	147,030	158,340	169,650
12,272	125	392.7	85,904	98,176	110,448	122,720	134,992	147,264	159,536	171,808	184,080
13,273	130	408.4	92,911	106,184	119,457	132,730	146,003	159,276	172,549	185,822	199,095
14,314	135	424.1	100,198	114,512	128,826	143,140	157,454	171,768	186,082	200,396	214,710
15,394	140	439.8	107,758	123,152	138,546	153,940	169,334	184,728	200,122	215,516	230,910
16,513	145	455.5	115,591	132,104	148,617	165,130	181,643	198,156	214,669	231,182	247,695
17,672	150	471.2	123,704	141,376	159,048	176,720	194,392	212,064	229,736	247,408	265,080
18,869	155	486.9	132,083	150,952	169,821	188,690	207,559	226,428	245,297	264,166	283,035
20,106	160	502.6	140,742	160,848	180,954	201,060	221,166	241,272	261,378	281,484	301,590
21,383	165	518.3	149,681	171,064	192,447	213,830	235,213	256,596	277,979	299,362	320,745
22,698	170	534.0	158,886	181,584	204,282	226,980	249,678	272,376	295,074	317,772	340,470
24,053	175	549.7	168,371	192,424	216,477	240,530	264,583	288,636	312,689	336,742	360,795
25,447	180	565.4	178,129	203,576	229,023	254,470	279,917	305,364	330,811	356,258	381,705
26,880	185	581.1	188,160	215,040	241,920	268,800	295,680	322,560	349,440	376,320	403,200
28,353	190	596.9	198,471	226,824	255,177	283,530	311,883	340,236	368,589	396,942	425,295
29,865	195	612.6	209,055	238,920	268,785	298,650	328,515	358,380	388,245	418,110	447,975
31,416	200	628.3	219,912	251,328	282,744	314,160	345,576	376,992	408,408	439,824	471,240

If weir overflow rate exceeds design requirements, use effluent trough with weirs on both sides.

Weir overflow rate (GPD/lin. ft. of weir) $= \dfrac{\text{flow (GPD)}}{\text{length of weir (ft.)}}$

Detention period (hours) $= \dfrac{\text{tank volume (cu. ft.)} \times 179.52}{\text{flow (GPD)}}$

Courtesy of Environmental Equipment Division, FMC Corporation.

that can be used to meet specific needs. Basically, the activated-sludge process is a system in which biologically active growths are continuously circulated with incoming biologically degradable waste in the presence of oxygen. The oxygen is provided by some means of aeration, and the process is aerobic except in the digesters. After a period of time, a large mass of floc or settleable solids is formed. These solids, called *activated sludge*, are the activated masses of microorganisms.

The ability of the floc to settle rapidly is affected by the compactness of the floc, the number of live organisms, and the mass of the floc (27). The compactness of the floc is dependent on the type of organisms in the floc. If the floc is predominately small bacilli or diplobacilli, it is compact. But if the floc is predominately long-chained, branching, or sheathed organisms, it tends to be loose. The latter sludge has predominantly poorer settling characteristics than the former, and the term *sludge bulking* is applied.

Sludge accumulation is influenced by the concentration of aeration-tank solids, the sewage strength in terms of BOD or suspended solids (SS), and the relationship between the sewage strength and the solids concentration in the aeration tank (28). The accumulation of suspended solids is less where the concentration of activated sludge is the greatest. The process may be visualized as a "fluid-bed" process in which microorganisms are distributed throughout the wastewater. The trickling filter is a fixed-bed process where the waste is trickled over an active biological mass adhering to the media comprising the filter bed. The activated sludge system generally requires more operational skill and attention than the other processes, but when properly designed and operated, some versions of the process are capable of the highest solids and BOD removals obtainable by biological action.

A diagram of one form of activated-sludge plant is indicated in Fig. 6-20. The portion of this diagram of immediate concern is the aeration tank and secondary settling tank (final clarifier). The re-aeration tank would be of interest in modifications of the process. The information necessary to design the primary sedimentation tank (if it is used at all) has been set forth in Sections 6-27 through 6-30. The same principle is applied to designing the secondary settling tank by using the surface settling rates specified by Ten-State Standards provided later in this chapter. In one form or another, the aeration tank, the return of sludge to the aeration tank, and where used, the supplemental aeration tanks which form the portion of the system, all make up the so-called *activated-sludge* system.

The presentation in this chapter is aimed primarily toward custom-built municipal plants. Chapter 8 contains more information on the process in connection with package treatment plants; Chapter 9 discusses the oxidation ditch variation of extended aeration in detail; and Chapter 10 provides a comprehensive discussion on mechanical aeration not covered in Chapter 9. Hence only diffused aeration is presented in this chapter.

a. Temperature. Thomas (29) made a review of low-temperature area treatment and indicated that activated sludge aeration units had less difficulty

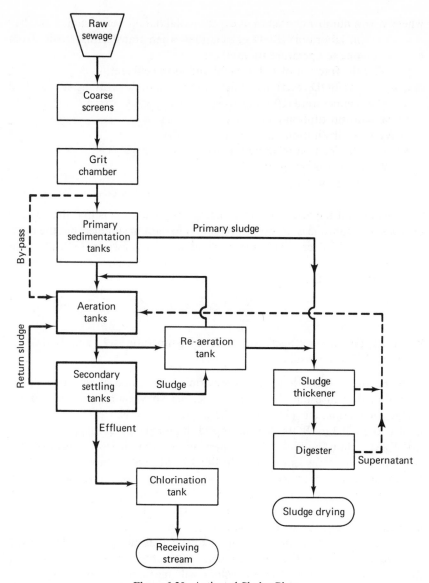

Figure 6-20. Activated-Sludge Plant.

with freezing than did trickling filters. He also indicated that the effect of low temperature can be counteracted, in a measure, by increasing either the aeration tank solids or the detention period.

Using material from Howland (30) and Bloodgood (31), the following expressions can be derived:

$$p = 1 - 10^{-t k_{20} M} \tag{6-18}$$

$$c = \left[\frac{t_1 \log f_2}{t_2 \log f_1} \right]^{N_\beta} \tag{6-19}$$

where c = a number chosen to fit experimental data (often assumed as 1.047 in laboratory BOD calculations when transferring results from one temperature to another,

f = the fraction of BOD remaining in the effluent,

k_{20} = the BOD reaction coefficient in common logarithms, at standard temperature (20°C) in 1/days,

M = a substitution factor = $c^{(T-20)}$,

N_β = a substitution factor = $1/(T_2 - T_1)$,

p = the fraction of BOD removed = $1 - f$,

T = temperature, in °C,

t = time, in days.

The effect of temperature on the biological population is that the number of types of organisms is lowest in winter and highest in summer, but the number of individuals decreases in summer and increases in winter (32). The temperature-respiration relationship for activated sludge in the temperature range between 0 and 25°C has been found to be (33):

$$\frac{R_{T_1}}{R_{T_2}} = e^{\beta(T_1 - T_2)} \tag{6-20}$$

where R_{T_1} and R_{T_2} are respiration rates at temperatures T_1 and T_2, respectively, and β is a constant.

b. Loading. BOD loading will determine the mode of activated-sludge operation. It will establish removal efficiency, oxygen requirement, excess sludge production, etc. (34). Domestic sewage cannot be treated at a loading higher than 0.3 lb BOD per lb mixed liquor volatile suspended solids (MLVSS) without bulking (35). Cannery and paper mill waste may be treated efficiently at loadings as high as 3 lb BOD/lb MLVSS without bulking (36). BOD removal decreases with an increase in loading of the same substrate but, at a given loading, varies with the chemical composition of the substrate (37). The chemical composition of the substrate affects the loading at which bulking will occur and the speed of bulking.

Table 6-10 indicates the aeration tank capacities and permissible loadings for several of the modifications of the activated-sludge process as well as for the conventional process. These values will meet most regulatory agency requirements.

c. Sludge Age. The sludge age is a measure of the time a particle of suspended solids has been undergoing aeration and is expressed in terms of days. $S = X_v/S_S = t_a S_a/24S_s$, where S is the Gould Sludge Age (GSA), X_v is the pounds of MLSS (total weight of suspended solids in the mixed liquor of the aerator tanks), S_S is pounds of incoming raw sewage suspended solids introduced from the primary clarifier or raw sewage if no clarifier is used, t_a is the aeration time in hours, S_a is the suspended solids in the mixed liquor of the aeration tank expressed in mg/l, and S_s is the suspended solids concentration

TABLE 6-10. Aeration Tank Capacities and Permissible Loadings (1)

Process	MGD-Plant Design Flow	Aeration Retention Period-Hours (Based on Design Flow)	Plant Design Lb BOD$_5$/Day	Aerator Loading Lb BOD$_5$/1000 ft^3	MLSS/Lb BOD$_5$*
Conventional	to 0.5	7.5	to 1,000	30	2/1 to
	0.5 to 1.5	7.5 to 6.0	1,000 to 3,000	30 to 40	4/1
	1.5 up	6.0	3,000 up	40	
Modified or high-rate	all	2.5 up	2,000 up	100	1/1 (or less)
Step aeration	0.5 to 1.5	7.5 to 5.0	1,000 to 3,000	30 to 50	2/1 to
	1.5 up	5.0	3,000 up	50	5/1
Contact stabilization	to 0.5	3.0 (in contact zone)†	to 1,000	30	
	0.5 to 1.5	3.0 to 2.0†	1,000 to 3,000	30 to 50	2/1 to
	1.5 up	1.5 to 2.0†	3,000 up	50	5/1
Extended aeration	all	24	all	12.5	as low as 10/1 to as high as 20/1

*MLSS/lb BOD$_5$—normally recommended values at ratio of mixed liquor suspended solids under aeration to BOD loading.

†Contact zone—30 to 35% of total aeration capacity. Re-aeration zone comprises the balance of the aeration capacity.

entering the aeration tank, mg/l from the primary clarifier or raw sewage entering the aerator.

According to Keefer and Meisel (38), the age of an activated sludge is indicated by its E_h (oxidation-reduction) potential. The E_h potential decreases rapidly during the early stage of its aging, and it takes from 15 hours to 3 days to reach an E_h of zero. When the potential is less than zero, it decreases at a much slower rate. Aged activated sludge with a low positive or negative E_h potential, when added to sewage and aerated 4 or 5 hr, produces a mixed liquor with a potential less than normal. The total 37°C bacteria count is greatest when the material is 1 hr old. There was no indication that the age of the activated sludge had any effect on the sludge index.

d. Sludge Concentration Factor (SCF). The process of reducing the water content present in a sludge is called *sludge concentration.* The concentration of sludge in general is the amount of sludge dissolved in a given body of liquid. The sludge concentration factor can be expressed as $\delta = S_R/S_a$, where S_R is the return sludge concentration (final clarifier to aerator) and S_a is the aerator sludge concentration. In Fig. 6-21 the plant $SCF = \delta = S_R/S_a = 10,000/2,000 = 5$.

e. Sludge Produced. Sludge produced from a waste can vary from 0.5 to 0.7 that of the BOD; however, for domestic waste 0.5 lb sludge per 1 lb BOD is frequently used.

f. Recycle Ratio. The recycle ratio is used to express the amount of sludge that is returned from the final clarifier to the aerator tank.

$$F = \frac{R_s}{I} \tag{6-21}$$

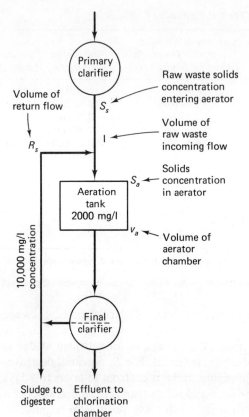

Figure 6-21
Sludge Return to Aerator.

where F is the recycle ratio, R_s is the volume of return flow from the secondary clarifier (Fig. 6-21), and I is the volume of incoming flow. The most common recycle ratio is $F = 0.25$ as a ratio, 25% as a percentage, or 1:4 proportion. The column "Sludge Recycle Ratio %" in Table 6-11 indicates the most common operating recycle ratios for several variations of the activated-sludge process. The values that are not in parentheses are the values provided as minimum and normal for activated-sludge plants by the Ten-State Standards (1). The values in parentheses are indicative of some plants that have been constructed but not operated in accordance with the Ten-States Standards. The "Max Recycle Rate %" column indicates the maximum pumping rate required for the return sludge pump by the Ten-State Standards. The values are particularly applicable to custom-built plants. Later in the text, with some of the proprietary units some deviations from these figures will be noted.

For activated-sludge plants (except extended aeration), the percentage value of the recycle ratio can be calculated

$$F \text{ (in \%)} = \frac{100}{120/\Psi S_a - 1} \qquad (6\text{-}22)$$

where Ψ is the Mohlman index.

TABLE 6-11. Range of Factors in Various Activated Sludge Processes

Type of Plant	OLF[5]	VLF[6]	Detent Time[7] (hr)	Sludge Recycle Ratio (%)	Max Recycle Rate (%)	Excess Sludge (%)	Max Excess Sludge (%)	Eff. BOD Removal (%)	MLSS (mg/l)
Conventional standard rate	0.2–0.5	30–40 (20–40)	6–7.5 (4–8)	15–30 (10–30)	75	6–15	25	85–95	2,000–3,000 (1,500–3,000)
Tapered aeration	0.2–0.5	30–40 (20–40)	6–7.5 (4–8)	15–30 (10–30)	75	6–15	25	85–95	1,000–3,000
Step aeration	0.2–0.5	30–50 (40–60)	5–7.5 (4–8)	20–50 (20–35)	75	6–15	25	85–95	2,000–3,000
Contact stabilization	0.2–0.5	30–50 (30–75)	1.5–3[1] (0.5–3[1]) 2–5[2]	50–100 (20–125)	150	6–15	25	85–95	2,500 (4,000–2,000)
Modified	0.5–5.0	100 (75–150)	2.5+ (0.5 Up)	10–20 (5–20)	50	3	25	50–75	200–500
Extended aeration	0.05–.20	12.5 (10–25)	24[3] (12–48)	50–100 (75–150)	200 300[4]	20–30		75–95	3,000–5,000
High-rate	0.5–5.0	100 (100–1,000)	2.5+ (1–5)	10–20	500	6–15	25	80–95	4,000–10,000

[1]Aerator.
[2]Stabilizer.
[3]Based on influent only.
[4]Without including recirculation.
[5]OLF = organic load factor (lb BOD/day.)/MLSS.
[6]VLF = volumetric load factor (lb BOD/day)/1,000 Ft[3]
[7]Detent time = detention time in hours.

$$\Psi = \frac{V}{X_v} \qquad (6\text{-}23)$$

where V is the volume of sludge settled in 30 minutes, in percent, and X_v is the mixed liquor suspended solids expressed as a percentage. S_a is the aerator solids concentration expressed as a percentage; thus an aerator solids concentration of 3000 mg/l = 0.4%.

$$S_a = \frac{R_s \times 10^6}{\Psi(1 + R_s)} \qquad (6\text{-}24)$$

Equation (6-24) was formulated by Kraus (39). R_s is the volume of return from the final clarifier and is in the same units as the incoming flow I (Fig. 6-21).

g. *Suspended Solids Control.*   The information available does not always fit a single equation or approach. Here are some additional approaches to computing the necessary data. The first three equations are based on the work of Torpey and Chasick (40):

$$S_R = \frac{(I + R_s)S_a - IS_s}{R_s} \qquad (6\text{-}25)$$

$$F = \frac{100R_s}{I} = \frac{(S_a - S_s)100}{S_R - S_a} \tag{6-26}$$

$$S_x = \frac{(R_s S_R + I_x S_s)}{(R_s + I_x)} \tag{6-27}$$

where F is a percentage, R_s, I and I_x are in MGD, and all other quantities are in mg/l. I_x is total primary effluent up to and including any point x, and S_x is the suspended solids to the section of channel between any two adjacent points of entrance of return sludge.

Another way to compute F is as follows:

$$F = \frac{S_a}{S_R - S_a} \tag{6-28}$$

where $S_a =$ MLSS in aerator, mg/l, and $S_R =$ settled sludge return, mg/l.

h. Hydrogen-Ion Concentration. Based on laboratory experiments (41), the best activated-sludge performance can be obtained at pH values from 7 to 7.5, although removals of oxygen consumed, suspended solids, and bacteria are good at pH values ranging from 6.0 to 9.0. Removals of oxygen consumed averaged about 43% at pH 4.0 and 54% at pH 10.0. At pH 5.0 to 5.5, the organisms seemed to become adjusted to the acidity, and after a few weeks the removal of oxygen consumed increased from 48.3 to 82.1%. Raising or lowering the pH on either side of neutrality greatly improved the sludge index.

i. Efficiency and Removal. Efficiency and settling characteristics of activated sludge differ with the chemical composition of the waste being treated (42, 43). Treatment efficiency can best be determined by a comparison of the BOD of the final effluent with that of the primary effluent.

BOD removal from a completely mixed basin for a relatively low-soluble effluent has been determined to be (44, 45)

$$kS_e = \frac{S_i - S_e}{X_a t} \tag{6-29}$$

where k is the removal rate coefficient (0.0012 for domestic sewage) (46), s_e is the soluble effluent BOD or COD, X_a is the average mixed liquor suspended solids, t is the aeration time, and S_i is the influent BOD or COD in mg/l.

$$\text{Percent efficiency} = e = \frac{100W_o}{W_i} \tag{6-30}$$

where e is the percentage efficiency of oxygen transfer, W_o is the weight of oxygen dissolved, and W_i is the weight of oxygen fed to the aeration tank. W_o and W_i are to be in same units.

The percentage efficiency of BOD removal is

$$e = \frac{100(\mathcal{L} - L)}{\mathcal{L}} \tag{6-31}$$

where L is the effluent BOD in pounds and $\mathcal{L}$ is the influent BOD in pounds.

Percent removal involves the pounds of BOD removed divided by the pounds BOD applied to the entire plant.

$$E = \frac{100(\mathcal{L} - L)}{\mathcal{L}_p} \qquad (6\text{-}32)$$

where **E** is the percent removal of the individual unit in a system and $\mathcal{L}_p$ is the total pounds of BOD applied to the plant. The last three equations can be used for quantities other than oxygen or BOD. Adding the percent removal of all the individual units should total 100% for the plant if the percent remaining in the effluent is included.

$$E = \frac{24L_a}{X \times t_a \times (1 + F)} \qquad (6\text{-}33)$$

where L_a is the applied BOD in mg/l, X is the aerator volatile suspended solids (MLVSS), t_a is the aerator retention time (hr), and F is the sludge recycle ratio as a decimal. This equation is normally used with pilot plant data.

j. Donaldson versus Mohlmann Index. The Donaldson index $= 100/\Psi$. A good settling sludge may have a Mohlmann index significantly lower than 100, whereas a sludge with a Mohlmann index of 200 is characterized by poor settling characteristics.

6-32. Aeration. There are two methods of achieving aeration. The first might be called a waterfall method. It includes cascades, multiple trays, spray nozzles, and any other means by which water is permitted to fall through the air or be tossed into the air by mechanical motion. Such methods are discussed in Chapters 9 and 10. The second method is injection (or bubble) aeration. It includes any method by which air is permitted to bubble through the water. The theoretical principles of oxidation adsorption have been summarized in the literature, references (47) through (70), and serve as background for some of the comments that will follow. See also Chapter 8.

a. Aeration Devices. There are a number of methods that have been used to inject air into water for aeration purposes. They include porous ceramic diffusers, nonceramic porous diffusers, Pemberthy ejectors, perforated pipes, slotted pipes, "Jet diffusers," spargers, swing-mounted plastic-wrapped tubes, and other schemes. These devices are adequately described in existing publications (71, 72). Spargers (Fig. 6-22), swing-mounted plastic-wrapped tubes, and porous ceramic diffusers are all in significant use. In a given application, the author would consider these devices in the order listed. These aeration devices have a dual purpose. They transfer oxygen from the air bubbles to the waste and serve as a means of achieving mixing in the aeration tank.

If air bubbles of 0.10-in. and 0.15-in. diameters are compared at 15-ft depth, zero D.O., 29.3 in. Hg, 20°C, the smaller bubble will have over 30% more absorption rate and requires only 55% as much power per pound of

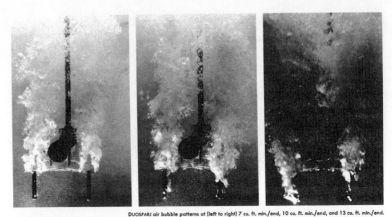

DUOSPARJ air bubble patterns at (left to right) 7 cu. ft. min./end, 10 cu. ft. min./end, and 13 cu. ft. min./end.

Figure 6-22. DUOSPARJ Air Bubble Patterns. (Courtesy Walker Process Equipment)

oxygen adsorbed. Large bubbles yield poorer gas transfer because of the smaller surface area available. The coefficient of oxygen absorption decreases as the 2.0 power of bubble radius, and smaller bubbles give a higher rate of transfer coefficient not only because of the larger surface contact area available but also because of the increase of frequency of collision between gas molecules and water molecules (68).

A wide-band diffuser arrangement gives a higher oxygen transfer as compared to a narrow-band diffuser arrangement at the same air-flow rates (60). Also, with Saran tubes in narrow-band arrangement, the oxygen absorption decreases as the air-flow rate per tube increases, whereas with the wide-band arrangement, the oxygen absorption decreases only slightly. With spargers, the air-flow per diffuser has little or no effect on the efficiency. Air-bubble entrainment increases with decreasing aeration tank width and results in an increase in oxygen absorption. Air diffusers can be clogged from the outside, but clogging from the inside by particulate matter is more critical. An air standard of 0.1 mg/1000 ft³ is recommended as a minimum (69). If porous ceramic diffusers are used, the author recommends not more than 0.04 mg of dirt per 1,000 ft³ is a desirable design standard if circumstances permit. The manufacturer of any given aeration device usually provides suggested standards on cleanness of the air required and may provide a range of design data.

Locate air diffusers 1-ft minimum and 2-ft maximum above the tank floors to aid in tank cleaning and reduce clogging during shutdowns. A discharge rate between 2 ft³/minute minimum and a maximum of 8 ft³/minute per square foot of effective diffuser area is suggested. Provide at least 3 ft³/minute of air per foot length of channel. A wide-band diffuser is recommended. The oxygen demand at the inlet is 4 to 5 times that at the outlet; hence a tapered aeration scheme is desirable if the process is not designed to switch to step aeration at a later point in the plant operation. Tapered and step operation will be discussed later in this chapter.

b. Air Application Rates. In a tank 15 ft deep (56),

$$r = K_1 A^{0.35} \tag{6-34}$$

$$r = K_2 A^{0.24} \tag{6-35}$$

where r is the air application rate in ft^3/minute per 1,000 ft^3 per ppm oxygen absorption per hour. Equation (6-34) is for a diffuser plate area of 5% and Eq. (6-35) is for a diffuser plate area of 10%. K_1 and K_2 are constants obtaining when all factors remain unchanged except air application rate, 4, and oxygen absorption rate, A, in ppm/hr. The air required per pound BOD removed from an equal volume of sewage, with diffuser plate area of 10% at equal flow rate, 15-ft-deep tank, with one sewage containing twice the BOD, would be $(2)^{0.24} = 1.18$ times that of the weaker sewage.

King (57) derived equations for bubble diameters of 0.1 to 0.2 in., depths of 1 to 2.5 ft, water temperature of 8° to 25°C, and D.O. less than 50% saturation.

$$A = \frac{0.0243 C H^{0.71} U^y R^x (1.024)^T}{d^{0.64}} \tag{6-36}$$

$$A = \frac{0.0175 C H^{0.75} U^y R^n (1.024)^T}{d^{0.70}} \tag{6-37}$$

Equation (6-36) applies to a diffuser plate, area of 5% and Eq. (6-37) to one of 10%. In these equations:

A = oxygen absorption in ppm/hr,
H = depth to porous plate diffusers in feet,
U = oxygen deficiency for water in ppm at atmospheric pressure,
R = air diffusion rate in ft^3/minute per 1,000 ft^3 of water (free air at 60°F and 29.3 in. Hg),
d = weighted average diameter of air bubbles in inches,
$y = 1/H^{0.1}$,
$n = 0.90/H^{0.04}$,
$x = 0.85/H^{0.05}$,
C = oxygen absorption coefficient (1 for clean water but lower values for sewage and mixed liquor).

$$\text{Air requirement} = \xi I L_a \tag{6-38}$$

where ξ = coefficient ≈ 0.005 ft^3 air/gallon, I = influent to aerator in gallons/day, and L_a = BOD of influent in ppm.

Air requirements are frequently obtained by running pilot plant studies. If pilot plant data is obtained, plot the lb 5-day BOD removed/day/lb MLVSS in the x direction and lbs oxygen/day/lb MLVSS in the y direction. The lbs oxygen required/day = slope of the curve x (lb BOD removed) + Y intercept × lbs MLVSS. The oxygen utilization rate = oxygen/day (lb) divided by (V × 8.34 × 24 hr), yielding the ppm/hr.

c. Total Flow.   The total flow in the aerator tank can be computed from:

$$Q_g = I + IF \tag{6-39}$$

where Q_g = total gallons/day flow, I = raw waste influent in gallons/day, and F = recycle ratio as a decimal.

In designing the size of a particular unit, remember that the actual flow through the unit will be the incoming raw waste flow plus all return flows that are adding to this flow.

d. Tank Volume.   The aerator tank volume (v_g) in gallons is obtained by

$$v_g = 60Q_g t_a \tag{6-40}$$

where Q_g = total flow through the aerator in gallons/minute, t_a is the aerator retention time in hours from Table 6-10, and 60 is minutes/hour used as a conversion factor. This is a simple hydraulic volume.

e. Freeboard.   Aeration tank freeboard as small as 8 in. has been used; however, a freeboard of 18 in. is recommended. This freeboard is in addition to the liquid depth of the tank that is normally meant when the term *tank depth* is used in this text.

f. Tank Length.   The tank length can be determined by several methods, one of which is as follows:

$$\text{Tank length (ft)} = \frac{Q_G t_a}{180WH} \tag{6-41}$$

where Q_G is the rate of flow in gallons/day, t_a is the period of aeration in hours, W is tank width in feet, and H is the liquid tank depth in feet.

g. Tank Depth.   Tank depth should not be less than 10 ft and not greater than 15 ft. If the size of the plant is large enough, the 15-ft depth will frequently be found the most economical depth compromise considering all factors.

h. Tank Width.   The tank width should be not less than 1.15 times the liquid depth and not greater than 2.25 times. Where a plant size warrants it, a compromise of 25-ft width is often convenient.

i. Tank Discussion.   The aeration tank calculation will yield a rather long length in many instances. This length can be achieved by connecting parallel tanks end to end at alternate ends of the tank. The aeration tank length should not be less than 100 ft as a minimum, and it should not ordinarily be longer than 300 ft in a single-section length of tank before doubling back. The *transverse velocity* at the bottom of the aeration tank should be a minimum of 1 ft/s, and a desirable design figure would be 1.2 ft/s in order to avoid solids settling. The *n* value of concrete at the recommended design figure will be approximately 0.013. The *horizontal velocity* should be approximately 5 ft/minute.

The inlets and outlets for each aeration tank unit should be suitably equipped with valves, gates, stop plates, weirs, or other devices to permit control of the flow to any unit and to maintain a reasonably constant liquid level (1). The hydraulic properties of the system should permit the maximum instantaneous hydraulic load to be carried with any single aeration tank unit out of service. Channels and pipes carrying liquids with solids in suspension should be designed to maintain self-cleansing velocities or should be agitated to keep such solids in suspension at all rates of flow within the design limits. Devices should be installed for indicating flow rates of raw sewage or primary effluent, return sludge, and air to each tank unit. For plants designed for sewage flows of 1.5 MGD or more, these devices should be totalized and should record, as well as indicate, flows.

j. Contact Stabilization. Contact stabilization is a modification of the conventional activated-sludge process that takes advantage of the adsorption characteristics of the activated sludge. Influent waste and stabilized activated sludge are aerated in the aeration tank for 20–60 minutes. The liquor and sludge is then passed into a clarifier where the sludge is separated from the liquor by sedimentation. The sludge is then pumped into a stabilization tank where it is aerated from 1.5 to 5 hours until oxidation is completed. Stabilized sludge is then mixed with the influent for another cycle. See Fig. 6-23.

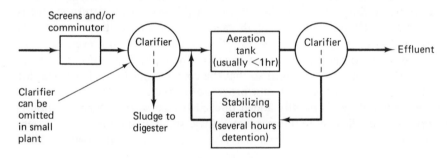

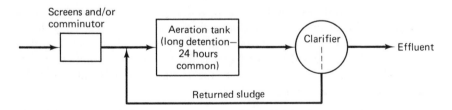

Figure 6-23. Comparison of Contact Stabilization and Extended Aeration Processes.

A small portion of the sludge drawn off the clarifier is wasted. In this process only the sludge is aerated for the extended period of time; therefore, less aeration basin capacity is required than in the conventional process. While in the stabilization basin, the organisms receive no additional food supply. It is convenient to visualize them as lean and hungry. Thus, when this stabilized sludge, with its hungry organisms, is placed in contact with the raw sewage influent of the aeration tank, the organisms have the capacity to remove large amounts of substrate. Refer to Table 6-11 for other details.

k. Extended Aeration. The extended aeration process is a modification of the activated sludge system in which the screened or comminuted raw waste is fed directly into an aeration tank. Waste is aerated much longer than in the conventional activated-sludge process. The effluent goes to a settling tank. Supernatant liquor from the settling tank is discharged as plant effluent, and all the settled activated sludge is returned to the aeration tank. This process is described in considerable detail in Chapters 8 and 9.

l. Tapered Aeration. There is a gradual decrease in the oxygen demand of the mixed liquor moving down a conventional activated-sludge aeration tank. A tapered aeration system merely spaces the air diffusors in the aeration tank to supply oxygen according to actual demand. Thus, in the first part of the aeration tank the diffusors are designed to supply more air than is supplied at a point further down the tank.

m. Step Aeration. Step aeration is merely another arrangement to even out the oxygen demand in the aerator. In this system the return sludge is introduced at the beginning of the aeration tank, as in the conventional activated-sludge system (73, 74, 75). The influent waste, however, is introduced at several points along the length of the tank. The tanks are usually divided into sections arranged so that the contents must pass from one side of the chamber to the other as it progresses to the effluent discharge end of the tank. Influent is usually added in each baffled section. Step aeration is primarily applicable to large plants. In a small plant the baffle system required tends to introduce head loss (76). Step aeration adds great flexibility to the handling of heavy organic loads and permits overloaded plants to function properly again (77).

n. Activated Aeration. The activated aeration provides substantial flexibility, but it is most likely to be suitable at a large municipal installation. It is essentially two conventional activated-sludge systems in parallel, with the return sludge from one unit fed back to the common influent waste distributor. The sludge from the second unit is fed to the same point; however, all the excess sludge is removed from the second unit. Effluents from the settling tank go to a common manifold.

o. Hatfield Process. The Hatfield process is essentially the same as contact stabilization. In addition, supernatant or digested sludge from the anaerobic digester is alternately fed to the stabilization (re-aeration tank). This process has greatest appeal to certain applications handling food industry waste.

p. Kraus Process. The Kraus and Hatfield processes are very similar. The essential difference is that part of the settled solids from the clarifier in the Kraus process is bypassed around the tank and fed back to the influent of the short-term aeration units (78, 79).

q. High-Rate Process. The BOD loadings in the aeration tank in the high-rate process are two or three times the maximum loading used in the conventional process; it requires a larger capacity in the secondary settling tank and very high rates of sludge return (80). A high-rate process usually involves a plant where substantial investment is incurred. Although this text contains a number of the fundamentals and details of the high-rate process, it is recommended that references (80) through (86) be consulted. Some of these are of primary value in leading to other references.

r. Complete Mixing. Kalinske and Busch (87) made pilot plant studies on domestic wastewater and found it possible to obtain 92 % BOD_5 reduction at organic loadings up to 400 lb BOD_5/day/1,000 ft³. McKinney et al. (88) provide a summary article and history of this activated-sludge deviation and provide a bibliography of material on the subject.

s. Ridge and Furrow Aeration. This system of placing diffusers in the bottom of the aeration tank is no longer used in new plants in the U.S. for economic reasons. The system consists of concrete channels with tops of the separators being inverted V shaped. The diffusers are mounted in the channels. This system requires more diffuser units than the spiral roll system.

t. Spiral Roll Aeration. In this system air diffusers are placed along one wall of the aeration tank and the rising air bubbles are used to provide a rolling transverse rotation of the tank contents of 1 ft/s or more. This is the most widely used system in the U.S.A, and it is shown in Fig. 6-24. Precautions must be taken to prevent short-circuiting of the flow through the center of the roll, i.e., center of the tank at mid-depth. The design suggestions in the text on tank width and depths, etc. were made partly for this purpose. Figure 6-25 shows spiral-roll aeration in progress.

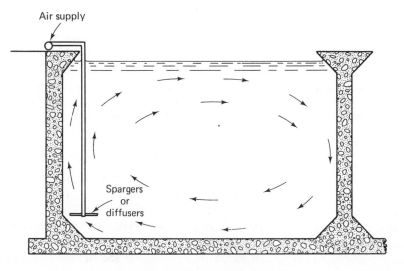

Figure 6-24. Spiral-Roll Aeration.

Figure 6-25
Spiral-Roll Aeration Tank in
Operation.

u. Swirl Mix Aeration. This is a patented proprietary process of the Walker Process Equipment, Inc. Information is available from the manufacturer. There is some evidence that the process is more efficient than the spiral-roll system; however, a cost study of the two systems is suggested before a decision is made. See Fig. 6-26.

6-33. Air System. Oxygen requirements generally depend on BOD loading, degree of treatment, and level of suspended solids concentration to be maintained in the aeration tank mixed liquor (1). Aeration equipment should be capable of maintaining a minimum of 2.0 mg/l of dissolved oxygen in the mixed liquor at all times and providing thorough mixing of the mixed liquor. Suitable protection from the elements should be provided for electrical controls.

a. Minimum Air. The minimum air permitted by the Ten-State Standards (1) is 2.0 mg/l or the value of air shown in Table 6-12. These require-

TABLE 6-12. Minimum Air

Process	Minimum Air To Be Applied Ft^3 Air/Lb BOD$_5$ Aeration Tank Load	
Conventional	1500	
Step aeration	1500	
Contact stabilization	1500	
Modified or High-Rate	400 to 1500	(depending on BOD$_5$ removal expected)
Extended aeration	2000	

ments assume equipment capable of transferring at least 1.0 lb of oxygen to the mixed liquor per pound of BOD$_5$ aeration tank loading. To these air volume requirements should be added air required for channels, pumps, or other air-use demand. The specified capacity of blowers or air compressors, particularly the centrifugal blower, should take into account that the air intake temperature may reach 40°C (104°F) or higher and the pressure may

Figure 6-26
Swirl Mix Aeration. (Courtesy
Walker Process Equipment)

be less than normal (1). The specified capacity of the motor drive should also take into account that the intake air may be $-30°C$ ($-22°F$) or less and may require oversizing of the motor or a means of reducing the rate of air delivery to prevent overheading or damage to the motor. The volume and weight of air versus temperature is shown in Fig. 6-27.

b. Blowers. Several schemes are available for compressing air. Centrifugal blowers are commonly used for air pressures between $\frac{1}{2}$ and 10 psi and $\frac{1}{2}$ to 500 horsepower (HP). Blower manufacturers publish charts showing pressure in psig versus ft³/minute of standard air, with the size of their most appropriate equipment noted. A more positive displacement-type blower, capable of slightly higher pressures, is shown in Chapter 8. These two types are the most commonly used.

The blowers should be provided in multiple units, so arranged and in such capacities as to meet the maximum air demand with the single largest unit out of service. The design should also provide for varying the volume of air delivered in proportion to the load demand of the plant (1).

Air volume changes in direct ratio to speed:

$$V_2 = V_1 \frac{N_2}{N_1} \qquad (6\text{-}42)$$

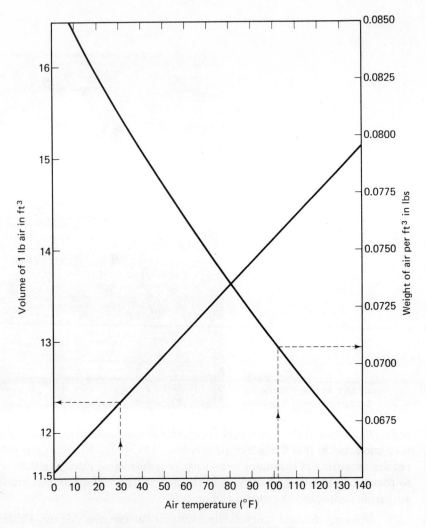

Figure 6-27. Volume and Weight of Air.

where V_1 is the original volume in ft³/minute, V_2 is the new volume in ft³/minute, N_1 is the original speed in rpm, and N_2 is the new speed in rpm. The pressure changes as the square of the speed ratio:

$$P_2 = P_1 \left(\frac{N_2}{N_1}\right)^2 \tag{6-43}$$

where P_1 is the original pressure in psig and P_2 is the new pressure in psig. The horsepower changes as the cube of the speed ratio:

$$HP_2 = HP_1 \times \left(\frac{N_2}{N_1}\right)^3 \tag{6-44}$$

where HP_1 is the original horsepower and HP_2 is the new horsepower. The pressure varies in direct proportion to the density:

$$P_a = \frac{P_g}{\Psi} \qquad (6\text{-}45)$$

where P_a is the pressure in psig, P_g is the gas pressure in psig, and Ψ is the specific gravity of the gas. Pressure varies in direct relation to barometric pressure:

$$P_{del} = P_{sl}\left(\frac{B_{elev}}{B_{sea}}\right) \qquad (6\text{-}46)$$

where P_{del} is the pressure delivered at a specified altitude in psig, P_{sl} is the pressure blower delivers at sea level in psig, B_{elev} is the local barometric pressure in inches of mercury, and B_{sea} is the barometric pressure at sea level in inches of mercury

$$P_{req} = P_{sl}\left(\frac{B_{sea}}{B_{elev}}\right) \qquad (6\text{-}47)$$

where P_{req} is the pressure required in order to deliver the same pressure that would have been delivered at sea level.

$$V_2 = V_1\left(\frac{B_{sea}}{B_{elev}}\right) \qquad (6\text{-}48)$$

and horsepower varies in direct proportion to the specific gravity (ratio of density of gas to density of air):

$$HP = HP(\text{for air})\Psi \qquad (6\text{-}49)$$

Lobe-type, constant-volume, variable-pressure blowers are built to deliver 800 to 23,000 ft^3/minute at pressures up to 12 psig. The principle of operation is shown in Fig. 8-7. In specifying these units, indicate the flow capacity in ft^3/minute, pounds of air per minute, the inlet air temperature in °F, inlet pressure in psia, discharge minimum and maximum pressure, and maximum discharge pressure, duty cycle, power supply, rpm, type of drive (electric, gas engine, diesel engine, or turbine), belt or direct drive, and type of coupling. The manufacturer usually publishes performance data indicating capacities in ft^3/minute and required HP at a specified discharge pressure and rpm.

c. Air Diffusers. The air diffusion piping and diffuser system should be capable of delivering 200% of the normal air requirements (1). The spacing of diffusers should be in accordance with the oxygenation requirements through the length of the channel or tank, and should be designed to facilitate spacing adjustments without major revision to air header piping. The arrangement of diffusers should also permit their removal for inspection, maintenance, and replacement without dewatering the tank and without shutting off the air supply to other diffusers in the tank (1).

Individual assembly units of diffusers should be equipped with control valves, preferably with indicator markings for throttling, or for complete shutoff. Diffusers in any single assembly should have substantially uniform pressure loss. Air filters should be provided in numbers, arrangement, and capacities to furnish at all times an air supply sufficiently free from dust to prevent clogging of the diffuser system used.

Air flow (ft³/minute per sq ft of diffuser) (89)

$$= \frac{12.5 \times \text{ratio of air to sewage (ft}^3/\text{gal}) \times \text{tank depth (ft)}}{\text{percentage ratio of diffuser area to tank area} \times \text{aeration period (hr)}}$$

(6-50)

$$\text{Air pressure at blower (psig)} = 0.54 \times \text{depth of sewage (ft)} \quad (6\text{-}51)$$

The wider a tank is constructed, the larger the amount of air required per linear foot of tank. Allow from 0.2 to 1.5 ft³ of free air per gallon of sewage treated. The proper amount of air depends on the waste characteristics and the method of process operation.

$$\frac{\text{ft}^3/\text{min}}{2.5 \text{ ft/min per sq ft diffuser plate}} = \text{sq ft diffuser} \quad (6\text{-}52)$$

$$\frac{\text{Sq ft diffuser}}{0.84} = \text{number of plates required} \quad (6\text{-}53)$$

$$\frac{\text{Sq ft diffuser}}{1.25} = \text{number of tubes required} \quad (6\text{-}54)$$

$$O_2 \text{ lb/min} \approx \text{ft}^3/\text{min air} \times \text{wt. air lb/ft}^3 \times \%O_2 \quad (6\text{-}55)$$
$$\approx \text{ft}^3/\text{min} \times 0.075 \times 0.23$$
$$\approx 0.017 \text{ cfm ft}^3/\text{min}$$

The blower design outlet pressure must equal the maximum pressure required in the system at any time. The worst condition is when the diffuser is partly clogged.

d. Air Piping. Figure 6-28 was developed by the author for preliminary estimation of piping for sewage treatment plants assuming standard conditions. The air velocity curve provides the maximum economical velocity in ft/min for a given pipe diameter and can be used as a guide in making design calculations. When air velocity exceeds 1400 ft/min, noise will become noticeable if the air passes over a sharp edge unless there is substantial background noise.

To keep air velocities under the values indicated, it will be found that the resistance drop must decrease as the pipe diameter increases. As a rule of thumb, the author endeavors not to exceed 0.4 in. of water pressure loss per 100 ft of pipe for pipe sizes under 36 in. This value drops to 0.15 in. for a pipe 80 in. diameter. The actual final design figure may exceed 0.4 in. of water for very small pipes. However, these figures at least give the reader some idea where to expect results. Even in small pipes it is not recommended that the resistance loss due to the air flowing in the pipe exceed 1.5 in. of water gauge, and one-third this value is more desirable.

It is suggested that Lemke (90) and Moody (91) be read in conjunction with air flow in pipes. A chart by Lemke (90) appears in the ASCE *Manual of Practice, No.* 36 (71). When other sources are used, it may be necessary to

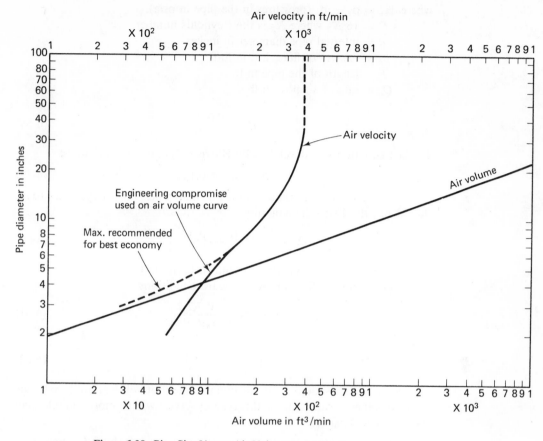

Figure 6-28. Pipe Size Versus Air Volume and Velocity.

use charts associated with ventilation systems instead of high-pressure air systems. There are various other sources having this information, three of which are (92, 93), and (94).

The Harris equation (95) indicates

$$f = \frac{0.1025LQ^2}{rd^5} \qquad (6\text{-}56)$$

where f = pressure drop in lbs/in^2,

$\quad L$ = length of pipe in ft,

$\quad Q$ = ft^3 of air per second (free air),

$\quad r$ = ratio of compression (from free air) at entrance of pipe,

$\quad d$ = actual internal diameter of pipe in inches.

$$H_f \approx 43.5\frac{(4f_a)LQ_a^2}{\rho_a d^5} \qquad (6\text{-}57)$$

where $H_f = p_1 - p_2$ (pressures in the pipe in psia),
$\quad f_a = 16/N_R$ where N_R is the Reynolds number,
$\quad d =$ internal pipe diameter in inches,
$\quad \rho_a =$ density of fluid, air, in lb/ft^3,
$\quad L =$ length of the pipe in ft,
$\quad Q_a =$ rate of air flow in lb/s.

$$4f_a = \frac{0.0313}{d^{0.31}} \tag{6-58}$$

The last equation was developed by Harris (95) and it is valid up to

$$p_1 - p_2 \leqq 0.1p_1 \tag{6-59}$$

Equation (6-56) is of primary value for small pipes. Pipes larger than 10 in. should use the Fritzsche equation

$$Q_a = \left[1.22 \frac{(p_1 - p_2)d^5}{v_s L} \right]^{1/1.35} \tag{6-60}$$

where v_s is the specific volume of fluid in ft^3/lb and
$\quad Q_a =$ flow in ft^3/min. At standard conditions

$$p_1 - p_2 = \frac{0.7Q_a L}{p_m d^5 1000} \tag{6-61}$$

where

$$p_m = \frac{p_1 + p_2}{2} \tag{6-62}$$

Due to the pulsations in an air piping system, it will be found that cone valves or other valves built for this type of service will be more suitable than standard check valves.

e. Air Cleaners. Air cleaners most often used include (a) two-stage electric precipitators, (b) viscous filters, and (c) dry filters.

The power rate in an electrostatic precipitator in a low-voltage two-stage filter ranges between 0.01 and 0.02 kW/1,000 ft^3 per minute, the velocity at the electrostatic units is approximately 200 ft/minute, and the resistance is less than 0.2 in. of water gauge. The resistance of a washable viscous type would be less than 0.1, and for a dry type, less than 0.5 (2–5 μ).

Additional information on filters and data on the various types of filter tests are readily available (92).

6-34. Final Settling Tanks. Inlets, sludge collection, and sludge withdrawal facilities should be designed to minimize density currents and assure rapid return of sludge to the aeration tanks. Multiple units capable of independent operation are desirable and should be provided in all plants where design flows exceed 100,000 gallons per day, unless other provision is made to assure continuity of treatment (1). Effective baffling and scum removal equipment should be provided for all final settling tanks.

Since the rate of recirculation of return sludge from the final settling tanks to the aeration or re-aeration tanks is quite high in activated-sludge processes, the detention time, surface settling rate, and weir overflow rate should be adjusted for sludge loadings, density currents, inlet hydraulic turbulence, and occasional poor sludge settleability (1). The design parameters in Table 6-13 should be observed in the design of final settling tanks for the activated-sludge processes indicated in the table. Due consideration must be given to the flow duration (e.g., school flows which may occur in a 6-hr period).

TABLE 6-13. Final Settling Tanks

Type of Process	Average Design Flow (MGD)	Detention Time (hours)	Surface Settling Rates (gal/day/ft²)
Conventional, modified, or high-rate and step aeration	to 0.5	3.0	600
	0.5 to 1.5	2.5	700
	1.5 up	2.0	800
Contact stabilization	to 0.5	3.6	500
	0.5 to 1.5	3.0	600
	1.5 up	2.5	700
Extended aeration	to 0.05	4.0	300
	0.05 to 0.15	3.6	300
	0.15 up	3.0	600

6-35. Trickling Filters. A trickling filter might be defined as a stationary biological bed upon which waste is periodically dosed and permitted to trickle through; the effluent is carried off underneath the bed by appropriate drain tiles. A cross section of a trickling filter is shown in Fig. 6-29 and a photograph of a trickling filter in operation is shown in Fig. 6-30. This unit, as generally constructed today, consists of a circular structure containing a bed of crushed stone or other insoluble medium placed to a suitable depth. Either rotary distributors or fixed nozzles are used to deposit waste on the bed. The

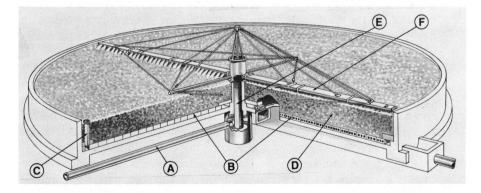

Figure 6-29. Cut-away View of Trickling Filter. (Courtesy W. S. Dickey Clay Mfg. Co.)

Figure 6-30. Trickling Filter Units in Operation. (Courtesy Lakeside Equipment Corporation)

term *filter* is probably a carryover from the intermittent sand filters from which the initial contact beds originated. An English engineer, Joseph Corbett, in 1893, permitted the liquid to flow through the bed instead of flooding the bed. The principle actually involved is one of adsorption rather than filtration. A gelatinous film of microorganisms forms over the stone or other material used to form the bed. This film is composed of bacteria, fungi, algae, protozoa, rotifera, chaetopoda, crustacea, nematoda, and insecta (96). The bacteria are the most important. The distribution of organisms varies with depth, type of waste, and other conditions that may prevail. Algae, for example, are found only on the upper surface where sunlight is present.

Stack (97) and Velz (98) visualized a trickling filter as a sort of self-regenerating adsorption tower. Each unit depth of the filter will remove a constant fraction of the removable BOD applied to that unit of depth. The removable BOD is the fraction of the observed BOD which can be removed by biosorption. It is also assumed that the quantity of BOD that can be adsorbed by one unit volume of a filter has a maximum limit.

Colloids are adsorbed by the film and are coagulated and hydrolyzed to smaller molecules. Soluble organics diffuse into the film and are oxidized by enzymatic action (99). Approximately 60% of the dissolved and colloidal organic matter removed from the waste stream via bio-oxidation goes into growth of microorganisms (sludge), with the remaining 40% going to carbon dioxide and water to provide the energy for the life processes. In addition, the sludge itself can undergo continuous auto-oxidation.

Trickling filter designs are calculated on the basis of persons served per cubic yard of filtering material (100). A surface loading of 30,000 to 40,000 persons per acre corresponds to a volume loading of 3 to 4 persons served per cubic yard of filtering material.

Hall (100) stated: "The factors which affect the performance of trickling filters are: the size, character, and depth of the media; the temperature of the air and sewage; the strength and volume of applied flow; the ventilation and cleanliness of media; the method of distribution; and seasonal variations."

The following comments are offered by the *Recommended Standards for Sewage Works* (1);

The sewage may be distributed over the filter by rotary distributors or other suitable devices which will permit reasonably uniform distribution to the surface area. At design average flow, the deviation from a calculated uniformly distributed volume per square foot of the filter surface shall not exceed $\pm 10\%$ at any point.

Sewage may be applied to the filters by siphons, pumps, or by gravity discharge from preceding treatment units when suitable flow characteristics have been developed. Application of the sewage shall be practically continuous. Consideration shall be given to a piping system which will permit recirculation.

All hydraulic factors involving proper distribution of sewage on the filters should be carefully calculated. For reaction-type distributors, a minimum head of 24 inches between the low water level in the siphon chamber and the center of the arms is generally desirable. A minimum clearance of 6 inches between media and distributor arms shall be provided. Greater clearance is essential when icing occurs.

The media may be crushed rock, slag, or specially manufactured material. The media shall be durable, resistant to spalling or flaking, and be relatively insoluble in sewage. The top 18 inches shall have a loss by the 20-cycle, sodium sulfate soundness test of not more than 10%, as prescribed by ASCE *Manual of Engineering Practice, No. 13*, the balance to pass a 10-cycle test using the same criteria. Slag media shall be free from iron. Manufactured media shall be structurally stable and chemically and biologically inert.

The filter media shall have a minimum depth of 5 feet above the underdrains and should not exceed 7 feet in depth, except where special construction is justified by studies.

Rock, slag, and similar media shall not contain more than 5% by weight of pieces whose longest dimension is 3 times the least dimension. They shall be free from thin, elongated, and flat pieces; dust; clay; sand; or fine material, and they shall conform to the following size and grading when mechanically graded over a vibrating screen with square openings:

> Passing 4½-inch screen—100% by weight
> Retained on 3-inch screen—95–100% by weight
> Passing 2-inch screen—0–2% by weight
> Passing 1-inch screen—0–1% by weight

Hand-packed field stone should have a maximum dimension of stone of 5 inches and a minimum dimension of stone of 3 inches.

Continuing with selected statements (1):

Underdrains with semicircular inlets or equivalent should be provided, and the underdrainage system shall cover the entire floor of the filter. Inlets opening into the underdrains shall have an unsubmerged gross combined area equal to at least 15% of the surface area of the filter.

The underdrains shall have a minimum slope of 1%. Effluent channels shall be designed to produce a minimum velocity of 2 ft/s at the average daily rate of application to the filter.

The expected reduction of BOD of the primary settling tank effluent by a single-stage filter packed with crushed rock, slag, or similar material and subsequent settling tank shall be determined from Fig. 6-31. In developing this curve, loading due to recirculated sewage was not considered. The expected performance of filters packed with manufactured media shall be determined from pilot plant and full-scale experience.

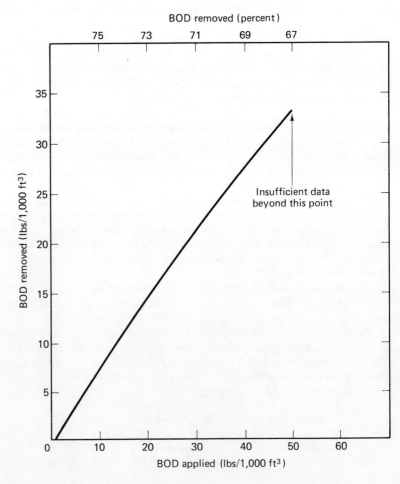

Figure 6-31. BOD Removal. Trickling Filter Units Including Final Settling Tanks.

6-36. Trickling Filter Design. A depth of 6 ft is suggested as a starting figure. In unusual cases, depths anywhere from 3 to 10 ft may become necessary. Eckenfelder (102) suggested the equation

$$\frac{L_e}{L_0} = \frac{100}{1 + 2.5(D^{0.67}/Q^{0.50})} \tag{6-63}$$

where L_e = BOD concentration in filter effluent, mg/l,
L_0 = BOD concentration applied to the filter, mg/l,
D = depth of the filter, ft,
Q = hydraulic loading, million gallons per acre per day (mgad).

When recirculation is involved, the influent BOD is diluted by the recirculated flow.

$$L_0 = \frac{L_a + NL_e}{N + 1} \tag{6-64}$$

where L_a = influent sewage BOD, mg/l,
N = recirculation ratio,
L_0 = BOD (mg/l) after admixture with recirculated flow, i.e., BOD concentration applied to the filter,
L_e = BOD concentration in filter effluent (mg/l).

Equations (6-63) and (6-64) can be combined to yield the estimate of filter performance of a recirculated flow

$$\frac{L_e}{L_0} = \frac{1}{(1 + N)[1 + 2.5(D^{0.67}/Q^{0.5})] - N} \tag{6-65}$$

which is applicable up to the point in a trickling filter where further removal cannot occur.

Figure 6-32 shows an aerial view of a complete trickling filter plant in operation.

Figure 6-32. Aerial View of Trickling Filter Plant in Operation. (Courtesy Lakeside Equipment Corporation)

The Sanitary Engineering Committee of the National Research Council developed the relations

$$F = \frac{(1 + R_s/I)}{(1 + 0.1R_s/I)^2} \tag{6-66}$$

in which F is the recirculation ratio, R_s is the volume of sewage recirculated in gal/day, and I is the volume of raw sewage in gal/day.

$$L_u = \frac{L_a}{v_t F} \tag{6-67}$$

where L_u is the unit loading applied in lb BOD/acre ft, L_a is the total BOD in lbs, v_t is the filter volume in acre ft, and F is the recirculation factor.

The percent removal ϵ_1 of BOD for a single-stage trickling filter and its associated following clarifier is:

$$\epsilon_1 = \frac{100}{1 + 0.0085\sqrt{L_u}} \tag{6-68}$$

The percentage removal, ϵ_2, by the second-stage filter may be found by

$$\epsilon_2 = \frac{100}{1 + \dfrac{0.0085}{1 - \epsilon_1}\sqrt{\dfrac{L'_u}{v'_t F'}}} \tag{6-69}$$

where L'_u is the BOD in effluent from the first-stage filter, in lb, v'_t is the volume of the second-stage filter, in acre ft, and F' is the recirculation factor for the second-stage filter.

PROBLEMS

Listed below are four sewage treatment plants located at 42°N, sea level elevation, and at an ambient air temperature of 20°C. When the necessary data cannot be derived from the facts given, make the necessary assumptions in order to complete the problems.

Plant	Flow MGD
A	0.5
B	1.0
C	10.0
D	100.0

1. Design a manually cleaned bar screen using sharp-edge rectangular bars, a velocity through the screen 1 ft/s, and a head loss of 1 ft minimum for plants A, B, C, and D.

2. Indicate all the information and data required on the plans for a mechanical screen for plants A, B, C, and D.

3. Assume that grit particle size of 0.25-mm diameter is to be removed. Design a grit chamber for plants A, B, C, and D.

4. Design a flocculator for plant C.

5. Design a rectangular primary sedimentation tank for plants A, B, C, and D.

6. Design a circular primary clarifier for plants A, B, C, and D.

7. Design a final sedimentation chamber (activated-sludge facility) for plants A, B, C, and D.

8. Design the aeration tank for plants A, B, C, and D.

9. Design the aeration tank for plants A, B, C, and D using the step aeration activated-sludge process.

10. Design the aeration tanks for plants A, B, C, and D using the contact stabilization activated-sludge process.

11. Design the aeration tank for plants A, B, C, and D using the extended aeration activated-sludge process.

12. Design the aeration tank for plants A, B, C, and D using the high rate activated-sludge process.

13. Design a trickling filter (high-rate) for plants A, B, C, and D with no recycling and with a 1:1 recycling ratio. Any time that a unit is so large, or good engineering sense indicates that its size is excessively large, use multiple units as required.

14. Design a conventional activated-sludge treatment plant for Jackson, Mississippi, using the latest census data. Include all treatment units from the sewer influent to the effluent of the final sedimentation tanks. Use multiple units as required to meet the Ten-State Standards.

15. Design a complete trickling filter treatment plant for Gulfport, Mississippi. (Exclude digesters and chlorination.)

16. Design a contact stabilization plant for Madison, Wisconsin. The plant is to be complete up to the point of digesters and chlorination.

REFERENCES

1. Great Lakes-Upper Mississippi River Board of State Sanitary Engineers, *Recommended Standards for Sewage Works*, 1971 rev. ed. Health Education Service, Albany, N.Y.

2. KIRSCHMER, O., "Untersuchungen über den Gefällsverlust an Rechen." *Trans. Hydraulic Inst.*, Munich, R. Oldenbourgh, 21 (1926).

3. METCALF, L., and EDDY, H. P., *American Sewage Practice*, Vol. III, 3rd ed. McGraw Hill Book Co., Inc., New York, 1935.

4. Data Sheet 315-2.101, Sept. 15, 1966, Product Manual, Sanitation Equipment Conveyor and Process Equipment Division, Rex Chain Belt Company, Milwaukee, Wisconsin.

5. Data Sheet 315-2.105, Aug. 1965, Product Manual Sanitation Equipment Conveyor and Process Equipment Division, Rex Chain Belt Company, Milwaukee, Wisconsin.

6. RETTGER, E. W., "A Proportional-Flow Weir." *Engineering News*, 71, 26, 1409 (1914).

7. PRATT, E. A., "Another Proportional-Flow Weir; Sutro Weir" (letter to the editor). *Engineering News*, 72, 9, 462 (1914).

8. SOUCEK, E., et al., "Sutro Weir Investigations." *Engineering News-Record*, 93, 679 (Nov. 12, 1936).

9. PARKER, HOMER W., *Environmental Engineering Handbook*. Prentice-Hall, Inc., Englewood Cliffs, N.J., 1973.

10. ROBINSON, A. R., *Parshall Measuring Flumes of Small Sizes*. Technical Bulletin 61, Colorado State University, Experiment Station, Fort Collins, Aug., 1960.

11. PARSHALL, R. L., *Parshall Flumes of Large Size*. Bulletin 426-A, Colorado State University, Fort Collins, June, 1961.

12. PARSHALL, R. L., *Measuring Water in Irrigation Channels with Parshall Flumes and Small Weirs.*, USDA, Soil Conservation Service Circ. 843.

13. ROTH, HELWIG, and HALL, cited in BEYCHOCK, MILTON R., *Aqueous Wastes from Petroleum and Petrochemical Plants*. John Wiley & Sons, London, 1967, p. 247.

14. SIMONSEN, cited in BEYCHOCK, MILTON R., *Aqueous Wastes from Petroleum and Petrochemical Plants*. John Wiley & Sons, London, 1967, p. 247.

15. ARGAMAN, YERACHMIEL, and KAUFMAN, WARREN J., "Turbulence and Flocculation." *Proc. ASCE*, SA2, 223 (1970).

16. CAMP, T. R., and STEIN, P. C., "Velocity Gradients and Internal Work in Fluid Motion." *Jour. Boston Society of Civil Engineers*, 30, 4, 219 (1943).

17. BEHN, VAUGHN C., "Settling Behavior of Waste Suspensions." *Proc. ASCE*, 83, SA5, 1423 (Oct., 1957).

18. HAZEN, ALLEN, "On Sedimentation." *Trans. ASCE* (1904), p. 45.

19. CAMP, THOMAS R., "A Study of the Rational Design of Settling Tanks." *Sewage Works J.*, 8, 5, 742 (1936).

20. MCLAUGHLIN, R. T., JR., "The Settling Property of Suspensions." *Proc. ASCE*, 85, HY12 (1959).

21. FREI, J. K., "Multiple Tray Clarification in a Modern Treatment Plant." *Sewage Works Eng.*, 12, 423 (1941).

22. CAMP, T. R., "Studies of Sedimentation Basin Design." *Sewage & Ind. Wastes*, 25, 1 (1953).

23. DRESSER, H. G., "Trays Nearly Triple Settling Tank Capacity." *Eng. News Record*, 147, 32 (1951).

24. SCHMITT, E. A., and VOIGT, O. D., "Two-Story Flocculation-Sedimentation Basin for the Washington Aqueduct." *Jour. AWWA*, 41, 837 (1949).

25. HANSEN, SIGURD P., and CULP, GORDON L., "Applying Shallow Depth Sedimentation Theory." *Jour. AWWA*, 59, 9, 1134 (1967).

26. DAGUE, R. R., and BAUMANN, E. R., "Hydraulics of Circular Settling Tanks Determined by Models." Presentation at the 1961 Annual Meeting Iowa Water Pollution Control Association, Lake Okoboji, Iowa, June 8, 1961. Published by Lakeside Engineering Corporation in bulletin *Spiraflo Clarifier Hydraulic Studies*.

27. McKinney, Ross E., "A Fundamental Approach to the Activated Sludge Process. II. A Proposed Theory of Floc Formation." *Sew. & Ind. Wastes.*, 24, 3, 280 (1952).

28. Heukelekian, H., et al., "Factors Affecting the Quantity of Sludge Production in the Activated Sludge Process." *Sew. & Ind. Wastes*, 23, 8, 945 (Aug., 1951).

29. Thomas, H. A., Jr., "Sewage Treatment in Low-Temperature Areas." *Sew. & Ind. Wastes*, 23, 1, 34 (1951).

30. Howland, W. E., "Effect of Temperature on Sewage Treatment Processes." *Sew. & Ind. Wastes*, 25, 2, 161 (1953).

31. Bloodgood, D. C., "The Effect of Temperature and Organic Loading in Activated Sludge Plant Operation." *Sewage Works J.*, 16, 5, 914 (Sept., 1944).

32. Horasawa, I. "Biological Studies on Activated Sludge in the Purification of Sewage." *Jour. Water Works & Sewage Assn.* (Japan), 148, 2 (1950).

33. Wuhrmann, K., "Factors Affecting Efficiency and Solids Production in the Activated Sludge Process." *Biological Treatment of Sewage and Industrial Wastes*, by McCabe, J., and Eckenfelder, W. W. Jr., Reinhold Publishing Corporation, New York, 1956.

34. Lesperance, Theodore W., "A Generalized Approach to Activated Sludge. Part I—Basic Biochemical Reaction." *W. W. & W. E.*, 2, 4, 44 (April, 1965).

35. Orford, H. E., and Isenberg, E., "Effect of Sludge Loading and Dissolved Oxygen on Performance of the Activated Sludge Process." Presented at the 3rd. Biol. Waste Trt. Conf., Manhattan College, New York (1960).

36. Eckenfelder, W. W. Jr., and Weston, R. F., "Kinetics of Biological Oxidation—Biological Treatment of Sewage and Industrial Wastes." *Aerobic Oxidation*, Vol. I, Reinhold Publishing Corporation, New York, 1956.

37. Genetelli, E. J., and Heukelekian, H., "The Influence of Loading Chemical Composition of Substrate on the Performance of Activated Sludge." *Jour. WPCF*, 36, 5, 643 (May, 1964).

38. Keefer, C. E., and Meisel, J., "Sludge Age and Its Effect on the Activated Sludge Process." *Sew. & Ind. Wastes*, 25, 8, 898 (1953).

39. Kraus, L. S., "Quantitative Relationships in the Activated Process." *Sewage Works J.*, 21, 4, 613 (July, 1949).

40. Torpey, W. N., and Chasick, A. H., "Principles of Activated Sludge Operation." *Sew. & Ind. Wastes*, 27, 11, 1217 (1955).

41. Keefer, C. E., and Meisel, J., "Activated Sludge Studies. III. Effect of pH of Sewage on the Activated Sludge Process." *Sew. & Ind. Wastes*, 23, 8, 982 (1951).

42. Dryden, F. E., et al., "High-Rate Activated Sludge Treatment of Fine Chemical Wastes." *Sew. & Ind. Wastes*, 28, 2, 183 (1956).

43. Eckenfelder, W. W., Jr., and Grich, E. R., "Cannery Waste Oxidation by Activated Sludge—Biological Treatment of Sewage and Industrial Wastes." *Aerobic Oxidation*, Vol. I, Reinhold Publishing Corporation, New York 1956.

44. ECKENFELDER, W. W., JR., "Theory of Biological Treatment of Trade Wastes." *Jour. WPCF*, 39, 2, 240 (1967).

45. ECKENFELDER, W. W., "Comparative Biological Waste Treatment Design." *Proc. ASCE*, 93, SA6, 157 (1967).

46. PARKER, HOMER W., "Activated Sludge Efficiency and Removal Rate Coefficient." Unpublished work.

47. HANEY, PAUL D., "Theoretical Principles of Aeration." *Jour. AWWA*, 46, 4, 353 (1954).

48. PASVEER, A., "Research on Activated Sludge. I. A Study of the Aeration of Water." *Sew. & Ind. Wastes*, 25, 11, 1253 (1953).

49. PASVEER, A., "Research on Activated Sludge. II. Experiments with Brush Aeration." *Sew. & Ind. Wastes*, 25, 12 1397 (1953).

50. PASVEER, A., "Research on Activated Sludge. III. Distribution of Oxygen in activated Sludge Floc." *Sew. & Ind. Wastes*, 26, 2, 28 (1954).

51. PASVEER, A., "Research on Activated Sludge, IV. Purification with Intense Aeration." *Sew. & Ind. Wastes*, 26, 2, 149 (1954).

52. PASVEER, A. "Research on Activated Sludge. VI. Oxygenation of Water with Air Bubbles." *Sew. & Ind. Wastes*, 27, 10, 1130 (Oct. 1955).

53. PASVEER, A., "Research on Activated Sludge. VII. Efficiency of Diffused Air System." *Sew. & Ind. Wastes*, 28, 1, 28 (1956).

54. ECKENFELDER, W. W., JR., and O'CONNOR, D. J., *Biological Waste Treatment*, Pergamon Press, 1961.

55. KING, H. R., "Mechanics of Oxygen Absorption in Spiral Flow Aeration Tanks. II. Experimental Work." *Sew. & Ind. Wastes*, 27, 9, 1007 (1955).

56. KING, H. R., "Mechanics of Oxygen Absorption in Spiral Flow Aeration Tanks. III. Application of Formulas." *Sew. & Ind. Wastes*, 27, 10, 1123 (1955).

57. KING, H. R., "Mechanics of Oxygen Absorption in Spiral Flow Aeration Tanks. I. Derivation of Formulas." *Sew. & Ind. Wastes*, 27, 8, 894 (1955).

58. DOBBINS, WILLIAM E., "Discussion." *Sew & Ind. Wastes*, 26, 7, 827 (1954).

59. DANCKWERTS, P. V., "Significance of Liquid-Film Coefficients in Gas Absorption." *Ind. & Eng. Chem.*, 43, 6, 1460 (1951).

60. BEWTRA, J. K., and NICHOLAS, W. R., "Oxygenation from Diffused Air in Aeration Tanks." *Jour. WPCF*, 36, 10, 1195 (1964).

61. GAMESON, A. L. H., and ROBERTSON, K. G., "The Solubility of Oxygen in Pure Water and Sea-Water." *J. Appl. Chem.*, 5, 502 (Sept., 1955).

62. HOWLAND, W. E., "Effect of Temperature on Sewage Treatment Processes." *Sew. & Ind. Wastes*, 25, 2, 101 (1953).

63. TRUESDALE, G. A., et al., "The Solubility of Oxygen in Pure Water and Sea-Water," *J. Appl. Chem.*, 5, 53 (Feb., 1955).

64. HABERMAN, W. L., and MORTON, ROSE K., "An Experimental Study of Bubbles Moving in Liquids." *Trans. ASCE*, 121, 227 (1956).

65. FISCHERSTROM, CLAES N. H., "Low Pressure Aeration of Water and Sewage." *Proc. ASCE*, 86, SA5, 21 (1960).

66. PASVEER, A., "Research on Activated Sludge. V. Rate of Biochemical Oxidation." *Sew. & Ind. Wastes*, 27, 7, 783 (1955).

67. ECKENFELDER, W. W. JR., and BARNHART, E. L., "The Effect of Organic Substances on the Transfer of Oxygen from Air Bubbles in Water." *A. I. Ch. E. Jour.*, 7, 4, 631 (1961).

68. POON, CALVIN P. C., "Theoretical Concept of Oxygen Transfer in Gas Bubble Aeration System." *W. & S. W.*, 113, R.N.-200 (1966).

69. MORGAN, P. F., "Maintenance of Fine Bubble Diffusion." *Proc. ASCE*, 84, SA2, 1609 (1958).

70. HANEY, PAUL D., "Theoretical Principles of Aeration." *Jour. AWWA*, 46, 4, 353 (1954).

71. American Society of Civil Engineers, *Manual of Engineering Practice No. 36.*

72. Water Pollution Control Federation, *Manual of Practice No. 5.*

73. GOULD, R. H., "Sewage Aeration Practice in New York City" *Proc. ASCE*, 79, Paper 307 (Oct., 1953).

74. GOULD, R. H., "Tallman's Island Works Opens for World's Fair." *Municipal Sanitation*, 10, 4, 185 (1939).

75. MCKEE, J. E., and FAIR, G. M., "Load Distribution in the Activated Sludge Process." *Sew. Works Jour.*, 14, 1, 121 (1942).

76. THOMAS, H. A., JR., and MCKEE, J. E., "Longitudinal Mixing in Aeration Tanks." *Sew. Works Jour.*, 16, 1, 42 (1944).

77. EDWARDS, G. P., "Discussion—Factors Affecting the Efficiency of Activated Sludge Plants." *Sew. Works Jour.*, 21, 640 (1949).

78. KRAUS, L. S., "Digested Sludge—An Aid to the Activated Sludge Process," *Sew. Works Jour.*, 18, 6, 1099 (1946).

79. KRAUS, L. S., "Dual Aeration as a Rugged Activated Sludge Process." *Sew. & Ind. Wastes*, 27, 12, 1347 (1955).

80. BLACK, S. A., "High-Rate, Combined Tank Activated-Sludge Process Evaluated." *Water and Poll. Control* (Canada) 105, 42 (1967).

81. SETTER, L. R., "Modified Sewage Aeration, Part I." *Sew. Works Jour.*, 15, 4, 629 (1943).

82. SETTER, L. R., and EDWARDS, G. P., "Modified Aeration. Part II." *Sew. Works Jour.*, 16, 2, 278 (1944).

83. CHASE, E. S., "High Rate Activated Sludge Treatment of Sewage." *Sew. Works Jour.*, 16, 5, 878 (1944).

84. BORROUGH, P. C., "Improved Activated Sludge Processes." *Chem. & Ind.* (Brit.), 36, 1507 (1967).

85. BARGMAN, ROBERT D., et al., "Aeration Requirements of a High Oxygen Demand Sewage." *Sew. & Ind. Wastes*, 29, 7, 768 (1957).

86. HIGGINS, P. M., "Waste Treatment by Aerobic Techniques." *Developments in Ind. Microbiol.*, 9, 146 (1968).

87. KALINSKE, A. A., and BUSCH, A. W., "New Equipment for the Activated Sludge Process." *W. & S. W.*, 103, 324 (1956).

88. MCKINNEY, ROSS E., et al., "Evaluation of a Complete Mixing Activated Sludge Plant." *Jour. WPCF*, 42, 5, 737 (1970).

89. IMHOFF, KARL, and FAIR, G. M., *Sewage Treatment*, John Wiley & Sons, Inc., New York, 1956.

90. LEMKE, A. A., "Flow of Air in Pipes." *Sew. & Ind. Wastes*, 24, 1, 24 (1952).

91. MOODY, L. F., "Friction Factors for Pipe Flow." *Trans. ASME*, 66, 671 (Nov., 1944).

92. BAUMEISTER, THEODORE, ed., *Standard Handbook for Mechanical Engineers*, 7th ed. McGraw-Hill Book Co., New York, 1967.

93. Chicago Pump, Hydrodynamics Division, bulletin, *Hydraulics and Useful Information*.

94. *Compressed Air Handbook*. Compressed Air and Gas Institute, New York, N.Y.

95. HARRIS, E. G., University of Missouri Bulletin 1, 4, (1912).

96. HOLTJE, R. H., "The Biology of Sprinkling Filters." *Sewage Works J.*, 15, 1, 14 (1943).

97. STACK, V. T., JR., "Theoretical Performance of the Trickling Filtration Process." *Sew. & Ind. Wastes*, 29, 9, 987 29, 9, 987 (1957).

98. VELZ, C. J., "A Basic Law for the Performance of Biological Filters." *Sewage Works J.*, 20, 4, 607 (1948).

99. ECKENFELDER, W. WESLEY, JR., and MOORE, T. L., "Biooxidation." *Chem. Eng.*, 62, 9, 189 (1955).

100. IMHOFF, KARL, "Highlights in the History of Biological Sewage Treatment." *W. & S. W.*, 101, 1, 28 (1954).

101. HALL, G. A., and HATCH, B. F., "Trickling Filters Loadings." *Sewage Works J.*, 9, 1, 50 (1937).

102. ECKENFELDER, W. W., JR., "Trickling Filtration Design and Performance." *Trans. ASCE*, 128, 3, 371 (1963).

Sewage
Treatment **7**
Part II

In the preceding chapter sedimentation, activated sludge, and trickling filter plants were discussed up to the point of sludge disposal problems. Since these plants traditionally have been used for municipal service, this chapter will first consider anaerobic sludge digestion and sludge disposal before considering other treatment methods. Chlorination may be applied to any of the processes and will be considered separately in Chapter 14.

Whether or not to use traditional methods is open to question. The engineer must acquaint himself with all the basic systems and make an economic analysis coupled with an engineering investigation before making that decision. Aerobic digestion will not achieve quite as great solids reduction as anaerobic digestion; however, a small plant cannot be relied upon to effectively use gas produced by digesters without costly design provisions. Even then, the quantity of production is usually too low to warrant the gas equipment investment involved. When interest rates are high and the trend is toward inflation, it may be that the municipality will have great difficulty floating the necessary bonds to design a plant using anaerobic digestion. A substantial amount of engineering judgment and investigation is the only sound way to make the decision.

ANAEROBIC DIGESTION

7-1. Sludge Digestion. The simplified explanation of sludge digestion that follows is based on some of the numerous articles in the literature (1, 2, 3, 4).

Sludge digestion is a biochemical phenomenon involving organisms, enzymes, food, and environment. A grossly simplified concept would be that complex organic materials are broken down by saprophytic bacteria into short-chain fatty acids. These volatile acids are, in turn, converted into gases by the methane-forming bacteria that must exist in the anaerobic digester. The success of operation of a digester depends on maintaining favorable conditions for the most efficient balance between the factors.

Mixing in some form is required for a reaction to occur. Digestion normally occurs over a range of temperatures encountered in practice; however, the rate at low temperatures may be too slow to be practical. Digester temperatures between 32 to 38°C promote the best reaction. Higher or lower temperatures decrease operating efficiency and may, in extreme cases, disrupt operation. There must be provision for sludge draw-off in a form that will not deplete the food supply or the buffer capacity of the system. Arrangements must also exist for properly feeding the digester and the withdrawal of supernatant. Gases are generated in the digester and must be removed by some appropriate method. The optimum pH range is 6.8–7.2.

The characteristics and color of sludge depend on its point of origin in the treatment process. Sewage sludge is a gray, dark, brown, or black smelly substance. The digestion process is designed to reduce the amount of organic and volatile solids in the sludge, reduce its volume, and make it less obnoxious and more readily disposable. The physical state of the sludge is altered during the process so that the water it contains can be more readily removed. The organics in sludge can be partly converted into gas which, in municipal plants, may be collected and utilized. A very important objective of sludge treatment is to destroy or effectively control microorganisms that may cause disease and infection. Additional information on sludge digestion is available in the literature (5, 6, 7).

Sludge is digested anaerobically to reduce its odor, volume, and putrescence and to produce gases which can be used to assist in lowering the cost of operating the plant by serving as fuel for the aerator-blower gas-powered engines.

7-2. High-Loading Rate Digesters. The principle of increased loading of sludge digesters has been investigated (8, 9, 10). The presently developed theories of digester operation do not adequately explain digester failures that occasionally occur. Buzzell and Sawyer (8) reported that digesting sewage sludge seems to be a pseudoplastic material possessing a relatively small yield value. It exhibits only slight thixotropic behavior. The viscosity of the digesting sludge is a function of both the total solids concentration and the volatile matter content. The viscous nature of digesting sludge is a possible limiting factor in the design and operation of high-rate digesters.

7-3. Detention Time. Due to the composition of the raw sludge, an excessive amount of time is required to bring about complete decomposition of the organic matter into CO_2 and H_2O (11). In sewage treatment practices currently employed, a more practical approach is taken: the organic matter is allowed to decompose to a point where it will no longer be objectionable.

Detention times (theoretical) are typically between 30 and 60 days. Detention times less than 30 days tend to lead to operating problems.

7-4. Sludge Concentration. Up to a limiting point of sludge concentration for stable operation, the degree of digestion during a specified detention time does not depend on the concentration of influent sludge to the digester. It is not recommended to concentrate feed sludges to more than 8% solids unless the fixed solids content of the raw sludge is less than 35% or the volatile matter content exceeds 65% (12). A dilute excess activated sludge might have a concentration around 0.5%, whereas a dense primary sludge may have a concentration varying from 8 to 12% solids. One way handle the problem is to combine primary and activated sludges in a separate thickening tank prior to injecting into the sludge digester (13).

7-5. Mixing. The contents of a high-rate digester should be continuously well mixed. Mixing accomplishes three major objectives, all of which contribute toward keeping the biological forces operating at or near peak capacity at all times (12). First, the active organisms are kept continuously in contact with the food supply; second, the food supply is uniformly distributed and made as available to the organisms as possible; and third, the concentration of inhibitory biological intermediates and end products are maintained at minimum levels. All of these factors serve as a means of keeping the working population of organisms performing at peak efficiency.

7-6. Input-Output Balance. For every pound of raw solids added, some 50 or 60% should be removed as digested sludge where long-term operation is considered. Some digesters are operated to produce a close balance between input and output of solids, while the substantial majority do not (14). The reduction of volatile matter, and hence the progress of digestion, is a function of the detention period of the daily sludge feed regardless of the type of sludge.

7-7. Digester Capacity. The net effective capacity of a sludge digester is the gross volume less that volume occupied by mineral deposition on the bottom, nominal trash accumulation at the top, and the zones of sludge liquor in the upper part and concentrating sludge in the lower part (15). For digestion systems providing for intimate and effective mixing of the digester contents, the system may be loaded up to 80 lb of volatile solids per 1000 ft³ of volume per day in the active digestion units (16). For digestion systems where mixing is accomplished only by circulating sludge through an external heat exchanger, the system may be loaded up to 40 lb of volatile solids per 1000 ft³ of volume per day in the active digestion units. This loading may be modified upward or downward, depending upon the degree of mixing provided. A single-stage anaerobic digester can treat raw sludge efficiently if the detention time is not much less than 30 days and the loading is in the range of 0.075 to 0.10 lb of volatile matter per ft³ of digester capacity per day (17).

Table 7-1 is a useful tool in estimating the size of the digester tank required. Multiple tanks are recommended. Where a single digestion tank is used, it is desirable to have a lagoon or storage tank for emergency use so that the tank may be taken out of service without unduly interrupting plant operations (16). Provision for sludge storage and supernatant separation in an additional unit may be required, depending on raw sludge concentration and disposal methods for sludge and supernatant. The proportion of depth to diameter should be such as to allow for the formation of a reasonable depth of supernatant liquor. The tank bottom should slope to drain toward the withdrawal pipe. At least two access manholes should be provided in the top of the tank in addition to the gas dome. One opening should be large enough to permit the use of mechanical equipment to remove grit and sand. A separate sidewall manhole should be provided (16).

In the use of Table 7-1, there are structural and economic considerations to the combination of depth versus tank diameter that are used to obtain a given digester volume. The total operating water depth at the center of the

TABLE 7-1. Sludge Digester Tank Volume

Tank Diameter, Ft	Depth Bottom Cone, Ft	Sidewall Depth in Ft of Water*				
		16	20	24	28	32
		Digester Volume, Ft³ × 1000				
20	2.50	5.28	6.54	7.80	9.05	10.31
25	3.12	8.36	10.33	12.29	14.26	16.22
30	3.74	12.19	15.02	17.85	20.67	23.50
35	4.37	16.79	20.64	24.49	27.74	32.18
40	5.00	22.20	27.23	32.26	37.29	42.28
45	5.62	28.42	34.78	41.14	47.50	53.86
50	6.25	35.49	53.35	51.21	59.06	66.92
55	6.87	43.44	52.94	62.44	71.95	81.45
60	7.50	52.30	63.60	74.91	86.23	97.52
65	8.12	62.09	72.36	88.63	98.91	112.17
70	8.75	72.79	88.18	103.38	118.98	134.37
75	9.37	84.49	102.16	116.83	136.50	155.17
80	10.00	97.18	117.29	137.40	157.41	177.61
85	10.62	110.93	133.63	156.33	179.03	201.73
90	11.25	125.64	151.09	176.54	201.98	227.43
95	11.87	141.41	169.76	198.11	226.46	254.81
100	12.50	156.36	189.78	221.39	252.61	284.03

*With floating cover, allow 30 in. above working water level to top of tank. Should allow freeboard of 18 to 24 in. between the liquid level outside the cover and the top of the tank wall. With a flat roof, a freeboard of 24 in. is desirable.

Total water depth = sidewall depth + depth of the bottom cone

This table has used a bottom slope of 3 in 12 and assumes that the tank contents will be mixed.

cone should ordinarily be between 20 ft and 45 ft. A single digester tank greater than 115 ft in diameter is not recommended. If anaerobic sludge digestion is used, it is recommended that a floating cover be used; however, there are other configurations in service.

Flat-bottom tanks have been used, and bottoms with a series of conical depressions, for example, could be used. A number of proprietary items are on the market. Figure 7-1 through Fig. 7-5 indicate typical arrangements of proprietary equipment.

7-8. Temperature. Unheated digesters are sometimes used in some locations where climatic conditions permit. A basic rule in digestion is to design the

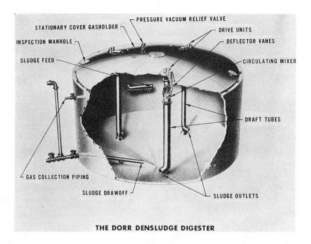

THE DORR DENSLUDGE DIGESTER

Figure 7-1
Single Stage Draft Tube Digester.
(Courtesy Dorr-Oliver, Inc.)

SYSTEM DESIGN

A modern two-stage digestion system is illustrated, including complete piping schematic. Solid lines represent our recommendations for a thorough job with good flexibility. A very few might be eliminated if a small cost saving is more important than operating flexibility and protection. Broken lines are optional and serve the following purposes:

(a) Recirculation return for heating Secondary Digester.

(b) Recirculation draw-off for heating secondary digester. (Supernatant lines in Secondary may also be used for this purpose but we do not recommend this application of a Supernatant Selector).

(c) Transfer from Secondary to Primary (this might occasionally help counteract the effects on the Primary of a shock load of toxic waste).

(d) Supernatant Selector: if used, some of the other supernatant lines may be omitted.

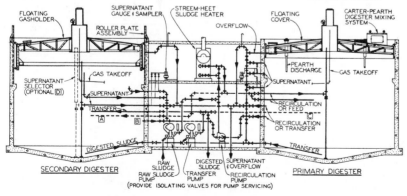

Figure 7-3. Digester System Design. (Courtesy Ralph B. Carter Company)

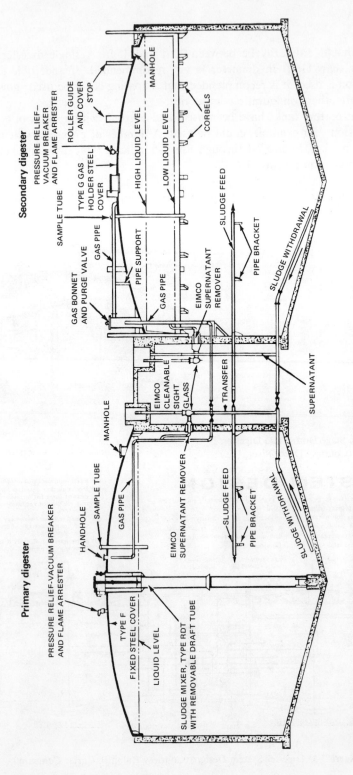

Primary digester

PRESSURE RELIEF-VACUUM BREAKER
AND FLAME ARRESTER

HANDHOLE

SAMPLE TUBE

MANHOLE

TYPE F
FIXED STEEL COVER

GAS PIPE

LIQUID LEVEL

EIMCO
CLEANABLE
SIGHT
GLASS

EIMCO
SUPERNATANT REMOVER

SLUDGE MIXER, TYPE RDT
WITH REMOVABLE DRAFT TUBE

SLUDGE FEED

TRANSFER

PIPE BRACKET

SLUDGE WITHDRAWAL

SUPERNATANT

Secondary digester

PRESSURE RELIEF—
VACUUM BREAKER
AND FLAME ARRESTER

ROLLER GUIDE
AND COVER
STOP

MANHOLE

SAMPLE TUBE

TYPE G GAS
HOLDER STEEL
COVER

HIGH LIQUID LEVEL

LOW LIQUID LEVEL

CORBELS

GAS PIPE

GAS BONNET
AND PURGE VALVE

PIPE SUPPORT

GAS PIPE

EIMCO
SUPERNATANT
REMOVER

SLUDGE FEED

PIPE BRACKET

SLUDGE WITHDRAWAL

Figure 7-2. Primary and Secondary Digester Arrangement. (Courtesy EIMCO)

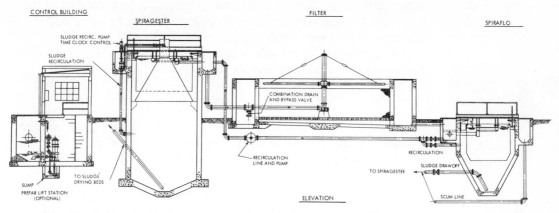

Figure 7-4. Spiragester Design. (Courtesy Lakeside Equipment Corporation)

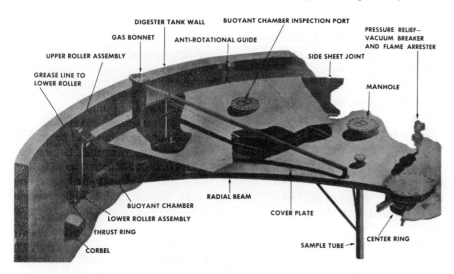

Figure 7-5. One Type Digester Floating Cover. (Courtesy EIMCO)

heating arrangement so that an operating temperature can be closely main-tained at a fixed value somewhere between 35°C and 60°C. The mesophilic range is 29–40°C (84–104°F) and the thermophilic range is 50–57°C (122–135°F.). A temperature in the range between 85 and 95°F is recommended. The higher thermophilic bacteria operation will accelerate digestion and in most cases provide better control on gas generation rate, but it can more easily incur operating problems.

7-9. *Garbage Digestion.* Ground garbage can be fed directly into a digester.

7-10. *Volatile Solids.* The total suspended solids in domestic sewage have in the past been frequently computed on the basis of 0.20 lb per capita per day.

Where garbage grinders are in use and, in general, a high standard of living is prevalent in the community, it is suggested that a figure of 0.32 lb per capita per day be used for design purposes.

The percentage of volatile matter destroyed during digestion can be expressed by the following formula (18):

$$E = \frac{100(\mho_r - \mho_D)}{100 - \mho_D} \tag{7-1}$$

where E is the percentage weight of total dry solids destroyed, $\mho_r$ is the percentage of weight of volatile matter, dry basis, in raw sludge, and $\mho_D$ is the percentage of weight of volatile matter, dry basis, in digested sludge.

$$\mho = \frac{100}{1.56 + 0.0456} S \tag{7-2}$$

where $\mho$ is the percentage of volatile solids remaining and S is the sludge age in days (19).

7-11. Fats and Oils. Fats and oils from plants and animals will decompose in digestion tanks if they are not present in quantities that form objectionable mats (20). Fats and oils, i.e., grease in general, present a constant problem in pumping raw sludge. Plugs of grease tend to build up in the line and restrict pipe flow capacity.

7-12. Gas Production. The major portion, 50 to 75%, of the gas produced in an anaerobic digester is methane gas. Gas production varies with temperature and ranges between approximately 6.0 and 7.2 ft³ of gas per lb volatile solids added or 7.0 to 15.1 ft³ gas per lb volatile solids digested. The heat value of sewage gas is 584 to 646 BTU/ft³ (21). Carbon dioxide is the second gas in quantity and it ranges from 15 to 20% of the gas production.

7-13. Gas Collection. If the digester is kept under a positive pressure, it is not possible for air to enter the system and is therefore impossible to produce an explosive hazard by an air-gas mixture.

It takes approximately 150 ft³ of sludge gas to equal 1 gallon of gasoline. An average heat value of about 640 BTU/ft³ of gas is characteristic of American practice, and a range of 500 to 700 BTU can be encountered on domestic sewage sludges. The yield of gas in the United States from plain sedimentation sludge is about 0.8 ft³ per capita per day (22). The range is from 0.5 to 1.3 ft³.

7-14. Digester Gas Lines. Insulate digester gas lines to prevent condensation in the lines. Meters should be kept warm, by location in control-room housing as an example, to prevent them from becoming gas condensation traps. The gas system should be sealed tight. Buildings where gas may collect should be provided with forced-draft fans. Provide flame traps, within 25 ft of the unit, on all gas-burning equipment. Pressure-relief and reducing valves should be used where appropriate. The pressure-relief valve on the roof of a digester should be set to release at some point between 8 to 12 in. of water gauge pressure. Use a bypass line around meters. Install a check valve and drip trap near the digester dome and a drip trap on the inlet side of the gas meters.

Pressure-control devices to gas engines are sized to function at pressures between 5 to 8 in. water gauge. Pressure range to a waste-gas burner is 6 to 10 in. water gauge pressure, and the burner should be at least 10 ft from the digester roofs or storage tanks. The Ten-State Standards require that waste-gas burners be readily accessible and be located at least 25 ft away from any plant structure if placed at ground level, or they may be located on the roof of the control building if sufficiently removed from the tank. In remote locations it may be permissible to discharge the gas to the atmosphere through a return-bend screened vent terminating at least 10 ft above the walking surface, provided the assembly incorporates a flame trap (16).

Slope gas line and size of line should be $\frac{1}{8}$-in. minimum with $\frac{1}{4}$-in. being desirable. Extra heavy wrought iron pipe can be used for gas piping and is most frequently used for gas lines $2\frac{1}{2}$ to 4 in. Cast iron is usually used in larger sizes.

7-15. Gas Disposal. In larger plants, sludge gas is used to power gas engines. Disposal of gas is normally by burning in small plants. Gas can be disposed of in a small trickling filter along with a large volume of air supplied by a blower (23). Gas can also be blown into a diffuser at the bottom of an activated sludge plant. Either method will remove the odor from the gas.

7-16. Heating Sludge. Wherever possible, digestion tanks should be constructed above groundwater level and should be suitably insulated to minimize heat loss. Sludge may be heated by circulation through external heaters or by units located inside the digestion tank (16). Piping shall be designed to provide for the preheating of feed sludge before the introduction of the digesters (Figs. 7-6 and 7-7). Provisions shall be made in the layout of the piping and valving to facilitate cleaning of these lines. Heat-exchanger sludge piping should be sized for heat transfer requirements (Fig. 7-8). Heating the interior of sludge digesters by circulating hot water in coils tends to build a steady accumulation of heat-insulating cake around the pipe (24, 25) and may stimulate corrosion (26, 27). Hot-water coils for heating digestion tanks should be at least 2 in. in diameter, and the coils as well as the support brackets and all fastenings should be made of corrosion-resistant materials (16). The use of dissimilar materials should be avoided to minimize galvanic action. The high point in the coils should be vented to avoid air lock.

Preheating cold, raw sludge by a submerged heating arrangement in a special heating unit before it enters the digester is said to yield approxi-

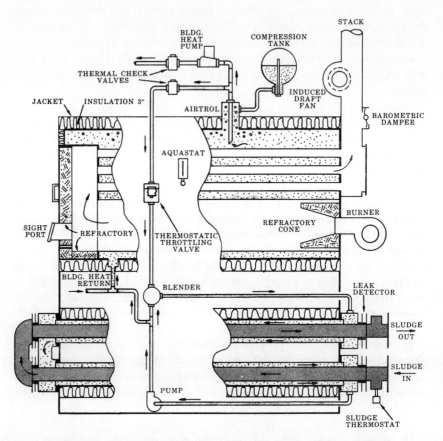

Figure 7-6. Sludge Heater and Heat Exchanger. (Courtesy Ralph B. Carter Company)

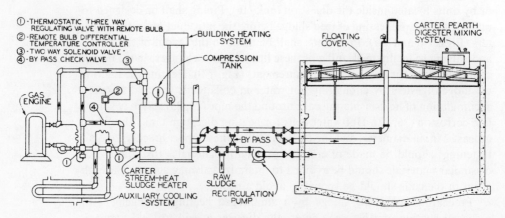

Figure 7-7. Basic Piping for External Heating of Sludge. (Courtesy Ralph B. Carter Company)

mately 85% thermal efficiency with few problems (28). A burner located in the top of the unit introduces a mixture of sludge gas and air. When ignited, the expanding gases are forced through a diffuser and their velocity creates a turbulence that rotates and mixes the sludge as it is heated to approximately 110°F.

7-17. Heat Transfer. The rate of heat flow can be determined as follows (29):

$$H_T = (T_3 - T_4) \times W_s \times c_s \qquad (7\text{-}3)$$

where H_T is the total rate of heat flow in BTU/hr, T_3 is the sludge outlet temperature in °F, T_4 is the sludge inlet temperature in °F, W_s is pounds of sludge flowing per hour, and c_s is the mean specific heat of sludge (usually taken as 1.0).

$$H_T = (T_1 - T_2) \times W_w \times c_w \qquad (7\text{-}4)$$

where T_1 is the hot water inlet temperature in °F, T_2 is the hot water outlet temperature in °F, W_w is pounds of water flowing per hour, and c_w the specific heat of water (taken as 1).

7-18. Sludge Piping. Based on published material (30) as well as personal experience, it is considered desirable to install all raw sludge piping in a man-

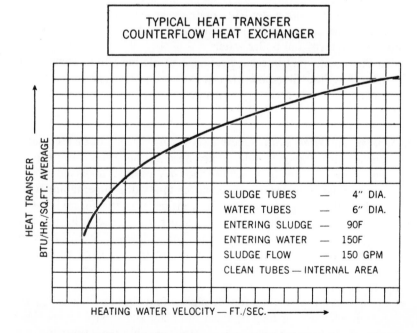

Figure 7-8 Typical Heat Transfer Counterflow Heat Exchanger. (Courtesy Ralph B. Carter Company)

ner that recirculation of hot liquor at 160°F (71°C), or slightly higher temperature, can be carried out any time the operator desires to clean grease accumulations out of the sludge pumps and piping. Otherwise, substantial operational problems may occur.

While comminutors are popular, they do complicate the problems of pumping solids by forming accumulations of stringy materials. Grindings of screenings are much worse. It is for this reason that the author does not recommend feeding grindings or screenings through the plant. It is better to have a well-designed screening operation to remove such materials and keep them out of the process. Excessive grease should also be removed by a suitable arrangement when anaerobic digesters are used and preferably when any type of plant is involved.

7-19. Sampling Pipes. It is recommended that sampling pipes be not less than 3 in. in diameter ($1\frac{1}{2}$ minimum required).

7-20. Controls. A suitable automatic mixing valve should be provided to temper the boiler water with return water so that the inlet water to the heat jacket or coils can be held below a temperature at which caking will be accentuated. Manual control should also be provided by suitable bypass valves (16). The boiler should be provided with suitable automatic controls to maintain the boiler temperature at approximately 180°F to minimize corrosion and to shut off the main gas supply in the event of pilot burner or electrical failure, low boiler-water level, or excessive temperature. Thermometers should be provided to show the temperatures of the sludge, hot water feed, hot water return, and boiler water.

Up to this point, only internal heating controls have been considered, following the Ten-State Standards (16). If external heater operating controls are involved, all controls necessary to insure effective and safe operation are required.

7-21. Supernatant Withdrawal. Supernatant piping should not be less than 6 in. in diameter (16). Piping should be arranged so that withdrawal can be made from three or more levels in the tank. A positive unvalved vented overflow should be provided. On fixed-cover tanks, the supernatant withdrawal level should preferably be selected by means of interchangeable extensions at the discharge end of the piping. If a supernatant selector is provided, provisions should be made for at least one other draw-off level located in the supernatant zone of the tank, in addition to the unvalved emergency supernatant draw-off pipe. High-pressure backwash facilities should be provided (16).

An alternate disposal method for the supernatant liquor such as a lagoon, an additional sand bed, or hauling from the plant site should be provided for use in case the supernatant is not suitable or other conditions make it advisable not to return it to the plant. Consideration should be given to supernatant conditioning where appropriate in relation to its effect on plant performance and effluent quality (16).

7-22. Ferric Chloride. Digestion of sludge treated with ferric chloride is considerably delayed, and gas yield is reduced by increased solids concentration (31). Ferric chloride is a strong poison that will kill bacteria. Digestion and methane production of chemically treated sludges does not appear feasible (32).

7-23. Heavy Metals. Heavy metals can be sufficiently toxic to digesters to put them out of service. A sewer monitoring program is desirable in a city having industry likely to release such contaminants. The application of some 25 to 50 lb (11 to 23 kg) of sodium sulfide to a primary digester of 1 million gal capacity, when the volatile acids approach the 600 mg/l level, will lower the concentration to an acceptable 200- to 300-mg/l level (33).

AEROBIC SLUDGE DIGESTION

7-24. Aerobic Digestion. Aerobic sludge digestion is a process that shows substantial promise as a means of sludge digestion in small plants. There is less sludge reduction in terms of volume than with anaerobic digesters, but the digestion time is reduced and the equipment is simpler and more practical for the small plant.

Aerobic sludge digestion is accomplished in a tank or tanks designed to provide effective air mixing, reduction of the organic matter, supernatant separation, and sludge concentration under controlled conditions (16). Multiple tanks are recommended. A single sludge-digestion tank may be used in the case of small treatment plants. The design of the facility should show that the single tank will not adversely affect normal plant operations.

Aerobic sludge digestion tanks should be designed for effective mixing by satisfactory aeration equipment. If diffusers are used, types shall be provided which are especially slow to clog and are designed to permit removal for inspection, maintenance, and replacement without dewatering the tanks (16).

7-25. Size and Number of Tanks. Chapter 9 describes aerobic digestion. Aerobic digestion in its simplest case consists of a bar screen, an aeration tank, and a clarifier in which the sludge can settle and be returned to the aeration chamber. See Chapters 8, 9, and 10 for further pertinent material on the subject.

SLUDGE

7-26. General. Some treatment processes aerate sludge until it nitrifies and breaks up into pinpoint floc. This is not particularly noticeable as it goes over the effluent weirs. However, the fact remains that sludge was produced, and in this instance it went into the receiving stream. The trend is toward harsher regulatory agency practice to prevent this wasting of sludge into receiving waters. Liquid sludge may be conveyed in tank trucks and spread on agricultural lands. It may be drained, dried, and then used as a soil conditioner on agricultural lands or any place a soil conditioner is needed. Sludge is not a competitive product with modern fertilizers; therefore it is a product that may require a subsidy for its use. It is the author's opinion, however, that in one form or another, the best disposal would be as a soil conditioner.

Sludge may be buried in landfill, or it may be partially consumed by incineration, and then used as landfill. In general, it is one of the biggest problems in the water carriage waste disposal process, and the manner in which sludge is handled may very well have substantial influence on process economics.

7-27. Solids Per Capita. It can be expected that the dry solids per capita per year in sludge will average at least 17 pounds, and a minimum design figure of 22 pounds is recommended.

7-28. pH of Sludge. Dewatering is usually in a pH range of 11.5 to 12.5.

7-29. Pathogen Reduction. Bacterial pathogens are significantly reduced by addition of sludge-conditioning chemicals such as lime.

7-30. Sludge-Conditioning Agents. Lime and ferric chloride have been widely used.

7-31. Sludge Compressibility. Most sewage sludges treated with ferric chloride have a compressibility of 0.85 or greater (34, 35).

7-32. Sludge Chemical Composition. Substances dissolved in sludge water (such as bicarbonate alkalinity) and the composition of the suspended solids (such as the ratio of volatile matter to ash) directly regulate the amount of chemicals required to condition a sludge (36).

7-33. Drainability of Sludge. One method of determining the drainability of sludge is that of Babbitt and Schlenz, which is described in the literature (37).

7-34. Flotation. Dissolved air-flotation can be used to concentrate a 1% activated sludge feed to a sludge blanket of greater than 5% solids (38). The air-flotation process was described earlier in this text, and additional information is available in the literature (38, 39).

7-35. Disc Centrifuge. The disc centrifuge is described in the literature (40, 41, 42). A problem has been reported of clogging of sludge discharge nozzles using this type of centrifuge (43).

7-36. Solid-Bowl Centrifuge. The solid-bowl centrifuge seems to offer a poorer effluent clarity than do the disc type centrifuges, but it also does not seem to be bothered with solids clogging. It is described in the literature (44).

7-37. *Elutriation.* In this process the sludge is washed by either fresh water or plant effluent to reduce the need for conditioning chemicals as well as to improve the settling or filtering characteristics of solids. The sludge solids concentration may be more than doubled if the process is properly conducted (45). Fraschina (46) reported that primary digester sludge washed by single-stage elutriation using a ratio of sludge to wash water of 1 to 3, resulted in a concentration of 5 to 6 % solids and a clear supernatant liquid not exceeding 1,300 mg/l of total solids.

7-38. *Sludge Concentration.* The sludge feed concentration to filter rate is linear (47).

7-39. *Filtration Rates.* Most of the sludge characteristics that determine filterability vary with the source of the sludge; therefore, the various processes of sewage treatment produce sludges of varying filterability, as indicated in Table 7-2.

TABLE 7-2. Minimum Filter Rates After Chemical Conditioning

Type of Sludge	Filter Rate (psf/hr)	
	Fresh	Digested
Primary	8	7
Primary and trickling filter	7	6
Primary and activated	6	5

7-40. *Vacuum Filtration.* On 1 ft^2 of continuous-vacuum drum filter area, the drainage of water from digested sludges is 2,000 times faster than is normally accomplished by gravity and air drying over 1 ft^2 of open sand area (49). Waste-activated sludge is virtually impossible to handle on metal media of any type, including the wire mesh or helical spring types, because of the particle size and low solids concentration, but it can be successfully dewatered on synthetic fabrics. Combination digested and undigested secondary sludge can be handled on metallic types of media but at the expense of rate loss, higher residual cake moisture, and excessive chemical costs. Filtrate solids are at least double that obtained when synthetic fibers are used.

7-41. *Sludge Tanks.* A sludge tank can be used as a settling tank in which to concentrate the sludge and/or as a holding tank until the sludge can be transported elsewhere or otherwise processed.

The author uses circular tanks of not less than 16-ft sidewalls and preferably 20-ft or more sidewalls. The absolute minimum base slope is 3 to 12 since mechanical equipment is not used. A bottom slope of 30° with respect to the horizontal is desirable. Use the same freeboard as digesters. If the locality will permit, leave the top open; however, if escape of gas is objectionable, use a floating cover. If tanks are covered, provide means of burning or using the gas and provide the same gas safety provisions on top of the tank as on a digester.

7-42. *Sludge Drying Beds.* In determining the area for sludge drying beds, consideration should be given to climatic conditions, the character and volume of the sludge to be dewatered, and other methods of sludge disposal (16). The lower course of gravel around the underdrains should be properly graded and should be 12 in. deep, extending at least 6 in. above the top of the underdrains. It is desirable to place this in two or more layers. The top layer of at least 3 in. should consist of gravel $\frac{1}{8}$ to $\frac{1}{4}$ inch in size. The top course should consist of at least 6 to 9 in. of clean, coarse sand. The finished sand surface should be level. Underdrains should be clay pipe or concrete tile, at least 4 in. in diameter, laid with open joints. Underdrains should be spaced not more than 20 ft apart. Walls should be watertight and extend 15 to 18 in. above and at least 6 in. below the surface. Outer walls should be curbed to prevent soil from washing onto the beds.

Not less than two beds should be provided, and they should be arranged to facilitate sludge removal. Concrete truck tracks should be provided for all percolation-type sludge beds. Pairs of tracks for percolation-type beds should be on 20-ft centers. The sludge pipe to the beds should terminate at least 12 in. above the surface and be so arranged that it will drain. Concrete splash plates for percolation-type beds should be provided at sludge discharge points.

Generally, washed sand with a uniformity coefficient of 4.0 or less and an effective size between 0.3 and 0.5 mm should be specified (50). Adequate freeboard should be allowed to use a depth of 12 in. of sand. The sludge discharge pipe must have provision to be drained during cold weather.

The bed should be designed so that the sludge can be applied at any desired depth up to 16 in. In practice, 7 to 9 in. is usually a suitable depth, but some sludge requires a greater depth (Figs. 7-9 and 7-10).

It has been observed that approximately 1 in. of sand will be removed with the sludge cake per year of 10 removals (51).

Hazeltine (52) developed the expressions

$$Y = 0.96X - 1.75 \qquad (7\text{-}5)$$

$$Z = 0.35X - 0.5 \qquad (7\text{-}6)$$

where X is the percentage of solids in the applied sludge, Y is the gross bed loading, and Z is the net bed loading. Gross bed loading is the pounds solids applied per 30 days of actual bed use. For example, if sludge containing 10% solids is applied 12 in. deep and removed after 40 days, the gross bed loading is $(62.5 \times 0.10 \times 30)/40$ or 4.69 lb ft²/day. Net bed loading is the gross bed loading multiplied by the percent solids in the sludge removed. Hazeltine considered temperature, solid content of the sludge applied to drying beds,

Figure 7-9
Sludge Drying Bed.

Figure 7-10. Sludge Drying Bed on Hillside.

and moisture content of the sludge as removed (in that order) the most important factors influencing drying. His 18-page paper is recommended reading.

A glass-covered drying bed is advisable in either wet climates or cold climates. The glass-covered structure should have frames constructed from aluminum. Wood frames tend to fail after 10 to 15 years' service. After the frame fails, water pours into the enclosure and defeats its purpose.

State regulatory agencies may spell out the drying bed area required. If not, an extremely rough rule might be 0.2 ft² per capita for an uncovered bed in the southwest, 0.3 ft² per capita for uncovered, or 0.2 ft² per capita in the southeast, 0.5 ft² per capita for a glass-covered bed in other portions of the continental U.S.A. excluding Alaska. These values are on the liberal side in most instances, but they are advisable.

7-43. Paved Sludge Drying Beds. It is feasible and permissible by most regulatory agencies to use a paved sludge drying bed with either a concrete or asphalt lining. This lining should have an unpaved area along each side or down the center of each bed to facilitate drainage. If the soil is fine textured, it would be necessary to resort to drain tile and graduated gravel below the sand. The bed would be covered by sand as before. Deerfield, Ill., had such a bed before 1964. South (53) describes such a bed. Some plants have been experimenting with spreading a very thin layer of sludge on beds of this type. In general, this type of bed seems to offer economic advantages over the conventional drying bed.

7-44. Intensity of Radiation. For an intensity of 1.00 cal/cm²/min, the evaporation rate was 0.91×10^{-3} g/cm²/min from a free water surface and 0.89×10^{-3} g/cm²/min from a sludge surface (54). Of the total energy incident on the sludge surface, approximately one-half is used in the vaporization of water. The evaporation rate from the sludge surface was depressed by 22% when both evaporation and drainage contributed to dewatering. One-half the energy incident to the sludge surface was found to be associated with the latent heat of evaporization. The emissivity was found to be dependent on the wavelength of radiation. The critical moisture of the sludge used ranged from 66 to 84%.

Under optimum conditions, the moisture content in 50 hr with evaporation only will be reduced 7 or 8%, but with a combination of evaporation and drainage, the moisture content may be reduced over 70%. Radiant energy has no major effect on the drying of the liquid sludge until it reaches a relatively solid state, at which time radiant energy plays a major part in the final dewatering of the sludge (55).

7-45. Sludge Lagoons. The use of shallow sludge-drying lagoons in lieu of drying beds is permissible subject to the following conditions: (a) The soil must be reasonably porous and the bottom of the lagoons must be at least 18 in. above the maximum groundwater table. Surrounding areas should be graded to prevent surface water from entering the lagoon; (b) Lagoons should not be more than 24 in. in depth; (c) The area required will depend on local climatic conditions. Not less than two lagoons should be provided; (d) Consideration shall be given to prevent pollution of ground and surface water. Adequate isolation shall be provided to avoid nuisance production (16).

7-46. Drainage and Filtrate Disposal. Drainage from beds or filtrate from dewatering units should be returned to the sewage treatment process at appropriate points. When disinfection is required, the filtrate should be returned to a point preceding disinfection (16).

PROBLEMS

1. Design a digester (or digesters) for each of the following treatment plants using conventional activated sludge process.

Plant	Flow Rate (MGD)
A	0.5
B	1.0
C	10.0
D	100.0

Assume that the plants are located in the vicinity of Chicago, Illinois. Determine the tank diameter, sidewall depth, depth of bottom cone, digester volume in cubic feet, bottom slope of the cone used, and all pipe sizes.

2. Design a heating system for each digester in Problem 1. Indicate pipe sizes and show all calculations pertinent to the design of a complete system.

3. Design a sludge drying bed for each of the plants in Problem 1.

4. Design a sludge drying lagoon for each of the above plants assuming that they are located in a climatic location comparable to that prevailing at El Paso, Texas.

5. Determine the gas production obtainable from each plant in Problem 1. Design the piping from the digester and size the gas engines for operating the aeration diffuser blowers. If there is an excess or deficiency of gas, indicate how much and what you propose to do with the surplus or how you will make up the deficiency.

REFERENCES

1. CHANIN, G., "Fundamentals of Sludge Digestion. Part II. Biology and Operation." *W & S. W.*, 108, 3, 85 (1961).

2. *Operation of Wastewater Treatment Plants.* WPCF Manual of Practice No. 11, WPCF (1966).

3. CHANIN, GERSCH, "Fundamentals of Sludge Digestion. Part I. Biochemical Background." *W & S. W.*, 108, 2, 55 (1961).

4. CHANIN, GERSCH, "Digestion is Not Really Digestion." *W & S. W.*, 113, 4, 117 (1966).

5. CLARK, JOHN W., and VIESSMAN, WARREN, JR., *Water Supply and Pollution Control.* International Textbook Co., Scranton, Pa., 1966.

6. FAIR, GORDON MASKEW, and GEYER, JOHN CHARLES, *Water Supply and Waste-Water Disposal.* John Wiley & Sons, Inc., New York, 1954.

7. *Anaerobic Sludge Digestion.* WPCF Manual of Practice No. 16, WPCF (1968).

8. BUZZELL, JAMES C., JR., and SAWYER, CLAIR N., "Biochemical Vs. Physical Factors." *Jour. WPCF*, 35, 2, 205 (1963).

9. LABOON, J. F., "Experimental Studies on the Concentration of Raw Sludge." *Sew. & Ind. Wastes*, 24, 4, 423 (1952).

10. SAWYER, CLAIR N., and ROY, H. K., "A Laboratory Evaluation of High-Rate Sludge Digestion." *Sew. & Ind. Wastes*, 27, 12, 1356 (1955).

11. HINDEN, ERVIN, and DUNSTAN, G. H., "Effect of Detention Time on Anaerobic Digestion." *Jour. WPCF*, 32, 9, 930 (1960).

12. SAWYER, CLAIR N., and GRUMBLING, JAY S., "Fundamental Considerations in High Rate Digestion." *Proc. ASCE*, 86, SA2, 49 (March, 1960).

13. TORPEY, W. N., "Concentration of Combined Primary and Activated Sludges in Separate Thickening Tanks." *Proc. ASCE*, 80, Paper 443 (May, 1954).

14. RANKIN, R. S., "Digester Capacity Requirements," *Sew. Works Jour.*, 20, 3, 478 (1948).

15. LANGFORD, L., "Digester Net Capacity." *W. & S. W.*, 105, 10, 435 (1958).

16. *Recommended Standards for Sewage Works*, rev. ed., Great Lakes–Upper Mississippi River Board of State Sanitary Engineers, Health Education Service, Albany, N.Y., 1971.

17. HINDIN, E., and DUNSTAN, G. H., "Some Aspects of Sludge Digestion," *W. & S. W.*, 106, 457 (Oct., 1959).

18. KEEFER, C. E., *Sewage Treatment Works*. McGraw-Hill, Inc. New York, 1940.

19. SPICKA, IVAN, "Design of Sludge Digestion Tanks." Discussion, *Proc. ASCE*, 95, SA4, 795 (1969).

20. BLOODGOOD, DON E., "Sludge Digestion." *W. & S. W.*, 101, 8, 376 (1954).

21. BLOODGOOD, DON E., "Sludge Digestion." Series 2–Part 8, *W. & S. W.*, 104, 1, (1957).

22. VAN KLEECK, LeROY W., "Sewage Sludge Gas—Its Collection and Use. Part 2." *Wastes Engineering*, 29, 2, 77 (1958).

23. POMEROY, RICHARD D., "Design of Sludge Digestion Tanks." *Proc. ASCE*, 94, SA5, 769 (1968).

24. CALIHAN, R. H., "Controlled Heating Solves Sludge Handling Problems." *Waste Eng.*, 25, 118 (1954).

25. "Penn State Operations Discussed." *Wastes Eng.*, 25, 11 (1954).

26. PARKS, G. A., "Electrolysis at Terminal Island." *Sew. Works Jour.*, 13, 48 (1941).

27. PARKS, G. A., "Operators Forum." *Sew. Works Jour.*, 21, 726 (1949).

28. BAXTER, S. S., and HORLACHER, H. F., "Submerged Sludge Heating at Philadelphia's Northeast Works." *Sew. & Ind. Wastes*, 26, 961 (1954).

29. BAFFA, JOHN J., "Forced Circulation Type Hot Water Heat Exchangers for Digestion Tank Heating." *W. & S. W.*, 96, 3, 85 (1949).

30. FISICHELLI, ANDREW P., "Raw Sludge Pumping—Problems and Interdisciplinary Solutions." *Jour. WPCF*, 42, 11, 1916 (1970).

31. KEEFER, C. E., "The Digestion of Sewage Sludge Containing Various Concentrations of Solids." *Sew. Works Jour.* 29, 9, 1117 (Sept. 1950).

32. SCHULZE, K. L., "Studies on Sludge Digestion and Methane Fermentation. I. Sludge Digestion at Increased Solids Concentration." *Sew. & Ind. Wastes*, 30, 1, 28 (1958).

33. REGAN, TERRY M., and PETERS, MERCER M., "Heavy Metals in Digesters: Failure and Cure." *Jour. WPCF*, 42, 10, 1832 (1970).

34. HALFF, A. H., "An Investigation of the Rotary Vacuum Filter Cycle as Applied to Sewage Sludges." *Sew. & Ind. Wastes*, 24, 8, 962 (1952).

35. JONES, B. R. S., "Vacuum Sludge Filtration. II. Prediction of Filter Performance." *Sew. & Ind. Wastes*, 28, 9, 1103 (1956).

36. GENTER, A., "Computing Coagulent Requirements in Sludge Conditioning." *Trans. ASCE*, 111, 637 (1946).

37. BABBITT, HAROLD, and SCHLENZ, HARRY, University of Illinois *Bulletin* 198, p. 21 (1930).

38. KATZ, WILLIAM J., "Sewage Sludge Thickening by Flotation." *Public Works*, 89, 12, 114 (1958).

39. CHASE, E. S., "Flotation," *Munic. Util.*, 95 (Dec., 1957).

40. AMBLER, C. M., "Evaluation of Centrifuge Performance." *Chem. Engr. Progr.*, 48, 3, 150 (1952).

41. BROWN, G. A., *Unit Operations*, John Wiley & Sons, Inc. New York, 1950.

42. HART, R. R., "General Considerations to Aid Centrifuge Selection." *Ind. Chem.* (Brit.), 38, 448 (1962).

43. ETTELT, GREGORY A., and KENNEDY, T. J., "Research and Operational Experience in Sludge Dewatering at Chicago." *Jour. WPCF*, 38, 2, 248 (1966).

44. WHITE, W. F., and BURNS, T. E., "Continuous Centrifugal Treatment of Sewage Sludge." *W. & S. W.*, 109, 384 (1962).

45. TORPEY, W. N., and LANG, M., "Elutriation as a Substitute for Secondary Digestion." *Sew. & Ind. Wastes*, 24, 813 (1952).

46. FRASCHINA, K., "Sludge Elutriation at the Richmond-Sunset Plant, San Francisco, California." *Sew. Works Jour.*, 22, 11, 1413 (1950).

47. SCHEPMAN, B. A., and CORNELL, C. F., "Fundamental Operating Variables in Sewage Sludge Filtration." *Sew. & Ind. Wastes*, 28, 12, 1443 (1956).

48. TRUBNICK, E. H., and MUELLER, P. K., "Sludge Dewatering Practice." *Sew. & Ind. Wastes*, 30, 11, 1364 (1958).

49. EIMCO Bulletin, *EIMCO—Process Clari-Thickener for Activated Sludge Plants.* EIMCO, Salt Lake City, Sept., 1965.

50. VAN KLEECK, LEROY W., "Conditioning Digested Sludge for Dewatering on Drying Beds." *Wastes Eng.*, 29, 3, 147 (1958).

51. RYAN, WILLIAM A., "Operations of Open Sludge Beds." *Sew. Works Jour.*, 10, 1, 153 (1938).

52. HAZELTINE, T. R., "Measurement of Sludge Drying Bed Performance." *Sew. & Ind. Wastes*, 23, 9, 1065 (1951).

53. SOUTH, W. T., "Asphalt Paved Sludge Beds." *W. & S. W.*, 106, R396 (1959).

54. QUON, J. E., and TAMBLYN, T. A., "Intensity of Radiation and Rate of Sludge Drying." *Proc. ASCE*, 91, SA2, 17 (1965).

55. JENNETT, J. C., and SANTRY, I. W., "Characteristics of Sludge Drying." *Proc. ASCE*, 95, SA5, 849 (1969).

8 Package Treatment Plants

A package sewage treatment plant consists of one or more prefabricated manufactured units that can be readily assembled, connected, and installed at the project site. Depending on the size of the facility, items such as heavy vertical electric motors or other special components may have to be installed at the site. Mechanically and electrically, the work at the site is comparatively insignificant compared to an on-site construction of a custom-designed conventional activated-sludge or trickling filter treatment plant. A package plant is usually a proprietary product of a manufacturer or engineer.

There are several treatment processes available in package form. These plants are usually designed for a domestic sewage loading cycle characteristic of a bedroom community or small town not having industrial waste. They are seldom designed for a BOD of the raw sewage greater than 300 mg/l, and many of them should be operated under 250 mg/l.

Package treatment plants seem to give either very good or extremely poor results. It appears that satisfactory results accompany well-designed equipment provided by reputable manufacturers, properly installed and operated. Entirely too often, a package treatment plant is procured and installed without the services of an appropriately trained consulting engineer. It is a fiercely competitive market and is literally one of buyer beware! Salesmen vary from extremes of highly reputable professional engineers to individuals with very little formal education.

A conventional treatment plant is generally less sensitive to poor operation than a package treatment plant. One reason for this is that the conventional plant is the product of a consulting engineer's work. The engineer has evaluated the receiving stream; the character of the organic load that will be placed on the plant and the variations in the sewage flow; the hydrographic and topographic condition of the site; the operation, supervision, and maintenance required; and he has worked with his client to insure that these factors are properly established. Also, there is usually more reserve margin in the conventional design than might be characteristic of the package plant to effect its economy. Another failing is that people are willing to operate and maintain a conventional plant but expect a package plant to perform miracles without attention. The two types of plants perform essentially the same—given the same attention!

8-1. National Sanitation Foundation. The National Sanitation Foundation, NSF, is a nonofficial and noncommercial agency incorporated under Michigan law. It is a nonprofit organization devoted to research, education, and service. It functions as a medium in which business, industry, official regulatory agencies, and the public may together confront issues of environmental quality. In scope, such problems range from the philosophical to the intensely practical, from matters of attitude all the way to the correct design for a restaurant dishwashing machine.

On January 1, 1965, the National Sanitation Foundation began a research project aimed at the development of criteria for the performance evaluation of package sewage treatment plants. This study was supported by the U.S. Public Health Service through a demonstration grant.

The NSF established a performance criteria and a standard performance evaluation method. A certificate of performance from the NSF indicates that a particular model of packaged waste treatment plant achieved a certain minimum level of performance. It does not in any way indicate the quality of construction of the plant. A plant may quite satisfactorily achieve a performance of acceptable level and yet be of poor quality construction. However, if the plant has at least been tested for performance, it is established that this phase of the plant complies with a minimum set of evaluation criteria.

8-2. Extended Aeration. The majority of extended aeration plants provide long-term aeration, complete-mixing activated-sludge systems. The system requires that microorganisms exist in the endogenous phase by maintaining a low food-to-microorganism ratio in the system. This ratio is produced by maintaining a low BOD loading, a high mixed liquor suspended solids, and a long retention time. The 24 hours used in these processes produces a BOD loading of 12 to 15 lbs of 5-day BOD/day/1000 ft^3. The BOD of the raw sewage must average 200 to 250 mg/l to obtain this load. The extended aeration process is better able to handle shock loadings or underloadings without detrimental aftereffects than the contact stabilization process. A typical extended aeration plant is shown in Fig. 8-1.

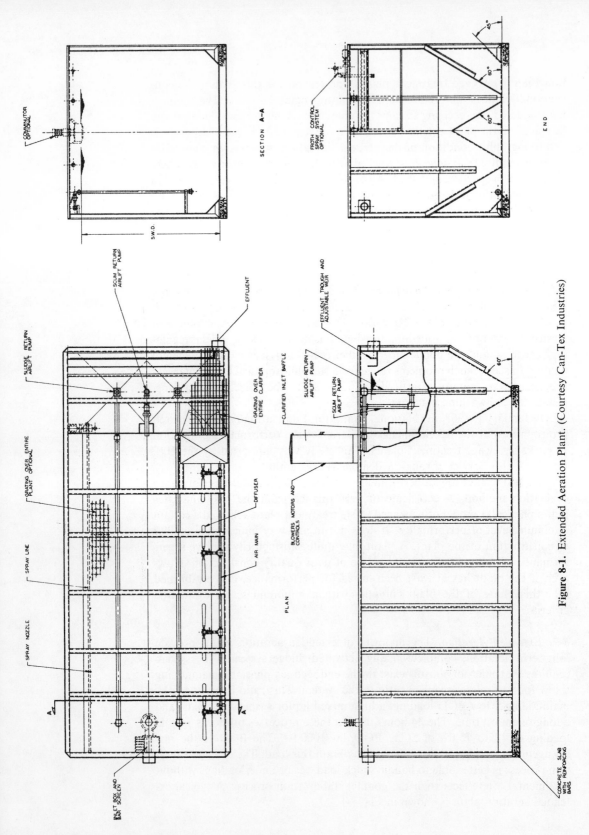

Figure 8-1. Extended Aeration Plant. (Courtesy Can-Tex Industries)

SECTION A-A

COMMINUTOR OPTIONAL

S.W.D.

FROTH CONTROL SPRAY SYSTEM OPTIONAL

END

45°

60°

60°

SCUM RETURN AIRLIFT PUMP

SLUDGE RETURN AIRLIFT PUMP

EFFLUENT

GRATING OVER ENTIRE CLARIFIER

CLARIFIER INLET BAFFLE

GRATING OVER ENTIRE PLANT OPTIONAL

EFFLUENT TROUGH AND ADJUSTABLE WEIR

SLUDGE RETURN AIRLIFT PUMP

SCUM RETURN AIRLIFT PUMP

60°

DIFFUSER

SPRAY LINE

AIR MAIN

BLOWERS MOTORS AND CONTROLS

SPRAY NOZZLE

PLAN

INLET BOX AND BAR SCREEN

CONCRETE SLAB WITH REINFORCING BARS

208

8-3. *Contact Stabilization.* Contact stabilization is a form of the activated-sludge process where aeration is carried out in two phases in two types of tanks, (1) the contact tank where raw sewage solids are adsorbed and absorbed on the microbio masses, and (2) the stabilization tank where the solids, which have been removed in a final settling tank, are partially stabilized by re-aeration before being recombined with the incoming raw sewage. An aerobic digestion tank may be included as a part of the process.

Contact stabilization processes may be upset by slug loadings which could result in a bad effluent for the period of times that the plant is overloaded. There are various thoughts regarding the type of operation to be expected from contact stabilization plants when they are underloaded. The contact stabilization process works best at design loading both hydraulically and organically. If a contact stabilization plant receives a higher-than-normal organic shock load or other biological toxic materials, the aeration zone or the contact zone may lose its effectiveness and upset the system.

An elevation view of a contact stabilization plant is shown in Fig. 8-2, and a plan view of the same plant is shown in Fig. 8-3.

8-4. *Trickling Filter.* A stationary biological bed is periodically dosed with waste, which is permitted to trickle through the bed. The effluent is carried off beneath the bed by means of drain tiles. One package plant arrangement on the market uses the arrangement shown in Fig. 8-4 ahead of the high-rate trickling filter. In this unit, the solids separate by gravity from the liquid in the upper cone section. The solid free liquid is removed at the weir trough, and the solids drop to the lower digester compartment. Figure 8-5 shows the complete plant.

Trickling filters are able to adjust to hydraulic flow variations but are generally less effective than the activated-sludge process for removing suspended solids, BOD, and nutrients. High-rate trickling filters have been used successfully in turnpike treatment plants where restaurant wastes are involved; however, the plants observed were custom-designed small units.

8-5. *Other Package Plant Processes.* There are plants available on the market that can be operated as extended aeration, conventional activated-sludge, step aeration, and contact stabilization. The plants can adapt to a variety of operating conditions. There are conventional activated-sludge plants, modified activated-sludge units, aerobic plants with sacks or bladders to hold the sludge, incinerator-type plants, plants designed to polish the effluents of other plants, and various other innovations.

Irrespective of whether it is an extended aeration, contact stabilization, trickling filter, activated-sludge, or other principle of operation, many of the problems in the small treatment plants are caused by the buyers that press for low bid without regard to the product. A package plant simply should not be procured on the basis of price. The services of an appropriately qualified consulting engineer should be secured, and the plants offered should

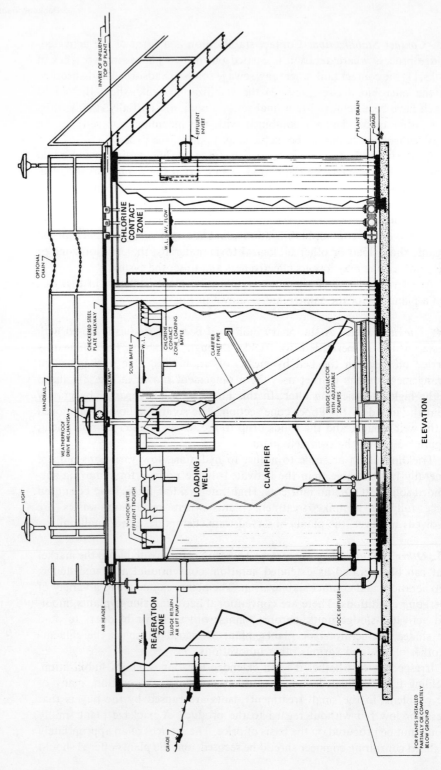

Figure 8-2. Elevation View Contact Stabilization Plant. (Courtesy Can-Tex Industries)

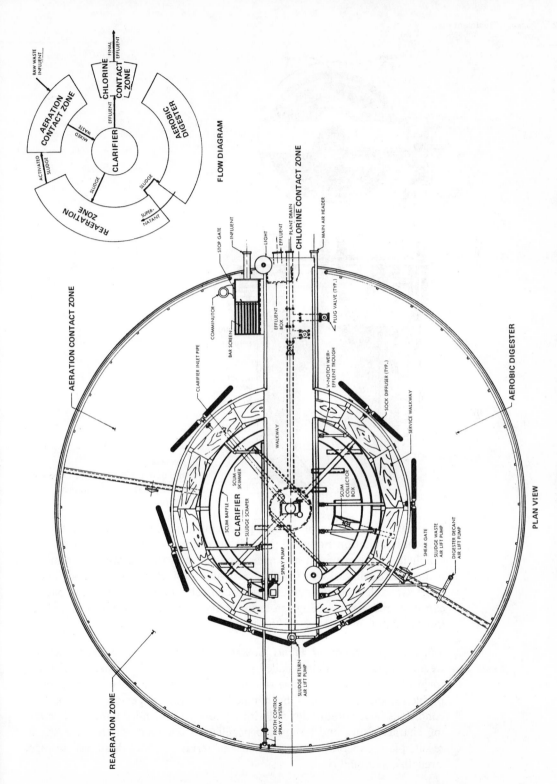

Figure 8-3. Plan View of Contact Stabilization Plant. (Courtesy Can-Tex Industries)

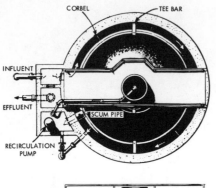

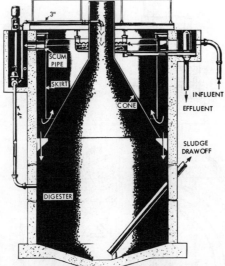

Figure 8-4
Combination Clarifier and Digester.
(Courtesy Lakeside Equipment
Corporation)

be evaluated in terms of effluent-producing capability as certified by the
National Sanitation Foundation, their operational controls and operational
simplicity, the instrument and measuring features, the maintenance requirements, the quality of construction, availability of spare parts, service assistance available from the manufacturer, controls to operate the unit at partial
capacity, and general design in terms of the application problem as well as
the suitability of the size of the unit.

8-6. Package Treatment Plants Compared to Other Systems. The package
treatment plant, commonly called package plant, was developed specifically
for domestic sewage use. It is suitable for certain types of industrial waste
problems, but this should be handled by a trained person. The loading
conditions versus time of day fall within known, well-defined limits. Its operating characteristics and design are peculiar to the needs of the user. As a
result, plant capabilities are restricted in terms of shock loadings, sewage
strengths, and quality.

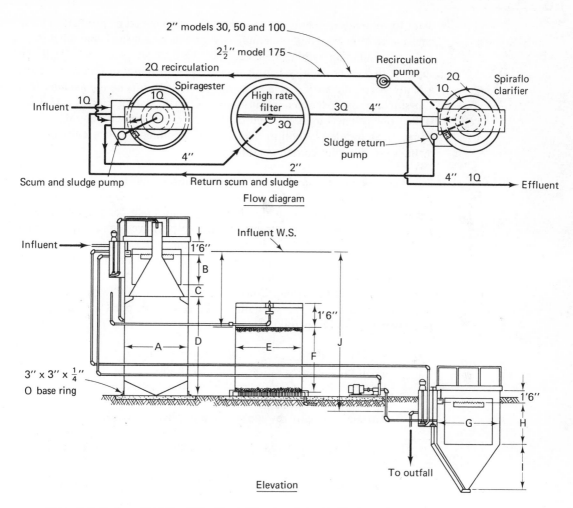

Figure 8-5. Package Trickling Filter Plant. (Courtesy Lakeside Equipment Corporation)

The author concluded that sources of the problems confronting package plants were: (a) lack of proper operation, (b) lack of proper maintenance, and (c) misconception by the owner as to what the plant could accomplish in the absence of (a) and (b). In addition, some of the designs suffered by one or more deficiencies, such as: (d) poor design (general), (e) inadequate quality of materials or thickness of materials, (f) unsuitable or no provision to handle foam and recirculate scum, (g) operating controls too complex or inadequate, (h) lack of timers or other provisions to regulate the amount of air according to loading, (i) inadequate provision for sludge handling, (j) inadequate provision for chlorination of effluent, (k) plant excessively oversized or undersized, and (l) inadequate measuring devices to properly control the plant.

If skilled operators are available and can spend the necessary time at the plant to control the operation, an acceptable effluent can be produced by the package treatment plants of a majority of the established manufacturers. Even with poor operation, the effluent of the package plant will usually be superior to a septic tank system. A package plant will not ordinarily produce a really good effluent *if it is not carrying at least half its design loading*. It is superior in most locations and instances to waste stabilization ponds when skilled operation is used. An aerated lagoon combined with a stabilization pond, or an aerated lagoon or aerated tank combined with a clarifier, has distinct merits under certain conditions.

8-7. Package Plant Calculations. The calculations that follow are made for two extremes of condition. The BOD of domestic waste is expressed in pounds per unit of time and is normally designated as 0.17 lb per person per day. Roadside rest areas, as an extreme, in this sample have been computed on 5 GPD per capita with 0.03 lb BOD per person. BOD is also designated in terms of parts per million (ppm) or in milligrams per liter (mg/l), both being approximately the same values. Normally sewage strengths are designated as between 220 and 300 ppm. Converting the amount of flow per person per day and relating it to the strength of the sewage in parts per million or pounds per person is as follows:

Hydraulic flow$_{(household)}$ = 100 gal/person/day.
Hydraulic flow$_{(rest\ area)}$ = 5 gal/person/day.
Organic loading$_{(household)}$ = 0.17 lb BOD/person/day.
Organic loading$_{(rest\ area)}$ = 0.03 lb BOD/person/day.
Normal strength sewage = 220 mg/l BOD (household).

Flow: at 100 gal/person/day we can determine the total flow per day by multiplying:

$$\text{Population} \times 100 \text{ gal/person/day} = \text{GPD or flow in gal/day} \quad (8\text{-}1)$$

that is,

$$750 \text{ persons} = 750 \times 100 \text{ GPD} = 75,000 \text{ GPD}.$$

The BOD is expressed as 0.17 lb/person/day. In sizing a plant for household use, we have to know the expected BOD in order to select the proper aeration tank volume or size.

The design flow is derived from Eq. (8-1). If the organic loading or raw waste BOD concentration is expressed in mg/l or ppm, we determine the pounds of BOD to be handled in aeration as

$$\text{MGD} \times \text{mg/l BOD} \times 8.34 = \text{lb BOD, or}$$

$$\text{Population} \times 0.17 = \text{lb BOD to be handled per day} \quad (8\text{-}2)$$

If the organic loading is expressed in pounds of BOD and we want to determine the BOD in mg/l:

$$\frac{\text{lb BOD}}{\text{MGD} \times 8.34} = \text{BOD mg/l} \quad (8\text{-}3)$$

If given the organic loadings in lb BOD as well as in mg/l of BOD, the flow in MGD is:

$$\frac{\text{lb BOD}}{8.34 \times \text{mg/l BOD}} = \text{flow MGD} \qquad (8\text{-}4)$$

Then

$$\frac{\text{MGD}}{100} = \text{population equivalent}$$

The establishment of the organic loading will now enable us to size the aeration tank for suitable operation. The most popular means of doing this for packaged sewage treatment plants is to calculate the lb BOD/1,000 ft^3 of aeration volume. Aeration tank volumes can be determined from the product of the length and width times the height of the tank, in ft. This volume in ft^3, when multiplied by 7.5, gives the capacity in gallons. Since we will be dealing with both values, we should know the capacity in gallons as well as the volume in cubic feet.

In current domestic practice, it is not uncommon to load extended aeration plants at 15 lb BOD/1,000 ft^3. When the lb BOD has been determined, we then have to establish what the loading permissible by state authorities would be. For the sake of making a specific determination, we will assume that a state regulatory body requires loadings of 15 lb BOD/1,000 ft^3 of aeration volume. Some states require loadings as low as 12.5 lb BOD/1,000 ft^3 of aeration volume.

8-8. *Extended Aeration Calculations.* Samples of calculations are as follows:
1. Determine the volume of the aeration tank.
    (a) Multiply:

    $$\text{aeration tank width} \times \text{length} \times \text{depth of water (ft)} \qquad (8\text{-}5)$$

    (b) Daily flow of 100 gal/person/day, e.g.,

    $$150 \text{ people} \times 100 \text{ gal/person} = 15,000 \text{ GPD flow}$$

Note: In extended aeration, the period of aeration is normally assumed to be 24 hours. Therefore, a 15,000 GPD unit would serve 150 people. (This is for normal, domestic household use and not for a rest area or other application. In a rest area, we would have 150 people $\times$ 5 = 750 GPD flow for (b).)
2. Determine organic loading of aeration tank. *Given:* maximum organic loading is set at 15 lb of BOD/1,000 ft^3 aeration volume. Sewage strength = 200 mg/l BOD, and daily flow = 15,000 GPD.

    (a)   Organic loading
    $$= 220 \text{ mg/l BOD} \times 8.34 \text{ lb/gal} \times 0.015,000 \text{ MGD}$$
    $$= 27.5 \text{ lb BOD/day} \qquad (8\text{-}6)$$

    (b) $\dfrac{27.5}{15} \times 1000 = \text{volume} = 1833 \text{ ft}^3 \text{ or } 13,748 \text{ gal} \qquad (8\text{-}7)$

(c)   Detention time $= \dfrac{13,748 \text{ gallons} \times 24 \text{ hr/day}}{15,000 \text{ GPD}}$ (8-8)

(d) Organic loading using 24-hr detention is

$$\frac{27.5 \text{ lb BOD/day}}{2,000} = 0.01375 \text{ lb} \qquad (8\text{-}9)$$

Therefore, 0.01375 lb BOD/ft³ or 13.75 lb BOD/1,000 ft³ aeration volume.

3. Determine the volume of the settling tank based on required detention time. *Given:* required detention time = 4 hr. Daily flow = 15,000 GPD.

$$\text{Volume} = \frac{15,000 \times 4}{24 \text{ hr/day}} = 2,500 \text{ gal} \qquad (8\text{-}10)$$

Given: required detention time = 2.5 hr based on runoff period. Runoff period = 16 hr, and daily flow = 15,000 GPD.

$$\text{Volume} = \frac{15,000 \text{ GPD} \times \frac{24}{16} \times 2.5 \text{ hr}}{24 \text{ hr/day}} = 1,927 \text{ gal} \qquad (8\text{-}11)$$

An important consideration in the proper sizing of a plant, especially the settling tank or clarifier, is the time in which flows are expected. This time may be designated as the runoff period or the number of hours in which the rated flow will run off. For schools, as an example, the runoff period would be 8 hours. The flow then can be designated as follows:

24-hr runoff = daily flow in MGD

16-hr runoff $= \dfrac{\text{daily flow}}{1.5}$ (normal housing development flow patterns)   (8-12)

12-hr runoff $= \dfrac{\text{daily flow}}{2}$ (8-13)

8-hr runoff $= \dfrac{\text{daily flow}}{3}$ (normal school flow pattern)   (8-14)

4. Determine the volume of the chlorine contact tank. *Given:* required detention time = 30 minutes and daily flow = 15,000 GPD.

$$\text{Volume} = \frac{15,000 \text{ GPD} \times 30 \text{ min}}{1,440 \text{ min/day}} = 312.5 \text{ gal} \qquad (8\text{-}15)$$

Given: required detention time = 15 minutes based on runoff period and a peak flow of 250%. Runoff period = 16 hr and daily flow = 15,000 GDP.

$$\text{Volume} = \frac{15,000 \times \frac{24}{16} \times 250\% \times 15 \text{ min}}{1,440 \text{ min/day}} = 586 \text{ gal} \qquad (8\text{-}16)$$

5. Determine the volume of the sludge holding tank. The sludge holding tank capacity is usually based on 1 to 3 ft³/100 GPD of design flow or per capita. *Given:* required volume to be based on 2.5 ft³/100 GPD of design daily flow. Daily flow = 15,000 GPD.

$$\text{Volume} = \frac{15,000 \text{ GPD} \times 2.5 \text{ ft}^3}{100 \text{ GPD}} = 375 \text{ ft}^3 \text{ or } 2,810 \text{ gal} \qquad (8\text{-}17)$$

6. Determine the surface settling rate of settling tank (SSR).

(a) Based on daily flow. *Given:* daily flow = 15,000 GPD and surface area of settling tank = 120 ft².

$$SSR = \frac{15,000 \text{ GPD}}{120 \text{ ft}^2} = \frac{125 \text{ GPD}}{\text{ft}^2} \qquad (8\text{-}18)$$

(b) Based on runoff period. *Given:* daily flow = 15,000 GPD, runoff period = 16 hr, and surface area of settling tank = 120 ft².

$$SSR = \frac{15,000 \text{ GPD} \times \frac{24}{16}}{120 \text{ ft}^2} = \frac{187 \text{ GPD}}{\text{ft}^2} \qquad (8\text{-}19)$$

8-9. *Contact Stabilization Calculations.*

1. Determine the volume of the aeration tank based on required detention time.

$$\text{Volume (gallons)} = \frac{\text{Daily flow (GPD)} \times \text{required detention time (hours)}}{24 \text{ hr/day}}$$

$$(8\text{-}20)$$

Given: required detention time = 3 hr and daily flow = 200,000 GPD.

$$\text{Volume} = \frac{200,000 \times 3}{24} = 25,000 \text{ gal} \qquad (8\text{-}21)$$

2. Determine the volume of the settling tank (use the formula set up for extended aeration plants).

3. Determine the volume of the re-aeration tank based on required detention time. *Given:* required detention time = 8 hr and daily flow = 200,000 GPD.

$$\text{Volume} = \frac{200,000 \text{ GPD} \times 8 \text{ hr}}{24 \text{ hr/day}} = 66,667 \text{ gal} \qquad (8\text{-}22)$$

4. Determine the volume of the aerobic digester. The capacity of the aerobic digester is generally based on 3.0 ft³/100 GPD of design flow or per capita. (See the formula for the sludge holding tank for extended aeration.)

8-10. *Other Useful Equations.*

$$\% \text{ BOD removed} = \frac{\text{mg/l BOD influent} - \text{mg/l BOD effluent}}{\text{mg/l BOD influent}} \qquad (8\text{-}23)$$

$$\text{Air requirements:} \quad \text{usually 1,500 ft}^3 \text{ air/lb BOD} \qquad (8\text{-}24)$$

1.04 ft³/min/lb BOD—This is derived by using 1500 ft³ of air per pound of BOD. Dividing 1,500 by 1,440, we get the 1.04 ft³/min/lb BOD. This will vary, depending on state requirements. Some states set air requirements as high as 3,500 ft³/lb BOD.

1.2 lb oxygen/lb BOD—This is derived by using a 16-hr flow period, a 3% aeration efficiency, 1,500 ft³ air/lb BOD and 0.0175 lb O_2/ft³ air.

BOD in lb/day may be reduced to BOD in lb/min by

$$\frac{\text{BOD lb/day}}{1,440} = \text{BOD lb/min} \qquad (8\text{-}25)$$

This formula will enable you to obtain the direct result of cubic feet per minute of air as related to BOD/lb/min.

In the event of shorter flow periods (which usually is the case in extended aeration plants) and the higher aeration efficiencies possible with larger (and therefore deeper) plants, the amount of oxygen supplied would be greater and therefore more than sufficient. In the steel plants, 2,800 ft³ of blower capacity, for example, may be provided per pound of BOD, with 3.0% aeration efficiency. The weight of oxygen supplied per pound of BOD would be:

$$2800 \times 0.03 \times 0.0175 = 1.5 \text{ lb } O_2/\text{lb BOD} \qquad (8\text{-}26)$$

Another value sometimes used in determining air requirements is 3 ft³/min of air per foot of length of aeration tank, or 2,100 ft³/lb BOD entering the tank daily, whichever is larger. Additional air is required if air lift is used for pumping return sludge from the settling tank.

Sludge holding tanks may be used from time to time. They are used when plants are overloaded and the mixed liquor suspended solids (MLSS) becomes excessively high. This would result in a high effluent suspended solids concentration. When insufficient air is transferred to the active solids, the process may become upset. In either case, withdrawing some of the sludge to the holding tank will relieve the undesirable condition and aid in returning the process to a good working order.

To maintain the extended aeration process in good working order, activated sludge must be returned to the aeration tank so that raw sewage can come contact with the return sludge for microorganism feed. The return sludge ratios must be maintained in proper proportions to develop good effluents. Ratios of 2,000 to 8,000 mg/l must be maintained as a mixture of activated sludge.

Additional equations which may be of use in working with aeration plants (not necessarily extended aeration only) are:

$$\text{Sludge age} = \frac{\text{lb solids under aeration}}{\text{lb solids applied/day in raw waste}} \qquad (8\text{-}27)$$

Mixed liquor suspended solids (MLSS), the quantity of solids to be maintained in the system:

$$\frac{\text{lb BOD to aeration daily}}{\text{BOD loading in lb applied/lb MLSS/day}} = \frac{\text{lb total MLSS to be}}{\text{maintained in system}} \qquad (8\text{-}28)$$

Using 0.03 lb BOD/lb MLSS/day as the loading, we take the total lb MLSS divided by 0.03, or

$$\frac{\text{lb total MLSS}}{0.03} = \text{total weight MLSS} \qquad (8\text{-}29)$$

To compute pounds of mixed liquor suspended solids under aeration:

$$\text{lb total MLSS} \times 0.08 = \text{lb MLSS in aeration tank} \qquad (8\text{-}30)$$

Select the MLSS concentration to be maintained in the aeration tank in mg/l. To compute aeration tank liquid capacity in million gallons:

$$\frac{\text{lb MLSS under aeration}}{8.34 \times \text{MLSS concentration}} = \text{mil gal capacity} \qquad (8\text{-}31)$$

To compute aeration liquid detention time based on raw waste flow only:

$$\frac{24 \text{ hr} \times \text{mil gal aeration capacity}}{\text{mil gal flow}} = \text{hours detention} \qquad (8\text{-}32)$$

We should check to see that the liquid detention time is within proper limits. If not, adjustments must be made on recomputed; if there is verification, then we can proceed to the next step. (If the detention time is fixed, the four preceding steps will be in reverse order. MLSS will then become a calculated result and not a selection.)

To compute aeration tank volume:

$$\frac{\text{gallons aeration capacity}}{7.5} = \text{ft}^3 \text{ aeration tank} \qquad (8\text{-}33)$$

To determine O_2/BOD ratio, use 1.2 lb O_2/lb BOD applied (based on 1,500 ft³/lb BOD); we must calculate pounds of oxygen dissolved each day.

$$\frac{O_2}{\text{BOD ratio}} \times \text{lb BOD to aeration}$$

$$1.2 \times \text{lb BOD in raw sewage} = \text{lb } O_2 \text{ dissolved each day} \qquad (8\text{-}34)$$

To compute the oxygen uptake in mg/l per hour:

$$\frac{\text{lb } O_2 \text{ dissolved}}{8.34 \times 24 \times \text{mil gal aeration capacity}} = \text{mg/l per hour} \qquad (8\text{-}35)$$

With the aeration tank volume, the O_2/BOD ratio, and the oxygen uptake rate, it is now possible to size the aeration tank and select aeration equipment.

8-11. Summary. The following guidelines can be derived from the foregoing information.

BOD Reduction Efficiency will be in the area of 95% when proper sludge control and proper sludge wasting is performed.

BOD LOADING. Use 0.03 lb BOD/lb MLSS/day for domestic wastes.

Design Flows. Use the average runoff flow rate in hours to determine the design flows in properly rating a plant.

$$\frac{\text{Flow in gal/day}}{\text{Hours of runoff flow}} \times 16 = \text{GPD design flow} \qquad (8\text{-}36)$$

Aeration Detention Time of Sewage. Use information provided by the design engineer or by state regulations. Usually from 12 to 24 hours.

Mixed Liquor Suspended Solids. The concentration is set by many factors and will range between 2000 and 8000 mg/l.

Mixed Liquor Suspended Solids Under Aeration. We can safely assume that 80% of MLSS will be under aeration and 20% will be in the clarifier at any given time.

Oxygen Requirements. Use 1.2 lb O_2/lb BOD at ambient temperature when 1,500 ft^3/lb BOD is used.

Excess Biological Sludge or Wastage. Use 0.15 lb sludge produced for each lb BOD reduced.

Return Sludge Flow. Use 50–100% of the raw waste flow.

For other typical data refer to Table 8-1.

8-12. Power Cost. The power cost of a package plant can be determined as follows.

a. Expense of Air Supply.

1. Obtain BOD loading in lb/day.
2. Determine air (ft^3/min) requirement.
3. Determine brake horsepower necessary to produce required air flow (factory-built blowers work under 3.5 psi).
4. Calculate input HP by dividing the brake HP by the belt drive efficiency and overall motor efficiency. The belt drive efficiency for fixed sheaves and standard V-belts is 0.96 to 0.98, and 0.80 if variable-pitch sheaves are used. Typical efficiencies for standard open motors are:

HP	Efficiency
5	0.87
3	0.85
2	0.83
1½	0.81
1	0.80

5. Calculate kilowatt input. Multiply the input HP by 0.746 kW per HP.
6. Calculate the daily kilowatt hours. Multiply kW input by 24 for daily kW input assuming that the blower operates continuously.
7. Calculate the blower power cost. Multiply kW input by the local power rate. The local power rate usually costs between one and five cents.
8. EXAMPLE.

$$\text{Organic load} = 33.3 \text{ lb BOD}_5/\text{day}$$

$$\text{Required blower cap.} = 33.3 \times \frac{1500}{1440} = 34.7 \text{ ft}^3/\text{min}$$

TABLE 8-1. Typical Values Used by Package Plant Manufacturers*

Parameter	Extended Aeration	Contact Stabilization
Lb BOD_5/1,000 ft^3 aeration	12 to 15	50 to 70
Lb BOD_5/100 lb M. L. solids	5 to 15	15 to 40
Air/lb BOD_5 removed (ft/min)	1500 to 2100	800 to 1200
Volume return sludge, % of flow	100 to 300	40 to 70
Waste sludge, % of flow	0	25
Waste sludge, % of flow	0	63
O_2/lb BOD_5	1.25	1.25

Notes:
    BOD_5 removal percentage $=$ BOD_5 removal in percentage of complete treatment times percentage of raw sewage flow in MGD as a decimal plus $(1 -$ percentage of raw sewage flow in MGD as a decimal times % BOD_5 removal in primary).
    *Note the values indicated in recommended design standards.

Use R-C #33 blower at 3.25 psi, BHP (from manufacturer's curve) $= 1.3$.

$$\text{Input blower HP} = \frac{1.3}{0.96 \times 0.81} = 1.67 \text{ HP}$$

$$\text{KWH/day} - 1.67 \times 0.746 \times 24 = 29.9$$

Blower power cost/month $= 29.9 \times \$0.03/\text{KWH} \times 30 = \$26.90$.

    b. Accessory Items. Obtain the nameplate HP of all the motors driving auxiliary equipment, e.g., foam spray pumps, comminutors, etc. Add the total auxiliary equipment HP to the input HP calculated in step 4 above, and determine the power cost for the total, as in steps 5, 6, and 7.

    c. Lighting and Miscellaneous Electrical Expenses. These expenses depend solely upon the facility (for example, number of lights used or laboratory equipment requiring electrical power).

8-13. Design Standards. The design standards are, for the most part, determined by regulatory agency requirements on the basis of experience and observation. In a fiercely competitive industry such as the package treatment plant industry, the consultant or specifying engineer is confronted with the problem of how to obtain a suitable quality plant. The design standard of the state regulatory agency, or the recommended standards that follow, whichever is more severe, should be used on all extended aeration treatment plants:

    a. Plant Influent.

    1. Comminutor required.
    2. Bar screen bypass with opening 1- to $1\frac{3}{4}$-in. opening required.
    3. Comminutor and bar screen to be located in a separate chamber similar to that in Fig. 8-6.

Figure 8-6. Comminutor and Bar Screen. (Courtesy Infilco Division, Westinghouse Electric Corp.)

b. *Aeration Zone.*

1. Minimum liquid depth = 10 ft.
2. Minimum freeboard = 18 in.
3. Detention 24 hr minimum based on design flow and/or 12.5 lb BOD_5/100 ft^3.
4. Aeration requirement exclusive of air lifts = 2,100 ft^3/lb BOD_5 or 3 ft^3/min/ft tank length, whichever is greater.
5. Spray system required for froth control.
6. Only positive return of sludge is acceptable.

c. *Settling Zone.*

1. Tank should not be less than 8 ft deep.
2. Four hours detention required, based on design flow.
3. Surface settling rate and weir loading should be not greater than indicated in Table 8-2.
4. Rate of return sludge = 10% to 200% of sludge design flow.
5. Sludge recirculation must be positive.
6. Method of flow measurement required.
7. Scum return to aeration tank.
8. For tanks with hopper bottoms, the upper third of depth of the hopper may be considered as effective settling capacity.
9. Two or more tanks should be provided on plants with capacities of 100,000 GPD or more.

d. *Sludge Tanks.*

1. Waste sludge holding tanks with at least 4 ft^3/capita or 25% of

TABLE 8-2. Surface Settling Rate and Weir Loading

Design Flow (GPD)	Surface Settling Rate (GPD/ft^2)	Weir Loading (GPD/ft)
2,000 to 5,000	200	600
6,000 to 7,000	250	800
8,000 to 15,000	300	1,200
16,000 to 20,000	300	1,600
25,000 to 90,000	300	2,100
100,000 to 250,000	300	2,600
250,000 to 1,500,000	300	10,000
1,500,000–Up	600	15,000

design sewage flow, whichever is greater. (A per capita is defined as a population equivalent.)

2. Sludge disposal facilities should be provided on all plants by digestion, vacuum filtration and incineration, provision for removal by tank truck to municipal facility, or other suitable disposal.

e. Chlorination.

1. Chlorination contact tank required. Separate tank required. Must not be integral within the basic plant.
2. Contact time 30 minutes minimum at design flow or 20 minutes at peak hourly flow. (A $2\frac{1}{2}$-hour contact time with a free residual chlorine level of 1ppm is actually desirable.)

f. Additional Provisions.

1. Scum-removal facilities are required on all final settling tanks.
2. Suitable froth spray control should be provided.
3. Multiple settling tanks required in plants larger than 100,000 gal/day unless other provisions are made to assure continuity of treatment. Double aeration tanks are actually desirable for any plant larger than 37,500 GPD.
4. Flow-measuring devices are required on all plants. For installations of 100,000 GPD or more, indicating, recording, and totalizing equipment is required.
5. Time clocks are required on blowers.
6. Duplicate equipment is required on all plants.
7. Mechanical sludge scrapers are required for treatment plants of 5,000 gal/day and larger.
8. Airlifts for sludge handling are not permitted for plants greater than 70,000 GPD capacity.
9. Continuous surface skimming is required.

g. Certification. Performance of package plants must be certified by the National Sanitation Foundation.

h. BOD and SS Removal. The plant should be capable of attaining 90% BOD and SS removal with a mixed liquor suspended solids concentration of 2,500 mg/l and when being operated at maximum design loading applied steady-state with respect to flow. Operation should be under the supervision of a properly trained and qualified operator and under spring and summer conditions.

8-14. *Diffused Aeration.* Figure 8-7 shows one type of blower used for diffused-aeration systems in package plants. Most of the package plants employ diffused-air systems which, in turn, require diffuser units to admit the air into the aeration chamber of the treatment plant.

The impellers rotate in opposite directions. As the left lobe passes the inlet, air is drawn into the case. The right hand impeller is starting to discharge.

Air is sealed in the left displacement pocket by close fit of impeller tips to case.

The left impeller is starting to discharge to the outlet. The displacement pocket on the right is just completing its charge.

Figure 8-7. One Type of Blower Construction. (Courtesy Fuller Company—Sutorbilt Products)

8-15. *Mechanical Aeration.* Since top speeds in mechanical aerators are generally limited to 10 to 15 ft/s for most efficient operation, depth becomes the variable factor (6). The submergence that provides the best combination of aeration and mixing depends on the diameter of the rotor. As the rotor diameter increases, the submergence also increases. The ratio of rotor diameter to rotor submergence is called the *submergence ratio*. The tank geometry is very important in the satisfactory and efficient performance of this type of plant.

The shape of the aerator blades, their rpm, and similar factors influence the oxygen transfer characteristics of the mechanical aerator (2). The ratio of transfer is dependent upon the deficit of oxygen existing in the water.

Figure 8-8 shows a vertical turbine mechanical aeration plant, and Fig. 8-9 shows a horizontal mechanical aeration plant. Figure 8-10 is a diagram of the plant in Fig. 8-9.

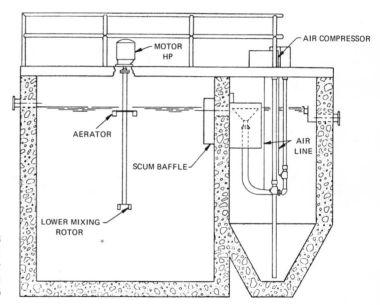

Figure 8-8
Vertical Turbine Mechanical
Aeration Plant. (Courtesy
INFILCO Division of Westinghouse
Corp.)

8-16. *Startup and Solids Accumulation.* The material in Sections 8-16 through 8-20 consists of selected passages from reference (3) and is used with the permission of the National Sanitation Foundation:

The startup period for extended aeration package plants is defined as the time between the placing of the plant in service and its attainment of the required level of treatment of the applied waste. Since, as it will be shown, treatment efficiency is, among other things, a function of the active solids mass present in the system, it follows that the rate at which such solids accumulate during startup is an important criterion of plant performance.

Solids will accumulate in an extended aeration system as long as the system is operating at less than its equilibrium value or separate sludge wasting

Figure 8-9
Horizontal Rotor Mechanical
Aeration Plant. (Courtesy Lakeside
Equipment Corporation)

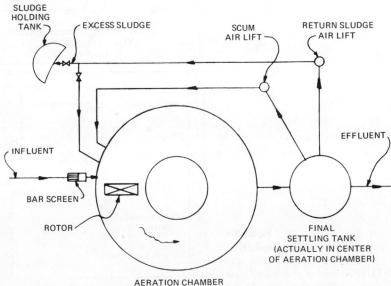

Figure 8-10. Diagram of a Horizontal Mechanical Aeration Plant.
(Courtesy Lakeside Equipment Corporation)

is employed to maintain the system at a solids level less than its equilibrium value. Otherwise, once the system reaches equilibrium, solids will be wasted from the system as they are formed.

The rate at which solids will accumulate in an extended aeration package plant beginning with the placing of the plant in service depends on:

1. The rate at which foodstuff enters the plant. Longer startup periods are therefore to be expected for underloaded plants than for those which receive their design load from the onset.

2. The rate at which solids are lost from the plant. Plants with poorly designed clarifiers or which are overloaded hydraulically will therefore have long startup periods.
3. The rate at which inert solids enter the system.
4. The rate at which solids are being oxidized in the system.

8-17. BOD Removal. For package sewage treatment plants of the extended aeration type, the BOD removal from the applied waste will depend on the MLSS concentration and the time the organisms contained in the MLSS are in contact with the waste, assuming that the concentration of oxygen in the aerator is not a limiting factor. It is further assumed here that the pH and temperature of the aerator contents are sufficiently near optimum values so as not to seriously impair biological activity, and that significant amounts of toxic substances are not present. It will be shown later, however, that temperature is not as significant a factor here as might otherwise be expected.

For extended aeration plants, it is not surprising that BOD removals of greater than 90% are not only common but indeed are to be expected. Substrate BOD is, for all practical purposes, nearly entirely removed by the system.

An MLSS concentration of 2,500 mg/l has been taken as evidence of process maturity based on the performance of the plants in the National Sanitation Foundation study in 1965–66.

8-18. Oxygen Requirements. Oxygen will be required for both synthesis and oxidation. In the study conducted in 1965–66 by the National Sanitation Foundation, all of the plants used proved capable of maintaining a residual dissolved oxygen concentration in their aeration compartments of 2.0 mg/l or better, except immediately in the area of raw waste introduction, at all times and under all conditions of loading. All but one of the plants used in the study utilized diffused aeration; the exception was a plant employing mechanical aeration. Judged on the basis of performance, the method of aeration is unimportant as long as it produces the required mixing effect and oxygen transfer in the aeration compartment. While there may very well be economic and other considerations, the types of diffusers employed are unimportant as long as the required mixing and transfer is achieved.

8-19. Nitrification. When sufficient oxygen is supplied, extended aeration package sewage treatment plants are capable of a high degree of nitrification. Ammonia nitrogen is a constituent of the waste fed these treatment plants.

The oxidation of 1 lb of degradable cell material requires a total of 1.98 lb of oxygen if the oxidation is to be carried completely through to nitrate. During one portion of the National Sanitation Foundation Studies, no particular attempt was made to control aerator dissolved-oxygen levels except to maintain them above 1.0 mg/l, with the result that levels of 5.0 mg/l or greater became common. At these dissolved-oxygen levels, a high degree of nitrification was experienced with subsequent denitrification in the settling portions of the plant and impairment of effluent quality due to the presence of increased amounts of solids and related BOD_5. All but two of the ten plants in operation at the time were originally equipped with a baffle just ahead of the effluent weir. The floated solids were largely trapped by the weir and held in the plant, so they were not being returned to the process. The food-to-microorganism ratio in the aerator rapidly changed, and the process began to fail due to the imbalance.

An attempt to combat this problem by the continuous operation of skimming mechanisms resulted in such high upflow rates in the settling compartment that in many cases the upflow potential exceeded the settling potential of the sludge and resulted in losses of solids. When the skimming mechanisms were not operated continuously, the floated solids became concentrated at the surface of the settling compartments, and in several severe cases nearly all the sludge in the plant was thus removed from the process. One control method was to reduce the intensity of aeration. In the plants having a baffle ahead of the effluent weir, symptomatic relief could be achieved by operating the skimmers once or twice a day while hosing down the surface of the settling compartment. The problem of denitrification can be illustrated by the following equation:

$$2NO_3 \longrightarrow 3O_2 + N_2 \uparrow$$

The reason for denitrification is the utilization of nitrate by the respiring organisms as a secondary source of oxygen in the absence of sufficient dissolved oxygen. This does not mean that during denitrification low dissolved oxygen values would occur in the plant effluent or the upper portions of the plant's clarification compartment. Indeed, during denitrification, such dissolved oxygen values were found to be only slightly lower than at other times. Denitrification, therefore, is a function of the respiration rate of the organisms contained in the settling sludge, the dissolved oxygen present in the immediate vicinity of the organisms, and the availability of nitrite and nitrate in the liquor surrounding the organisms.

For these reasons, another control method would be to reduce the residence time of the organisms in the settling compartment so that they are returned to the aeration compartment while they still have an adequate supply of dissolved oxygen. A third method involved the maintenance of very high dissolved-oxygen levels in the aeration compartment to permit a more normal sludge residence time in the clarification compartment. The economics of the situation does not seem to favor such an approach.

8-20. Solids Separation. Nearly all applied BOD is removed from the raw waste in the aeration compartment of a package sewage treatment plant of the extended aeration type. The effluent BOD from such plants is almost entirely due to the endogenous respiration of the organisms in the effluent. Therefore, the overall plant efficiency is dependent on the efficiency of the plant's solid-liquid separation or settling compartment. The efficiency of this compartment is, of course, design-dependent.

Periodic losses of solids with reduction in overall process efficiency have frequently been noted with the operation of both package and conventional activated sludge plants. Such losses are a function of the settling rate of the solids and the hydraulic load applied to the settling compartment.

When mixed liquor is allowed to settle, the contained solids settling rate is at first constant and uniform, creating a rather discrete solids-liquid interface. See Fig. 8-11. If the height of this interface is measured and plotted versus time, the zone settling rate for the particular activated sludge can be computed. Activated sludge from domestic waste treatment plants commonly displays zone settling rates of from 20 to 30 ft/hr at a concentration of 1,000 mg/l MLSS down to 5 to 6 ft/hr at a concentration of 4,000 mg/l MLSS. See Fig. 8-12. Therefore, as the MLSS concentration increases, the zone settling rate decreases until a point is reached where, for all practical purposes, no sedimentation occurs. Aside from the matter of concentration dependency, a particular sludge's zone settling rate is further influenced by

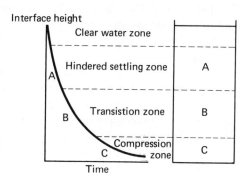

Figure 8-11
Zone Settling Rate, Activated Sludge. (Courtesy National Sanitation Foundation)

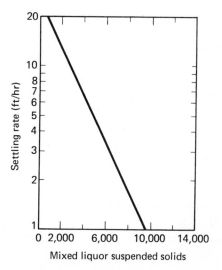

Figure 8-12
Zone Settling Rate, Activated Sludge. (Courtesy National Sanitation Foundation)

the biological and physiochemical characteristics of the particular system, such as the food-to-microorganism ratio, intensity of aeration, and the period of aeration.

Proper design of the clarifier portion of an extended aeration package plant must take into consideration, among other things:

1. The adjusted influent flow rate.
2. Peak flow rates expected.
3. MLSS concentration to be maintained.
4. Use of foam control and scum return equipment.
5. Biological and physiochemical characteristics of the system expected.

8-21. Specifications. Sections 8-22 through 8-31 are written in abbreviated specification form to provide samples of items that should be considered in the preparation of package plant contracts. A few are not applicable in all cases.

8-22. Established Manufacturer. The sewage plant manufacturer must be an established organization employing engineers. It must have an established record of building sewage treatment plants of the general type offered to meet these specifications, or provide adequate proof to the engineer that it has the capacity to build such plants and the necessary technical know-how.

The plant should include all internal and integrally mounted miscellaneous piping, fittings, and equipment necessary to make an operational unit unless specifically exempted under *Exceptions and Exclusions*.

Submission of alternative variations in configuration certified capable of achieving equal or superior treatment results is subject to approval by the engineer.

8-23. Blowers. If used in a plant, blowers should not turn at a speed greater than 1,800 rpm. Each blower unit should have two flexible connectors, one check valve, one pressure-reducing valve, one blast gate, one inlet snubber, one set of spare sheaves, one set of spare filters for outdoor mounting, one spare set of Vbelts, and one horizontal motor of not more than 1,800 rpm. Motors that will operate indoors may be of the ODP type. Motors that operate outside should be of the TEFC or TENV type. Motors should be adequately sized to have 15% overload margin at the maximum operating condition that can exist with the blower. Blowers must have timing gears, running in an oil bath or grease, which prevent the internal parts of the blower from combing into frictional contact with each other.

8-24. Service Area on Plant. When a plant is located above ground level, it should be provided with a 60° ship ladder from ground level to plant floor. An access bridge should be provided if necessary to properly service the plant, unless the plant is small enough that all servicing can be done from the sides and all areas can be safely and easily reached by the operator or maintenance man. When an access bridge is required, it should be complete with stiffeners, railing, and either checkered walkway or suitable grating.

8-25. Grit-Removal Channel. A grit-removal channel is not required unless stipulated on the drawings or in the specifications. When a grit-removal channel is stipulated, it may be installed on top of the plant if the plant is large enough, or it may be a separate unit. It may consist of an inlet channel for grit decanting located between the outer and inner shells adjacent to the bridge. The grit-removal channel should have a proportioning weir to regulate the velocity with bypass gates to the distributing channel, or it should be of an alternative design that has been approved by the engineer.

8-26. Comminutor. A comminutor should be provided in a chamber separated from the treatment plant. A comminutor mounted integrally with the plant itself is not acceptable. The comminutor chamber may be a shop-fabricated unit complete with a separate bar screen bypass channel, or the comminutor chamber may be constructed from concrete at the site with the comminutor mounted in one channel and a bar screen mounted in an adjacent channel. Schemes with a comminutor and bar screen built together in a single channel should not be accepted. The bar screen channel should have an overflow weir such that liquid will not flow through the bar screen unless the comminutor fails or the flow rate exceeds its capacity.

The bars of the bar screen should be constructed from 6061 alloy aluminum or galvanized coated steel (hot-dipped). Hydraulically, the bar screen should be capable of passing the peak design flow of the plant. The comminutor must be designed so that it is adequate to handle the variety of wastes characteristic of the specific application and so that, if the comminutor becomes jammed, protective devices will prevent the comminutor motor from burning out until the operator visits the plant and clears the jam.

8-27. Froth Control. Each plant should be provided with suitable sprays to provide froth control.

8-28. Aeration Regulation. Package treatment plants must be capable of operating under extremes of loading conditions. The minimum flow at time of startup may be as low as 10% of the design flow of the plant. It is therefore necessary that provisions exist, through timers or other control devices, that the aeration of the plant can be reduced. It must furthermore be assumed that an operator will not be in continuous attendance at the plant. A low skill level must be assumed upon the part of the operator. Extremely complex aeration regulation devices or controls are unacceptable.

8-29. Chlorinator. Liquid-type chlorinators are not acceptable under any condition. The chlorinator must use the true vacuum principle of operation from the chlorine gas cylinder to the ejector. No portion of the chlorine piping or tubing between the chlorine cylinder and the ejector can be under pressure. The chlorinator must either be capable of being mounting directly on 150-lb chlorine cylinders with vacuum shutoff valve built into the chlorinator, or it must have the vacuum shutoff valve mounted on the chlorine cylinder so that if the tubing between the chlorine cylinder and the ejector is broken, accidently removed, or acquires an air leak, the valve on the cylinder will automatically close and prevent any danger from chlorine. None of the parts inside the chlorinator or the vacuum valve assembly may be made from materials subject to corrosion by chlorine even in the presence of moisture and/or oxygen.

8-30. Structural Design. The equipment manufacturer should furnish all fabricated tanks. The walls and partitions of the tanks should be fabricated from steel plate of a thickness not less than indicated in Table 8-3 and from steel of specification according to Table 8-4. Small treatment plants with tanks not over 16 ft high and less than 30 ft in diameter may substitute sheet and/or plate of 3003, 5052, 5454, 5456, or 5083 aluminum alloy not less than $\frac{1}{4}$-in thick when the coldest one-day mean ambient temperature is 5°F or warmer. Reinforced concrete tanks of adequate structural strength may be cast in place. (Concrete tanks should be designed specifically for each installation.)

It should be the responsibility of the treatment plant manufacturer to provide a design that has adequate structural strength to withstand earth pressures (if submerged), wind, installation, operation, and maintenance demands.

8-31. Corrosion Protection. The Steel Structures Painting Council (SSPC), Mellon Institute, 4400 Fifth Ave., Pittsburgh, Pa., 15213 has publications for sale that treat coatings in detail. Where feasible, SSPC surface preparation specification no. 8, "Pickling," and specification no. 10 are the most satisfactory preparations followed by thorough removal of any residue or dust. The degree of surface preparation must be a function of end use (exposure) and, to a limited extent, of the coating material. In some cases, for immersion service, a coating would require white metal blast preparation by specification SSPC-SP-5-63. Cost is a function of surface preparation, the coating used, the number of coats required, and the thickness of the coating material.

PROBLEMS

1. A small housing development in a middle-class neighborhood has 3,000 people. Calculate the gallons per day, the organic loading, the volume of aeration tank required on an extended aeration plant, the detention time, and the volume of the settling tank, all on the assumption of a runoff period of 16 hr.

TABLE 8-3. Tank Shell Plate Thickness

Diameter (*ft*)	Tank Height in Feet				Tank Height in Feet		
	8	16	24	32	8	16	24
	Shell Plate Thickness (in.)				Capacity in Gallons × 1,000		
5	0.25	0.25	0.25	0.25	1.17	2.34	3.52
10	0.25	0.25	0.25	0.25	4.69	9.41	14.08
15	0.25	0.25	0.25	0.25	10.58	21.19	31.78
20	0.25	0.25	0.25	0.25	18.8	37.8	57.5
24	0.25	0.25	0.25	0.25	27.1	54.2	81.3
26	0.25	0.25	0.25	0.25	31.7	63.5	95.2
30	0.25	0.25	0.25	0.25	42.4	84.8	127.0
36	0.25	0.25	0.25	0.25	60.9	122.0	182.5
42	0.25	0.25	0.25	0.27	82.9	165.5	248.2
48	0.25	0.25	0.25	0.28	108	217	325
50	0.25	0.25	0.25	0.32	117	235	352
55	0.25	0.25	0.27	0.33	141	282	423
60	0.25	0.25	0.30	0.36	169	339	508
65	0.25	0.25	0.32	0.40	196	391	587
70	0.25	0.25	0.35	0.43	228	456	684
75	0.25	0.25	0.37	0.46	264	528	794
80	0.25	0.27	0.40	0.49	297	595	892
85	0.25	0.28	0.42	0.53	340	680	1019
90	0.25	0.30	0.45	0.56	380	760	1140
95	0.25	0.31	0.47	0.59	425	850	1275
100	0.25	0.33	0.49	0.62	470	940	1410
110	0.25	0.36	0.54	0.66	564	1130	1695
120	0.31	0.40	0.59	0.72	675	1355	2035

Note: Tank capacities are only approximate.

TABLE 8-4. Steel Specification Versus Temperature
(Permissible Minimum Specifications)

Temperature†	Max. Plate Thickness* (in.)	Specification No.	Grade
5°F or warmer	$\frac{1}{2}$	A36, A131, A441	A
	1	A131	B
−25°F or warmer	$\frac{1}{2}$	A131	B
		A516	60 and 70
	1	A131	C
		A516	60 and 70
−55°F or warmer	$\frac{1}{2}$	A131	C
	$1\frac{1}{4}$	A514, A517	All

*Plates shall not under-run calculated thickness by more than 0.01 in.
†Temperatures are the coldest one-day mean ambient temperature.

2. Determine the volume of a sludge holding tank for a mobile home trailer park having 210 trailers. Make any assumptions necessary.

3. Determine the volume of a sludge holding tank for a picnic park in which 7,000 people per day are expected to use the rest rooms (no showers).

4. Provide all calculations necessary to specify a contact stabilization plant for a mill town of 1,960 population equivalent in domestic waste.

5. If the influent BOD is 250 mg/l and the effluent is 25 mg/l, what percentage of BOD was removed?

6. How many pounds of MLSS are to be maintained in a plant serving an economy multi-dwelling row house development having 9,000 persons?

7. How many cubic feet of aeration tank are required on a plant to serve a motel having 110 rooms? The restaurant has 80 seats.

8. How many pounds of oxygen are required per day in Problem 1? Determine the air required in ft^3/min.

9. Calculate the HP necessary to produce the air required for the plant in Problem 6. Calculate the kilowatt input.

10. Assume the local power rate is 2 cents per kilowatt hour, and calculate the power blower cost on a plant serving a penal institution having 3,500 inmates.

11. Explain what would be the result of startup of a plant having a poorly designed clarifier.

12. On what is the plant efficiency dependent?

REFERENCES

1. SMITH, ARNOLD, "Surface Aerators Spin to a Comeback." *Water and Wastes Engineering*, 3, 12 (1966).

2. SMITH, ARNOLD., "Testing and Rating Surface Aerators." *W. & S. W.*, 111-R.N. (1964).

3. *Package Plant Criteria Development, Part 1: Extended Aeration*. National Sanitation Foundation, Ann Arbor, Michigan, Sept., 1966.

Oxidation Ditch Sewage Waste Treatment Process 9

The oxidation ditch is an economical, highly efficient, and simple waste treatment process for a wide variety of applications where adequate space of suitable terrain is available to construct the facility. The basic process was developed by the Institute of Public Health Engineering, T. N. O., Holland. The first full-scale plant was installed at Voorschoten, Holland, in 1954. The configuration and design details discussed in this chapter have many substantial improvements compared to the earlier plants.

The most commonly used oxidation ditch configuration for municipal sewage treatment is a trapezoidal cross section of relatively shallow depth forming a continuous circuit. Other cross-sectional configurations can be used in certain circumstances. This ditch serves as an aeration reactor. The continuous circuit provides an excellent, complete mix basin and provides simple excavation and economic construction. The heart of the process is the cage rotor, which is an efficient, mechanical surface aerator. The principle of operation differs from the vertical turbine mechanical aerators now on the market.

The oxidation ditch process is a modified form of the activated-sludge process and may be classified in the complete-mix, long-term aeration group. This is not a mechanically aerated lagoon because the settled solids produced by the process are continually returned to the ditch to provide a high solids concentration in the aeration basin. The process is able to take shock loadings without upset, and this is one of its advantages for roadside rest areas or other installations having an uneven loading cycle. The rotor operation affords easy reduction or change in oxygenation input by adjustment of immersion in the configuration described herein.

9-1. Process Flow Sheet. There is normally no primary tank used in the oxidation ditch process. See Fig. 9-1. Raw sewage passes directly through a bar screen to the ditch. A comminutor is not needed unless required by state law. The oxidation ditch forms the aeration basin where the raw sewage is mixed with the active organisms in the ditch. The cage rotor entrains the necessary oxygen into the liquid and keeps the contents of the ditch mixed and moving. The mean velocity of the liquid in the ditch must be kept above a minimum of approximately 1 ft/s to prevent settling of solids. The ends of the ditch must be rounded to prevent eddying and dead areas, and the outside edges of the curves must always be given erosion protection.

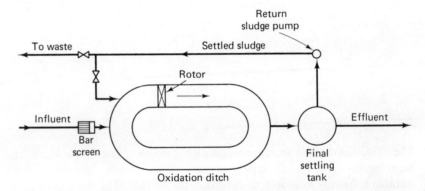

Figure 9-1. Oxidation Ditch Plant Diagram.

The mixed liquor in the ditch flows to the clarifier for separation. Quiescent conditions in the clarifier allow the separation of the solids from the liquid. The clarified liquid passes over the effluent weir, through the chlorination tanks, and then to the receiving stream.

The settled sludge is removed from the bottom of the clarifier by an air lift or pump and is returned to the ditch. All sludge formed by the process and settled in the clarifier is returned to the ditch. Scum which floats to the surface of the clarifier is also returned to the oxidation ditch.

The oxidation ditch is operated as a closed system, and the net growth of the volatile suspended solids will require periodic removal of some sludge from the process. Wasting of excess sludge, through a connection to the return sludge line, lowers the concentration in the ditch and keeps the metabolism more active. Sludge concentration control by wasting excess sludge is one of the reasons for the high reductions possible by this process. Excess sludge may be dried directly on sludge drying beds or stored in a holding tank or sludge lagoon for later disposal.

9-2. Process Theory. The oxidation ditch entails a physical and biological process. A small portion of the organic matter undergoes direct chemical oxidation, but the bulk of the organic matter must be stabilized by the biochemical activities of the microorganisms previously formed in the system. The substrate level or food value of the used water as measured by the biochemical oxygen demand must provide the proper diet for the microorganisms.

The oxidation ditch process, with its long-term aeration basin, is designed to carry mixed liquor suspended solid concentrations of 3000 to 8000 mg/l. This provides a large organism weight in the system. The food-to-organism ratio or loading factor is low, ranging from 0.03 to 0.1 $\#$BOD/day/$\#$VSS. This low factor produces a system that can absorb loading without upsetting the operation. Also, because of the low food-to-organism ratio, the growth of volatile sludge because of endogenous respiration is relatively low. The volatile sludge is less reactive and has a lower BOD than an equivalent weight of sludge produced by processes having larger loading factors. Therefore, the solids lost from the system over the effluent weir in the final settling tank would have a lower BOD than the same quantity lost from the system with other processes. Excessive aeration with high D.O. levels in the aeration basin can cause the formation of pinpoint floc which will not flocculate and will be lost over the effluent weir, carrying with it quantities of BOD. Continual BOD removals of 90 to 97% are possible with this system.

9-3. The Ditch. The ditch forms the reactor or aeration tank. The shape of the ditch is generally an elongated oval as shown in Fig. 9-1, but it may be bent at one end, both ends, circular, or any shape, as long as it forms a complete circuit. The depth of the ditch ranges from 3 to 5 ft for a small rotor ditch, and this shallow structure is an easy and economical excavation. The immersion depth for a 42-in. rotor would normally be designed at 7 in. The depth from the surface depth to the bottom of the ditch would be 8 ft for the latter case. The compacted earth ditch should preferably be lined with poured concrete or shot-crete. A variety of other materials such as asphalt, wood, preformed materials, and clay have been used.

For normal domestic sewage, the volume of the ditch is sized based on 74 ft³/lb of BOD applied per day. This is equal to a daily loading of 13$\frac{1}{2}$ lb of BOD per 1,000 ft³ of ditch volume. The ditch volume for weak wastes may be sized for a loading as low as 10 lb of BOD per 1,000 ft³, and for strong industrial waste, loadings as high as 40 have been used. Loadings used for strong industrial wastes depend on the strength type and amenability of the waste to aerobic degradation.

Roadside rest areas, recreation areas, amusement parks, and similar facilities have a varied usage over the year in most states. In such a case, it may be desirable to use a design as shown in Fig. 9-2. During the winter, the outer ditch would normally be out of operation and only the inner oxidation ditch would function. The photograph in Fig. 9-3 shows a plant designed for 220 population equivalent. As a comparison, Fig. 9-4 shows a plant suitable for serving 33,750 persons.

9-4. The Rotor. The cage rotor is placed across the ditch to carry out a two-fold function of (a) supplying the necessary propulsion for complete mixing of the ditch contents and (b) introducing the necessary oxygen to support the biological activity. The rotor rotates on an axis in a plane parallel to the liquid surface, and perpendicular to the direction of flow. The small Lakeside Equipment Company rotor consists of twelve blades mounted on the periphery of circular end plates. The blades have teeth on either side with the teeth on each successive blade occupying the void in the preceding blade.

Rotors are manufactured in 1-ft intervals up to 15 ft in length for the small rotors and 30 ft for the large 42-in.-diameter rotors. A rotor assembly can be of multiple lengths, but it must be supported by intermediate bearings. Rotors may be mounted directly on concrete pads or suspended from an overhead support structure. Small rotor installations normally use the concrete pad mounting.

The small cage rotor is $27\frac{1}{2}$ in. in diameter, and operating immersions are 2 through 10 in. The large rotor would usually be kept within the range of 4 to 13 in. immersion. Rotor operation has shown that it produces some comminution of raw sewage solids and is self-cleaning. Rags and other stringy material do not attach themselves to the rotor.

The velocity of the liquid contents of the ditch is an important factor and normally dictates the length of the rotor required for a given project. Velocity criteria based on tests and actual operating experience, both here and abroad, have shown that, for normal designs, with a lined ditch, 16,000 gallons of actual ditch volume per foot of small rotor will produce a velocity that will maintain all solids in suspension. The normal movement of liquid in the ditch is at an average velocity of 1 ft/s. Excessive rotor length can produce a velocity in excess of $2\frac{1}{2}$ ft/s which is not an economical design because of reduced rotor oxygenation.

The selection of the rotor assembly length dictates the approximate cross section of the ditch. The width of the ditch divided by the rotor length should give a ratio between 1.5 and 2.8. The larger ratios are normally used only with the smaller rotor assemblies or short length (3 or 4 ft). The rotor imparts a velocity in the ditch which is fairly uniform throughout its cross section and may be classified as a push-type flow.

With the rotor assembly length setting a relatively constant ditch cross section, the total length around the ditch, or between rotor assemblies, falls within a fairly narrow range. For a 1 ft/s velocity, the ditch contents will travel the complete circuit of the ditch every 3 to 5 minutes. Or restated, the ditch contents are re-aerated every 3 to 5 minutes. This is complete mixing.

To suitably apply mechanical aerators, it is essential that the oxygenation

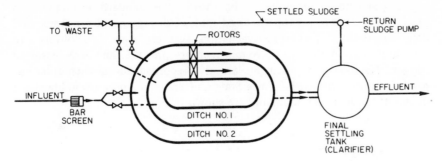

Figure 9-2. Oxidation Ditch Plant for Extreme Load Variation Summer to Winter Where Dual Tanks Are Required.

Figure 9-3. Oxidation Ditch for 220 P.E. (Courtesy Lakeside Equipment Corporation)

Figure 9-4. Oxidation Ditch for 33, 750 P.E. (Courtesy Lakeside Equipment Corporation)

capacity and power requirements be known. The standard technique is to place the aerator in a properly sized tank of deoxygenated tap water and determine the rate of water reoxygenation. Test results are then converted to standard conditions of 20°C, 76 mm mercury pressure, and zero dissolved oxygen. Net power requirements are also recorded during each run. The results of tests conducted at Iowa State University at Ames, Iowa, under the direction of Dr. E. R. Baumann, are shown in Fig. 9-5.

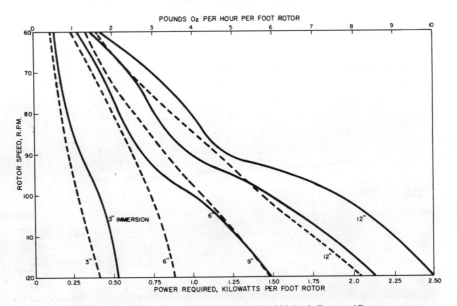

Figure 9-5. Oxidation and Power Data for $27\frac{1}{2}$ Inch Rotor. (Courtesy Lakeside Equipment Corporation)

The small cage rotor was tested at various immersions and speeds ranging from 60 to 120 rpm output. At 100 rpm and with the tips of the blades immersed 3 in, one can introduce 1.63 lb oxygen/hr per foot of rotor. Increasing the immersion to 12 in. at the same speed raises the oxygen introduced to 7.62 lb. Sufficient power must be provided to allow for the maximum immersion considered for a particular design. Under optimum conditions, it is possible to obtain 6.8 lb of oxygen input per kilowatt.

Figure 9-6 shows the oxidation capacity of a 42-in. rotor, and Fig. 9-7 shows the power requirements for the same rotor. These are not the only rotors available on the market, but they do represent two widely used sizes.

The curves give the rotor oxygenation and power requirements in tap water under standard conditions. An equation has been developed to convert oxygen input under standard conditions to the oxygen input required for various forms of activated sludge and different wastes that are amenable to aerobic degradation. The basic formula is

$$N_0 = \frac{1.5 \times N}{\alpha \left(\dfrac{\beta C_{sw} - C_L}{C_{st}} \right) (1.024^{(T-20)})} \qquad (9\text{-}1)$$

where N_0 = lb O_2/day transferred to water at zero D.O. and 20°C,

α = oxygen transfer ratio,

β = ratio of saturation of waste to saturation of water,

C_{sw} = saturation value of oxygen in waste at operating temperature,

C_{st} = saturation value of oxygen used in test operation,

C_L = operating D.O. level,

T = temperature of waste in °C,

1.5 = conversion from five-day BOD to ultimate BOD,

N = lb O_2/day transferred to the waste mixture.

Suitable figures that can be substituted into the equation have been developed for the alpha and beta factors for domestic waste. Solving Eq. (9-1) for N_0 in terms of N results in a conversion factor. The normal domestic-waste conversion factor for oxidation ditch application is 2.35. Adjustment of this conversion factor must be considered for industrial wastes. Correction must also be taken into account for sewage treatment plants that are located at elevations greater than 2,000 ft. The equipment manufacturer of the rotor can provide the necessary correction for higher altitudes.

With the conversion factor, the BOD can be converted to pounds of oxygen required at standard conditions. This is then divided by the total length of rotor derived by the velocity criteria and results in pounds of O_2 required per hour per foot of rotor. The normal procedure is to use 6 in. of immersion and select the rotor speed to give the required oxygen input. This practice permits the use of one of the advantages of the rotor. The plant design is a projected figure for loading several years in the future. For operation under present conditions, it is necessary to lower the rotor immersion. This feature allows for operational flexibility, and thus better operation, because the waste is not over aerated and because less power is required at the lower immersion. It also can be used for adjustments from summer to winter loadings in recreation areas, roadside rest areas, and similar installations having seasonal loads. Figure 9-8 indicates an oxidation ditch located at a 5,000-ft altitude.

9-5. *Clarifier*. The oxidation ditch operates as a closed system in that all solids formed by the process should be retained within the process. Operation must be such that a suitable size of flocculated solid is formed so it can be separated from the liquid when subjected to the near-quiescent area in the settling tank. This settling tank is commonly called the clarifier.

In the oxidation ditch, the raw sewage and the returned sludge should be injected just upstream of the rotor so the material can be immediately mixed with the ditch contents. The effluent from the ditch should be removed some-

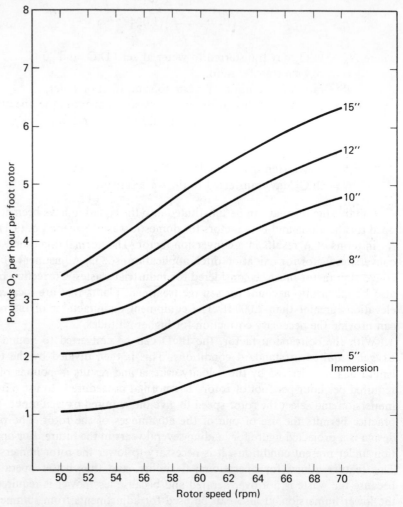

Figure 9-6. Oxidation Capacity 42 in. Magna, Zero D.O., 20°C, 760 mm. (Courtesy Lakeside Equipment Corporation)

what further upstream. The removal point should be far enough away from the injected flow so that no short-circuiting occurs.

Ditch contents at and directly downstream of the rotor are highly turbulent. Further downstream from the rotor, flow lines in the ditch straighten out to form a fairly even slow-rolling motion. In a ditch carrying a healthy activated sludge, it is possible to observe heavy sludge flocculation with crevices opening up and showing clear liquid. The operation in the ditch provides distinct flocculation of solids before removal to the final tank for settling.

Sizes of final settling tanks are generally based on a surface settling rate of 600 gallons per square foot per day. To allow for sufficient depth in the

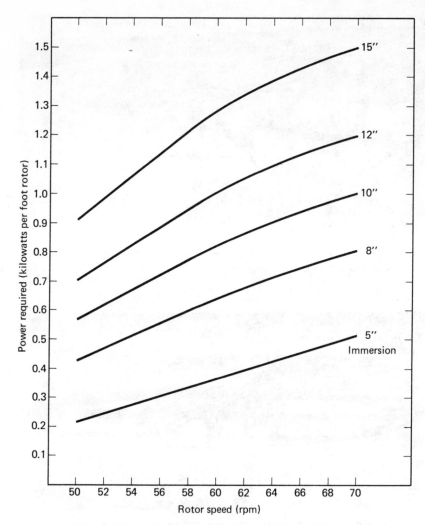

Figure 9-7. Power Requirements, 42 in. ϕ Magna. (Courtesy Lakeside Equipment Corporation)

tank and to handle the high concentrations of sludge carried in the ditch, 3 hours' detention time is needed. Since a primary settling tank is normally not included in this process, the final tank must be equipped with surface skimming. Figure 9-9 shows a large mechanized clarifier, although a much smaller nonmechanized unit will be sufficient in many small plants.

9-6. The Activated Sludge. All activated sludge settled in the clarifier is returned to the ditch. This sludge normally has a solids concentration of between 1 and 2 %. For a normal mixed liquor suspended solids concentration of 6000 mg/l in the ditch and an average concentration of solids in the recirculated flow, the return sludge air lift or pump should be rated at approxi-

Figure 9-8. Oxidation Ditch Located at 5,000 Ft. Altitude. (Courtesy Lakeside Equipment Corporation)

Figure 9-9. Large Mechanized Clarifiers. (Courtesy Lakeside Equipment Corporation)

mately 67% of the average incoming flow. The return sludge mechanism usually should be operated somewhat above this point to insure that solids are returned. The required capacity of the sludge return pump can vary, depending on the solids concentration carried in the ditch and the concentration of the solids in the returned flow.

The return sludge mechanism should be sized to handle at least 100% of the design flow. Adjustment of the rate of return should be provided for any needed operational change in the process.

Because the oxidation ditch process operates as a closed system, with efficient operation a slow growth or increase in suspended solids occurs. The process operates at the lower end of the endogenous respiration curve, and the actual growth rate is low. Normally the growth rate for domestic sewage varies from 1.6 to 2.23 ft^3 per 100 population equivalent per day.

Normal operation is to allow the mixed liquor suspended solids to accumulate in the ditch to an 8000 mg/l concentration or to where the D.O. in the ditch directly upstream of the rotor falls between 0 and 0.5 mg/l. At this time solids must be removed from the process. They may be wasted directly to sludge drying beds, or stored in the holding tank or excess sludge lagoon for later disposal.

9-7. Oxidation Ditch Calculations. When small-sized ditches are used, there are some complications in following the design procedure. The reason is that the regular procedure results in a velocity of flow that is too high in small ditches. The high velocity can result in uneconomical operations and undesirable hydraulic flow patterns. To correct for these conditions in small plants, the necessary sizes have been computed for a series of plants and tabulated in Table 9-1, which indicates the dimensions of plants suitable for various population equivalents. It is necessary to refer to Fig. 9-10 to determine the significance of the dimensions shown.

The next problem is to come up with a procedure that will enable any size oxidation ditch to be calculated when the ditch is larger than the sizes shown in Table 9-1. To illustrate the principles involved, two cases will be taken in which some of the quantities are identical. Case 1 involves a town of 3,500 people that requires an oxidation ditch to replace a malfunctioning lagoon. It is assumed that there is no industrial waste contribution in this problem and that the water consumption of 100 gal/capita/day and 0.17 lb BOD/capita/day will be used. Case 2 involves a ditch designed for 20,000 people/day using a roadside rest area (population equivalent of 3,500 people = 20,000 people using a roadside rest area). Rest area usage currently assumes 5 gal/day/capita water usage and 0.03 lb BOD/capita/day.

a. Design Data.
Case 1. Population or PE: 3,500.

Design flow:

$$3500 \times 100 = 350,000 \text{ gal/day} = 243 \text{ gal/min} \qquad (9\text{-}2)$$

Design strength:

$$3500 \times 0.17 = 595 \text{ lb BOD/day} \qquad (9\text{-}3)$$

Design based on 90% of better BOD reduction.
Case 2. Population or PE: 3,500 PE or 20,000 facility users.
Design flow:

$$20,000 \times 5 \text{ gal/day/cap} = 100,000 \text{ gal/day} = 69.5 \text{ gal/min} \qquad (9\text{-}4)$$

Design strength:

$$20,000 \times 0.03 = 600 \text{ lb BOD/day} \qquad (9\text{-}5)$$

The discrepancy between the 595 lb for domestic sewage use and 600 lb

TABLE 9-1. Typical Oxidation Ditch Designs

Persons Design[1]	Ditch Dimensions[2]							Final Tank Dimensions[3]		Cage Rotor[4]			Sludge Return[5] gal/min
	A	B	C	D	E	F	G	H	J	Length	RPM	HP	
150	50' 0"	28' 0"	6' 0"	3' 0"	8' 0"	3' 0"	7' 0"	4' 4"	5' 5"	3'	65a	1½	15
200	68' 0"	28' 0"	6' 0"	3' 0"	8' 0"	3' 0"	8' 0"	5' 3"	5' 4"	3'	75a	1½	18
250	86' 0"	28' 0"	6' 0"	3' 0"	8' 0"	3' 0"	8' 0"	5' 3"	7' 0"	3'	75b	2	22
300	105' 0"	28' 0"	6' 0"	3' 0"	8' 0"	3' 0"	9' 0"	6' 1"	6' 4"	3'	70	2	25
350	94' 0"	32' 0"	7' 0"	3' 6"	9' 0"	3' 0"	10' 0"	6' 11"	5' 9"	3'	75	3	26
400	108' 0"	32' 0"	7' 0"	3' 6"	9' 0"	3' 0"	10' 0"	6' 11"	6' 10"	3'	80	3	28
450	106' 0"	34' 0"	7' 0"	3' 6"	10' 0"	4' 0"	11' 0"	7' 10"	6' 0"	4'	74	3	32
500	118' 0"	34' 0"	7' 0"	3' 6"	10' 0"	4' 0"	11' 0"	7' 10"	6' 10"	4'	78	3	35
600	140' 0"	36' 0"	8' 0"	3' 6"	10' 0"	4' 0"	12' 0"	8' 8"	6' 9"	4'	86	5	42
700	107' 0"	42' 0"	8' 0"	4' 0"	13' 0"	6' 0"	13' 0"	9' 7"	6' 5"	5'	80	5	49
800	123' 0"	42' 0"	8' 0"	4' 0"	13' 0"	6' 0"	13' 0"	9' 7"	7' 8"	5'	89	5	56
900	127' 0"	44' 0"	8' 0"	4' 0"	14' 0"	7' 0"	14' 0"	10' 5"	7' 3"	6'	86	7½	63
1,000	131' 0"	44' 0"	8' 0"	4' 6"	14' 0"	6' 0"	15' 0"	11' 3"	6' 8"	6'	90	7½	70
1,100	121' 0"	48' 0"	8' 0"	4' 6"	16' 0"	8' 0"	16' 0"	12' 2"	6' 2"	7'	88	7½	77
1,200	123' 0"	50' 0"	8' 0"	4' 6"	17' 0"	9' 0"	16' 0"	12' 2"	7' 0"	8'	86	7½	84
1,300	134' 0"	50' 0"	8' 0"	4' 6"	17' 0"	9' 0"	17' 0"	13' 0"	6' 5"	8'	89	7½	91
1,400	134' 0"	52' 0"	8' 0"	4' 6"	18' 0"	10' 0"	18' 0"	13' 10"	5' 10"	9'	88	10	98
1,500	144' 0"	52' 0"	8' 0"	4' 6"	18' 0"	10' 0"	18' 0"	13' 10"	6' 5"	9'	91	10	105
1,600	144' 0"	54' 0"	8' 0"	4' 6"	19' 0"	11' 0"	19' 0"	14' 9"	5' 10"	10'	89	10	111
1,700	153' 0"	54' 0"	8' 0"	4' 6"	19' 0"	11' 0"	19' 0"	14' 9"	6' 5"	10'	91	10	118
1,800	145' 0"	58' 0"	8' 0"	4' 6"	21' 0"	13' 0"	20' 0"	15' 7"	5' 10"	11'	90	10	125
1,900	145' 0"	60' 0"	8' 0"	4' 6"	22' 0"	14' 0"	21' 0"	16' 6"	5' 3"	12'	89	15	132
2,000	152' 0"	60' 0"	8' 0"	4' 6"	22' 0"	14' 0"	21' 0"	16' 6"	5' 8"	12'	90	15	139

(Courtesy Lakeside Equipment Corporation)

[1] Design criteria for number of persons per day using is based on sewage only at 0.17 lb BOD/capita/days and 100 gal/capita/day.

[2] For letter location and general layout, refer to Fig. 9-10.

[3] Designs shown are all for hopper-bottom final tanks equipped with nonmechanized Spiraflo clarifiers. Design can be altered to accommodate flat-bottom tanks with mechanized, Spiraflo clarifiers or other manufacturers' clarifiers if proper sizing is employed.

[4] All units are 27½-in.-diameter cage rotors operating at 6-in. immersion, except (a) which is 4-in. immersion and (b) which is 4½-in. immersion. Rotor lengths are based on a lined ditch.

[5] Sludge return is by 3-in. air lift, except (c) which is a 4-in. air lift and (d) which is a 5-in. air lift. A pump can be used instead of an air lift.

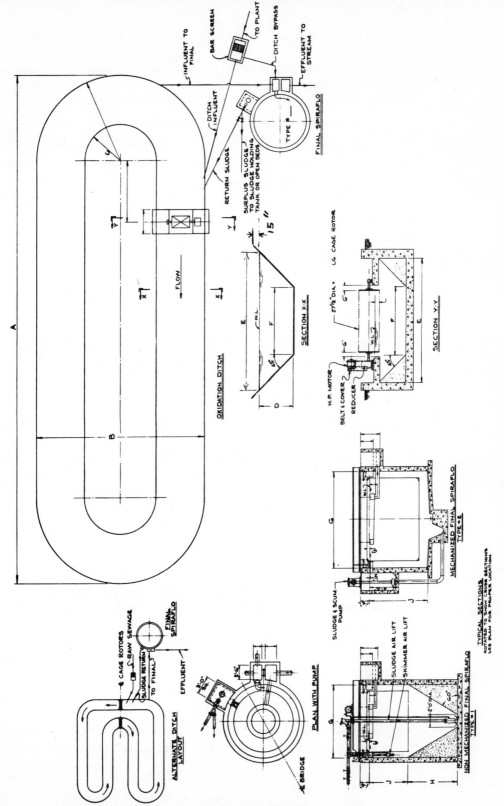

Figure 9-10. Dimension Identification for Oxidation Ditch Plants in Table I. (Courtesy Lakeside Equipment Corporation)

for the roadside rest area is due to the fact that 3,500 population equivalent is actually equal to a number slightly less than 20,000. For simplicity of calculation, we will use the value 595 lb BOD/day for both cases.

b. Oxidation Ditch Size.

$$\text{Ditch volume} = 74 \text{ ft}^3/\text{lb BOD} \times 595 \text{ lb BOD} = 44,030 \text{ ft}^3 \quad (9\text{-}6)$$

$$44,030 \text{ ft}^3 \times 7.48 \text{ gal/ft}^3 = 330,000 \text{ gal} \quad (9\text{-}7)$$

$$\text{Loading} = \frac{595 \text{ lb BOD} \times 1000}{44,030 \text{ ft}^3} = 13.5 \text{ lb BOD}/1000 \text{ ft}^3 \quad (9\text{-}8)$$

In aeration-ditch volume calculations with separate settling, use 74 ft³/lb BOD. This figure was developed as a result of the experience from more than 100 oxidation ditch plants.

$$\text{Total design BOD} \times 74 = \text{ditch volume in ft}^3 \quad (9\text{-}9)$$

c. Rotor Length. In smaller-sized plants like this, the $27\frac{1}{2}$-in. rotor is used. Standard operating conditions are expressed in RPM and depth of immersion.

The conversion factor of water-to-sewage used in ditch applications is 2.35. Therefore, 2.35 pounds of oxygen must be entrained for each pound of design five-day BOD. The oxygenation capacity/hour needed is

$$\text{Total design BOD} \times 2.35 \text{ lb} \div 24 \text{ hr} = \text{lb oxygen/hr} \quad (9\text{-}10)$$

Lakeside test data gives $O_2/\text{hr/ft}$ of rotor at rpm and immersion. The length of rotor needed is:

$$\text{lb } O_2/\text{hr} \div \text{value from Fig. 9-5} = \text{feet of rotor} \quad (9\text{-}11)$$

To maintain a 1-ft/s velocity in a lined ditch, the maximum volume per foot of the $27\frac{1}{2}$-in. cage rotor must not exceed 16,000 gal.

Rotor needed for velocity:

$$\text{Ditch volume in ft}^3 \times 7.48 \text{ gal/ft}^3 \div 16,000 \text{ gal/ft} = \text{rotor length in ft} \quad (9\text{-}12)$$

The actual rotor length would be the size obtained from the larger figure of the two methods. Two or more rotors of equal length can be used to meet the requirement if a single rotor is inadequate or if the dual rotor arrangement will make a better design. A recommended ditch liner would be 4-in.-thick concrete.

We can move 16,000 gallons of water at a velocity of 1 ft/s required to keep the solids in suspension. This requires

$$\frac{330,000 \text{ gal}}{16,000 \text{ gal/ft}} = 20.6 \text{ ft of rotor} \quad (9\text{-}13)$$

In this case we will use two 11-ft rotors. The conversion factor of water to waste for this type of waste under 2,000 ft altitude is 2.35.

$$\frac{2.35 \times 595}{24} = 58.3 \text{ lb } O_2/\text{hr} \quad (9\text{-}14)$$

$$\frac{58.3 \text{ lb } O_2/\text{hr}}{22 \text{ ft of rotor}} = 2.65 \text{ lb } O_2/\text{hr/ft of rotor} \qquad (9\text{-}15)$$

A $27\frac{1}{2}$-in.-diameter cage rotor at 6-in. immersion and 89 rpm output speed gives 2.72 lb O_2/hr/ft of rotor at 0.600 kW/ft of rotor (Fig. 9-5).

$$\text{Power} = 11 \times 0.600 \times 1.34 = 8.84 \text{ BHP at rotor/unit} \qquad (9\text{-}16)$$

Use two 11-ft rotor assemblies, each driven by a 15-HP motor (permits operating at maximum depth of immersion if more oxygen is required).

d. Ditch Dimensions. To calculate ditch dimensions, refer to Fig. 9-10. The ditch cross-sectional area from section X-X is the area in ft²:

$$\text{Cross section} = D^2 + D \times F \qquad (9\text{-}17)$$

Volume in ends of ditch:

$$\text{Circumference} = \pi D \qquad (9\text{-}18)$$

$$\text{Volume in ends} = \text{ditch cross section} \times \text{circumference} \qquad (9\text{-}19)$$

Total ditch volume needed (Eq. (9-9))	_____ ft³
Volume in ends (Eq. (9-19))	_____ ft³
Volume needed in middle	_____ ft³

$$\frac{\text{Volume needed in middle}}{\text{Cross-sectional area}} = \text{total linear feet} \qquad (9\text{-}20)$$

$$\text{Linear feet} \div 2 = \frac{\text{ft}}{\text{side}} \text{ (length of side)} \qquad (9\text{-}21)$$

$$\text{Overall length of ditch } (A) = \text{straight side} + \text{ends} = \_\_\_ \text{ ft} \qquad (9\text{-}22)$$

e. Final Settling Tank. Design basis: 600 gal/ft²/day surface settling rate and 3 hours' detention time. With domestic sewage application, the required surface area is:

$$\frac{\text{Design flow in gal/day}}{600} = \text{surface area in ft}^2 \qquad (9\text{-}23)$$

$$\text{Surface area} = \frac{350{,}000 \text{ gal/day}}{600 \text{ gal/ft}^2} = 583 \text{ ft}^2 \qquad (9\text{-}24)$$

$$\text{Volume} = 243 \times 3 \times 60 \times 0.1334 = 5840 \text{ ft}^3 \qquad (9\text{-}25)$$

$$\frac{\text{Design flow in gal/day} \times 3 \text{ hr}}{24 \times 7.48 \text{ gal/ft}^3} = \text{volume in ft}^3 \qquad (9\text{-}26)$$

In the nonmechanized clarifier:

Less volume in upper one-third of hopper in ft³.
Straight-wall volume needed = Eq. (9-26) in ft³

$$\frac{\text{Straight-wall volume needed}}{\text{Tank area}} = \text{total SWD}$$

where SWD is side wall depth.

A 28-ft-diameter clarifier (mechanized) is 616 ft².

$$\frac{5840 \text{ ft}^3}{616} = 9.46 \text{ ft SWD} \tag{9-27}$$

We can use one 28-ft diameter by 9-ft 6-in. SWD final settling tank, mechanized type in this example.

Figure 9-11 shows the principle of operation of the clarifier used in these calculations. The unit shown is the mechanized type.

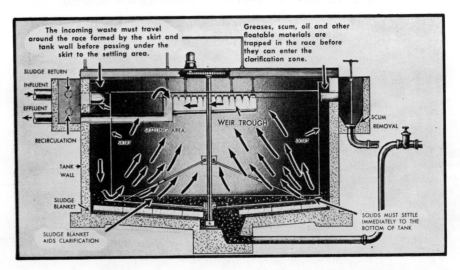

Figure 9-11. Clarifier Operation. (Courtesy Lakeside Equipment Corporation)

Now consider the same problem for the roadside rest area case using 600 gal/ft²/day surface settling rate and 3 hours' detention time. Here the quantity of flow is 100,000 gal/day. The problem that occurs at this point is that we have a higher peak hour usage for rest and recreation areas than is characteristic for domestic sewage. Based on available data, the correction factor has been found to be 2.3 times the gal/day rate.

$$2.3 \times 100,000 \text{ gal/day} = 230,000 \text{ gal/day} \tag{9-28}$$

This is the figure we actually used in our subsequent calculations.

$$\frac{230,000 \text{ gal/day}}{1440 \text{ min/day}} = 160 \text{ gal/min} \tag{9-29}$$

$$\text{Surface area} = \frac{230,000 \text{ gal/day}}{600 \text{ gal/ft}^2\text{/day}} = 383 \text{ ft}^2 \tag{9-29}$$

$$\text{Volume} = 160 \text{ gal/min} \times 3 \text{ hr} \times 60 \text{ min/hr} \times 0.1334 = 3850 \text{ ft}^3 \tag{9-30}$$

A 22-ft diameter = 380 ft². (Columns 1 and 2 of Table 9-2 can be used. This is close enough in this case.)

TABLE 9-2. Mechanized Spiroflo Size

Tank Dia. (ft)	Area (ft²)	Based on 900 Surface Rate Recommended G for weir troughs*
8	50	3' 9"
9	63	4' 0"
10	78	4' 6"
11	95	4' 9"
12	113	5' 3"
13	133	6' 0"
14	154	6' 6"
15	177	7' 0"
16	201	7' 6"
17	227	8' 0"
18	255	8' 3"
19	284	8' 9"
20	314	9' 3"
21	346	9' 9"
22	380	10' 3"
23	415	10' 6"
24	452	11' 0"
25	491	11' 6"
26	531	12' 0"
27	573	12' 6"
28	616	13' 0"
29	661	10' 6"
30	707	11' 0"
31	755	11' 3"
32	804	11' 9"
33	855	12' 0"
34	908	12' 3"
35	962	12' 9"
36	1,018	13' 0"
37	1,075	13' 6"
38	1,134	13' 9"
39	1,195	14' 3"
40	1,257	14' 6"
41	1,320	14' 9"
42	1,385	15' 3"
43	1,452	15' 6"
44	1,520	16' 0"
45	1,590	16' 3"

*The weir length G is found in Figs. 9-28, 9-29, and 9-30.

Now, referring to Table 9-3, opposite 22' 0" diameter, we find that in column 5 we have the value 1601 ft³. In other words, the upper one-third of the hopper may be used as settling volume.

$$3850 - 1601 = 2249 \text{ ft}^3 \quad \text{(which is less than } V \text{ in column 4)}$$

$$(9\text{-}31)$$

TABLE 9-3. Hopper Bottom Spiraflo Data

Tank Dia. (ft)	Tank Area (ft²)	Hopper Depth 60° Slope	Volume of V	Volume Upper 1/3 V
6	28.3	3′ 5½″	47	26.0
7	38.5	4′ 4″	74	43.4
8	50.2	5′ 2½″	114	67.0
9	63.6	6′ 1″	164	98.5
10	78.5	6′ 11″	228	137.0
11	95.0	7′ 9½″	299	186.0
12	113.0	8′ 8″	390	245.0
13	133.0	9′ 6½″	496	314.0
14	154.0	10′ 5″	620	395.0
15	177.0	11′ 3″	762	489.0
16	201.0	12′ 2″	926	599.0
17	227.0	13′ 0″	1156	721.0
18	255.0	13′ 10″	1321	861.0
19	284.0	14′ 9″	1553	1019.0
20	314.0	15′ 7″	1812	1192.0
21	346.0	16′ 5½″	2097	1380.0
22	380.0	17′ 4″	2412	1601.0

$$\frac{2249}{380} = 5.92 \text{ ft SWD or } 5' \ 11'' \text{ SWD} \tag{9-32}$$

Thus we can use a nonmechanized unit 22′ 0″ diameter by 5′ 11″ SWD and 17′ 4″ hopper bottom. Now, consider an alternative using a mechanized unit. Again, using 22′ 0″ = 380 ft²,

$$\frac{3,850 \text{ ft}^3}{380 \text{ ft}^2} = 10.13 \text{ ft SWD} \tag{9-33}$$

Thus we can use one 22′ 0″ diameter by 10′ 2″ SWD final settling tank equipped with a mechanized Spiraflo.

If some other make of clarifier is used, the calculations must be altered accordingly.

f. Return Sludge (from final tank to ditch). Pump or air lift for 40 to 150% of design flow. From Section 9-7a, take the design flow figure in gal/min.

Maximum return sludge pump rate = 1.5 × design flow = 240 gal/min

$$\tag{9-34}$$

Minimum return sludge pump rate = 0.4 × design flow = 64 gal/min

$$\tag{9-35}$$

This variation in flow must be obtained by a variable-speed pump, a splitter box, a time clock, a bypass, or an air lift.

g. Weirs. In Table 9-2, column 3, through 29′ in diameter the dimension is for a round trough, and above this dimension, for a square trough. It should also be noted that the total feet of weir shown in the table is approximately equal to the length of the skirt. This is done so that the theoretical velocity coming under the skirt is equal to the velocities leaving the weir trough.

h. Sludge Drying Beds. If a sludge drying bed is planned, it is necessary to first consult applicable regulatory agencies. Then refer back to Chapter 7. In an uncovered sludge drying bed for a small plant like this, it would not be unreasonable to use 1 ft² per population equivalent (PE) or 3500 × 1 ft² = 3500 ft² sludge drying bed area. In this case, four sludge drying beds, each 18′ 0″ × 49′ 0″, are used.

i. Excess-Sludge Holding Tank. Use 2.23 ft³/100 PE/day or, in the example being used,

$$(2.23 \times 3500)/(100) = 78.1 \text{ ft}^3/\text{day} \qquad (9\text{-}36)$$

is the holding tank capacity required. Unless circumstances dictate otherwise, it is usual to design on the basis of 20-day holding time:

$$20 \times 78.1 = 1562 \text{ ft}^3 \text{ total tank capacity required} \qquad (9\text{-}37)$$

j. Excess-Sludge Lagoon. One way to calculate sludge lagoons is to take the 78.1 ft³/day figure in Section 9-7i above and use 100 days of storage time; thus 100 × 78.1 = 7800 ft³ sludge lagoon volume required. In that case, we use two excess-sludge lagoons, each 3,905 ft³.

k. Plant Photo. Figure 9-12 is a photograph of the configuration for the plant calculations just completed. It is equipped with sludge-holding

Figure 9-12. Oxidation Ditch for 3500 Population Equivalent. (Courtesy Lakeside Equipment Corporation)

lagoons. The sludge coming from an oxidation ditch plant is aerobically digested and is not a smelly sludge as in some of the other processes. The sludge has very little odor. It can be applied directly on agricultural land or pasture land, either in the liquid or dry form.

9-8. Oxidation Ditch Design Details. After calculations have established the general dimensions, design details are resolved. Either under oxidation ditch size in the calculations or at this point, the cross-sectional area must be determined. Since 11-ft rotors are being used, a cross section such as section X-X in Fig. 9-10 is appropriate. The cross-sectional area is 80 ft, and the total length is 550 ft. The 550 ft, however, is the distance around the complete circular path of the ditch and not distance A. Distance $A = 275$ ft. Use distance $M = 15$ ft, $B = 58$ ft, and radius from point $C = 8$ ft to the inner channel wall and 29 ft to the outer channel wall. Refer to Fig. 9-13 for these and other dimensions.

Dimension F in section X-X and section Y-Y (Fig. 9-13) is the same width as the rotor. Distance D is normally kept between 3 and 5 ft. In a plant of this size, 5 ft should be used. The last two statements assume a $27\frac{1}{2}$-in. rotor. If the 42-in. rotor is used, the dimension D would be 5 and 8 ft. Measurement E is determined from the 45° angles shown on the cross section X-X. The rotor assemblies are spaced halfway around the plant.

9-9. Effluent Weirs. There are a number of different designs that might be used to remove the effluent from the oxidation ditch. The lowest-cost unit is shown in Fig. 9-14. This design requires the operator to lie down with his face virtually in the effluent section in order to reach below the water surface to loosen and adjust nuts. It is sometimes difficult to properly adjust this type of weir for the desired height under the conditions indicated. This design is not recommended because the operator will not make changes that should be made, thus leading to improper operation.

Figure 9-15 shows a design that is highly recommended for roadside rest areas or any of the domestic sewage treatment plants whose flow conditions are within its range of construction. Figure 9-16 shows two oxidation ditch water-level control designs that might be used. The main advantage of types 3 and 4 over type 2 (Fig. 9-15) is in the length of weir that may be constructed. Thus, they are primarily for plants too large to be served by type 2.

9-10. Rotor Cross Section. Section Y-Y in Fig. 9-17 represents the most desirable cross section to use where a single rotor section is mounted across the ditch at a given location. Drain connections and drains are sized according to weather conditions. Piping must be carried far enough to empty in an appropriate location, such as an effluent sewer.

A single rotor length is sufficient in most small plants and even in a number of medium-sized plants. If the plant requires a wider ditch, then it might

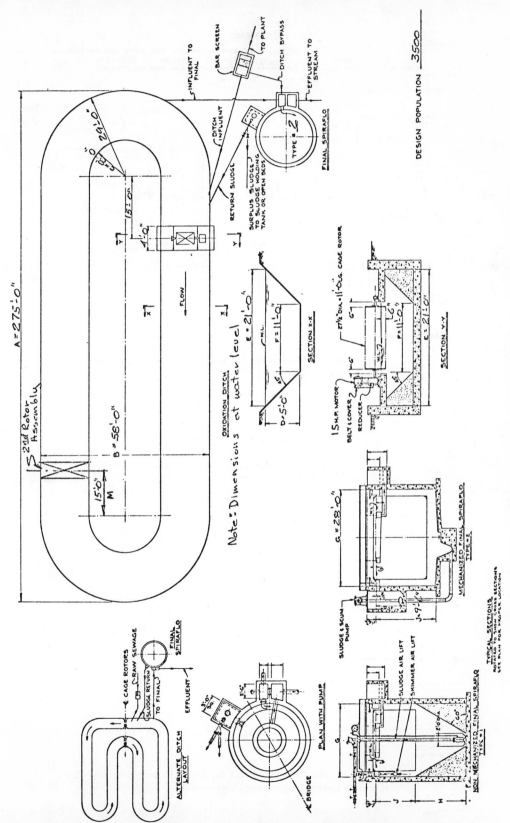

Figure 9-13. Oxidation Ditch Dimensional Detail. (Courtesy Lakeside Equipment Corporation)

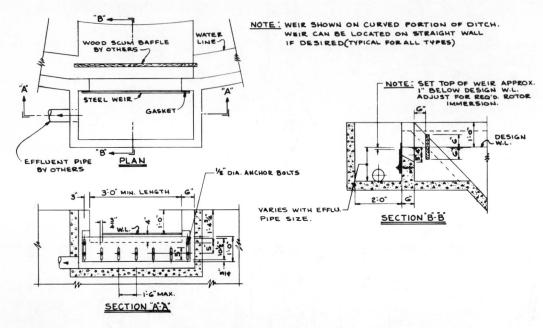

Figure 9-14. Adjustable Effluent Weir Type 1. (Courtesy Lakeside Equipment Corporation)

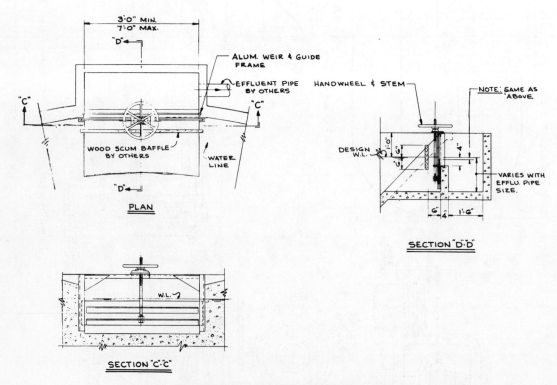

Figure 9-15. Handwheel Operated Adjustable Effluent Weir Type 2. (Courtesy Lakeside Equipment Corporation)

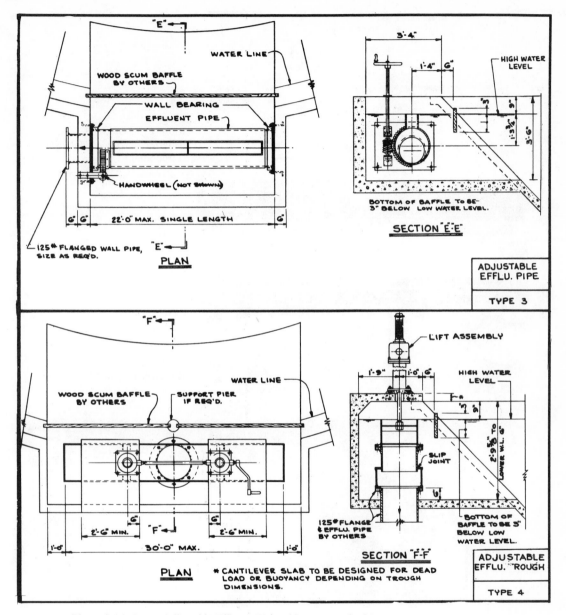

Figure 9-16. Large Adjustable Effluent Weirs. (Courtesy Lakeside Equipment Corporation)

be necessary to use a design like that shown in Fig. 9-18. Figure 9-19 shows the same two rotors in operation.

Ice and winter operation have not created problems with rotors provided that the plant is not excessively sized for the flow of wastewater available. Figure 9-20 shows a rotor operating at Woodstock, New Brunswick, Canada,

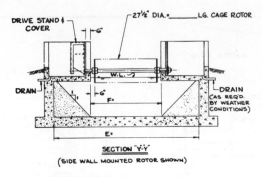

Figure 9-17
Rotor Cross-Section Construction.
(Courtesy Lakeside Equipment
Corporation)

Figure 9-18
Tandem Rotor Construction.
(Courtesy Lakeside Equipment
Corporation)

Figure 9-19. Tandem Rotors in Operation. (Courtesy Lakeside Equipment
Corporation)

during the winter. Figure 9-21 shows a protective housing that has been
built over rotors on a plant in Maine. A number of the oxidation ditch plants
are operating in Canada.

For trouble-free operation, use multiple V-belt drives, not gear-driven
rotors. In the initial design, try to keep the motor speed under 90 rpm and

Figure 9-20. Winter Operation of Rotor in Canada. (Courtesy Lakeside Equipment Corporation)

Figure 9-21. Protective Housings for Rotors in Severe Cold. (Courtesy Lakeside Equipment Corporation)

the depth of immersion at a design value around 6 in. If it becomes necessary to deviate from these values, call the equipment manufacturer to discuss the problem. The values given above are for the $27\frac{1}{2}$-in. rotor. When any other size rotor is involved, the general rule is to keep the initial design speed slightly under one-half the available speed range and the design immersion halfway between the minimum and maximum immersion points. Thus, a 42-in. rotor installation would be designed for 10-in. immersion and not to exceed 60 rpm.

Several construction techniques are recommended to help prevent vandalism. One is to place the electric motor on the closed interior side of the ditch. Another technique is to place a wire fence door and protective fencing at the entrance to the bridge walkway that crosses to the center of the ditch. Finally, use underground wiring from the motors to the control building.

The electric motors should have epoxy windings and should be sized on the basis of maximum rotor immersion and rotor operation at a speed faster than specified in the initial design. There should be a switch on the rotor that will enable the maintenance man to turn it on and off at that location. It should be a type that can be locked to prevent vandalism.

9-11. *Oxidation Ditch Lining.* A number of oxidation ditch plants have been constructed without linings; however, this is not being advocated or recommended unless the ditch is cut out of rock or other material where the movement of the ditch contents (the liquid wastewater), at velocities of 1 to 2.5 ft/s, is not likely to cause erosion problems. Erosion may occur at all points where the liquid changes direction if the channel is not lined. It is usually a rather poor-looking job and is often subject to problems later when an attempt is made to put lining at only the critical points. It is always desirable to use concrete in the immediate vicinity of the rotors. One might then switch to asphalt for the remainder of the ditch. Plastics are not recommended. Plastic has been used, but it tends to fail. Failure usually occurs in the vicinity of the rotors, after which it progressively occurs around the ditch because the liquid works under it and lifts it out (contents are moving at 1 ft/s or higher and, hence, have some lifting force).

A gunnite lining is satisfactory. In most locations, concrete may be more economical. Figure 9-22 should assist the contractor who has never built

Figure 9-22. Oxidation Ditch in Process of Construction. (Courtesy Lakeside Equipment Corporation)

an oxidation ditch. The 4-in.-thick concrete ditch lining should be placed just like a sidewalk. Pour the bottom first, with provision for waterstops, using wire mesh reinforcement, except in the area of rotors where structural requirements are higher. More steel and structural support is, of course, required for the rotors. Heavier bars can be used in place of the wire mesh.

A ditch should not be placed in a location where surface water can drain into it. Exterior drainage water, in any significant quantity, could easily upset the biological balance in the ditch or exceed the capacity of the clarifier. A protective embankment around the ditch or some other protective scheme should be used if it must be located where excess water could seep in. Whether the area in the center is large enough to cause concern is a matter of engineering judgment. How the ditch is laid out and the amount of rainfall in the area are influencing factors.

A number of patterns can be used in laying out an oxidation ditch. The sloping side wall, as shown, is usually the most economical configuration; but other types of side walls can be used. Figure 9-23 shows one form of design with a vertical center wall requiring cast concrete using forms. Figure 9-24 illustrates another vertical wall unit having provision for better hydraulic properties at the ends of the ditch. The sloping walls of the typical plant are constructed without the use of forms or by the use of a very simple slip form.

A design where one ditch is constructed inside another is shown in Fig. 9-25. This is only one scheme of several that might be utilized. Ditches may be constructed in concentric rings for any number of ditches. They may be built in a variety of shapes to fit the contour of the land or to fit the space available. Figures 9-26 and 9-27 indicate two of the schemes used to adapt a plant to the available site. One plant uses a boomerang shape.

Placing the clarifier in the center section of the ditch aids in preventing vandalism. The ditch itself serves as a barrier, provided fencing is placed across the access bridge entrance.

Figure 9-23. Vertical Center Wall with Sloping Side Walls. (Courtesy Lakeside Equipment Corporation)

Figure 9-24. Vertical Wall Construction. (Courtesy Lakeside Equipment Corporation)

Figure 9-25
Oxidation Ditch Constructed Inside Larger Ditch. (Courtesy Lakeside Equipment Corporation)

Figure 9-26. "L" Shape Ditch. (Courtesy Lakeside Equipment Corporation)

Figure 9-27. "U" Shape Construction. (Courtesy Lakeside Equipment Corporation)

9-12. Clarifier. The particular clarifier discussed in this chapter is arranged so that the liquid enters the space between the skirt and the wall. The liquid goes down and under a skirt and then has to rise to the center to reach the effluent weir.

Concrete-wall tanks can be built on the site according to the manufacturer's drawings, and all steel work can be furnished to complete the unit. Steel walls can also be purchased, and only the base has to be constructed, particularly with the mechanized unit. Either design seems to work satisfactorily. As long as the clarifier and rotor are properly matched, it is not mandatory that they be purchased from the same manufacturer.

Figure 9-28 shows some of the combinations of weir troughs and effluent pipes that are available. Figure 9-29 contains typical arrangements of boxes and effluent pipes. Figure 9-30 shows an elevation cross section of a mechanized peripheral feed clarifier, and Fig. 9-31 indicates the details of a non-mechanized clarifier.

9-13. Excess Sludge. The need for wasting excess sludge is governed by the actual plant loading and the quality of the final effluent required for a particular project. The oxidation ditch can be operated in a balanced state with no excess sludge buildup, but the ultimate plant efficiency is lowered with the loss of suspended solids (mostly ash) and some BOD over the effluent weirs. When a plant is fully loaded and a highly polished effluent is required, excess sludge will build up.

The rate of sludge wasting depends on the inert solids in the raw waste and the BOD removed daily, since these determine the rate at which the mixed liquor suspended solids build up in the oxidation ditch. Solids should be wasted before difficulties arise in the treatment plant or before the effluent is

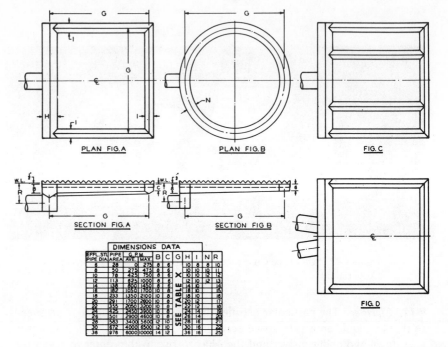

Figure 9-28. Weir Troughs and Effluent Pipes. (Courtesy Lakeside Equipment Corporation)

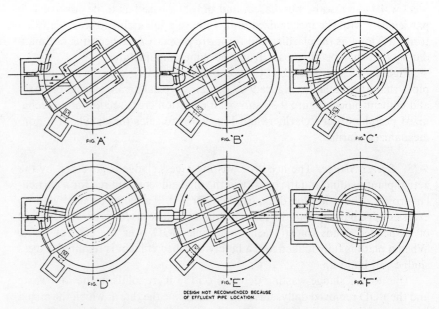

Figure 9-29. Typical Arrangements of Boxes and Effluent Pipes. (Courtesy Lakeside Equipment Corporation)

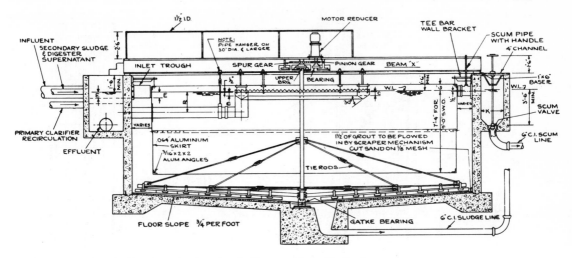

Figure 9-30. Clarifier Cross Section Detail. (Courtesy Lakeside Equipment Corporation)

adversely affected. When the suspended solids test shows 55 to 65% sludge volume and/or the D.O. level is reduced to 0.5 mg/l, solids must be wasted.

Depending on the plant design, excess sludge is discharged onto sand drying beds or pumped into a storage tank or sludge lagoon. The preferred operation is to reduce the sludge concentration 35 to 50% in a single day to enable rapid solids regrowth. Removal of only a small portion results in rapid return to full solids capacity and necessitates frequent sludge removals. Experience indicates that, with heavy sludge removal, excess sludge can be discharged at intervals of 1 to 4 weeks.

9-14. Maximum Size. The two oxidation ditches shown in Fig. 9-32 are over 1,000 ft long; however, they are less than half the length of ditches that are in operation. If suitable land is available, an oxidation ditch over a half mile long is feasible under some circumstances, and this is not necessarily the maximum size that might be constructed. Each case would have to be studied in terms of the cost of land, how much grading would be involved, etc. In the large oxidation ditches it is recommended that the $27\frac{1}{2}$-in. rotor be operated at speeds of 60 to 95 rpm. The 42-in. rotor should be operated at speeds of 50 to 72 rpm.

Based on the rotor immersion and output speed, at design conditions, the actual brake horsepower can be calculated from the power curves supplied by the manufacturer for either size rotor. This calculation is made using the following equation and gives the actual power required at the rotor for design conditions.

$$\text{BHP} = \text{rotor length} \times \text{kW/ft of rotor} \times 1.34 \qquad (9\text{-}38)$$

When small oxidation ditches are involved, the motor should be sized for maximum rotor immersion. In many large installations, this would require

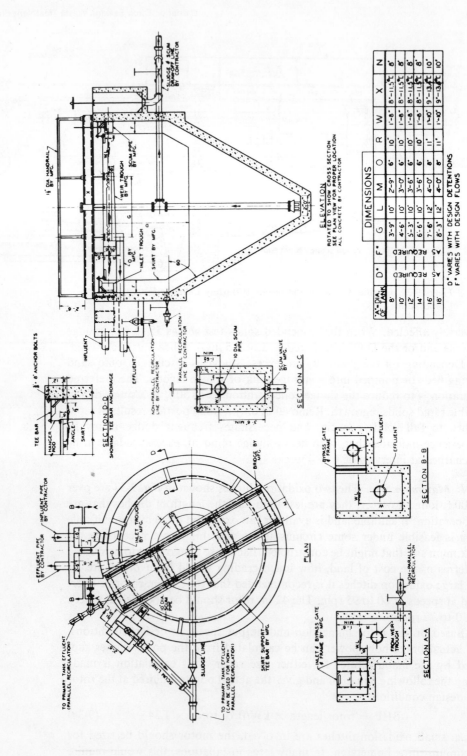

Figure 9-31. Non-Mechanized Hoppered Clarifier. (Courtesy Lakeside Equipment Corporation)

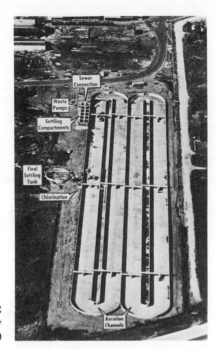

Figure 9-32
Large Oxidation Ditch. (Courtesy
Lakeside Equipment Corporation)

an excessively large motor. In the latter case, the motor selected for a given application must take into account the drive efficiency and must afford an increase in rotor immersion above design immersion.

1. For domestic waste a $1\frac{1}{2}$-in. increase in rotor immersion above the design immersion is the minimum allowance.
2. For industrial waste a 2-in. increase in rotor immersion above the design immersion is the minimum allowance.
3. Reducer and drive efficiency for standard drive is 0.95.

$$\text{Drive HP} = \frac{\text{rotor length} \times \text{kW/ft at increased immersion} \times 1.34}{\text{rotor and drive efficiency (0.95)}}$$

$$(9\text{-}39)$$

It is preferable to use a single rotor to a maximum single length and then go to two rotor assemblies located at alternate ends of the aeration reactor. After the two rotors have reached the maximum length, dual rotor assemblies may be used.

Mountings available include (a) fixed concrete mount, (b) side-wall mount, and (c) suspended from overhead support structure. The last mounting is not recommended for rotors over 15 ft or for dual rotor assemblies.

Preferred design is to use V-belt drive. Special drives are available, but the general construction is as recorded below.

1. $1\frac{1}{2}$- to 40-HP motor, shaft-mounted reducer with belt drive to motor.

2. 40-HP or larger motor, fixed-mounted reducer with belt drive to motor.
3. $1\frac{1}{2}$ through 5 HP may be suspended from an overhead support, and gear motor with belt drive to rotor. This is not a popular arrangement with the author.
4. Counter-rotating reducer is available for fixed reducer mounted on center dividing wall with belt drive between reducer and motor. Author prefers to use separate drives.

9-15. New Schemes. A number of schemes that are related to the oxidation ditch are in existence. One consists of rectangular basins with rounded corners and guide vanes at each end to provide the necessary hydraulic change of direction of the flow. Multiple-section construction using this principle is feasible. Figure 9-33 illustrates a plot plan of this scheme. Many of the design factors are established, but more data needs to be accumulated. Table 9-4 contains a number of design factors and information that can be used concerning oxidation ditches.

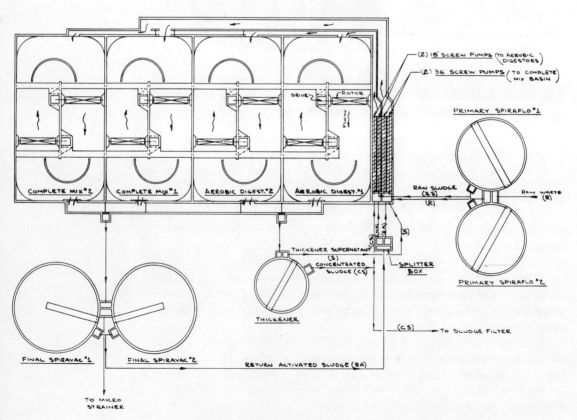

Figure 9-33. Variation of Oxidation Ditch Process. (Courtesy Lakeside Equipment Corporation)

TABLE 9-4. Design Factors

Rotor Diameter[6] (in.)	Rotor Length[5,7] (ft)	Ditch Width[1] Rotor Length	Minimum Center Island Width Without Baffles[2,4] (ft)	Minimum Length Straight Section[3] (ft)
$27\frac{1}{2}$	3 to 5	3 to 2.5	14	40
$27\frac{1}{2}$	6 to 15	2.5 to 2	16	40
42	16 to 30	2.0 to 1.3	20	40
42	>30	<1.5	24	Note 3

[1]Ratio of ditch width to rotor length. Use higher ratio with smaller rotor length.

[2]Minimum island without using return baffles at ends of aid hydraulics.

[3]Length of straight section of ditch should be a minimum of 40 ft or at least two times the width of the ditch at liquid level.

[4]If a center dividing wall is used in place of a center island, then return flow baffles are required at each end. The channel width, W, is calculated as flat bottom plus $\frac{1}{3}$ of sloping side wall. Baffle radius is $W/2$. Baffles should be offset with a larger opening in the direct of flow by $W/8$ for Lakeside rotor. Baffle design is slightly different for other makes of rotors.

[5]Large contractions in ditch width at rotor mounts should be avoided. If desired, support posts can be used, leaving open space at ends. Where contraction is made, special baffles, critically placed, are required.

[6]Liquid depth with $27\frac{1}{2}$-in. rotor is 5 ft up to 1 MGD, and 6 ft deep above 1 MGD. With 42-in. rotor, 10 ft is normal. Special designs using other depths usually require special baffles. Use baffles when 6-ft depth is used with $27\frac{1}{2}$-in. rotor.

[7]With $27\frac{1}{2}$-in. rotor, if tank volume less than 60,000 gal use 13,000 gal/ft of rotor. Over 60,000 gal, use 16,000 gal/ft of rotor. With 42-in. rotor, use 21,000 gal/ft of rotor.

There is a modification of the oxidation ditch principle in which conventional vertical aerators, located near the end baffles that serve as center islands, provide both aeration and movement. This scheme currently involves a royalty arrangement; therefore, it is not included in this book. There is a question of what can and cannot be patented in the process.

a. Nitrates and Nitrites. In oxidation ditches using wastes encountered up to now, the ammonia compounds will be converted to nitrates and nitrites as long as the D.O. level in the ditch is maintained above 0.5 mg/l dissolved oxygen. If the sludge is not required for by-product usage, there is a scheme that can be used to reduce the nitrates and nitrites in the effluent. The recirculated sludge is not pumped back to the ditch. Instead, it is pumped back to an appropriate point in the raw sewage influent line. If the incoming sewage is septic or near septic, adequate mixing and reaction may take place before the combined sludge and sewage enters the oxidation ditch. If the sewage is not at the septic point, a wet well of appropriate size can be constructed and the returned sludge pumped into this wet well in such a way that mixing of the raw sewage and recirculated sludge takes place. From a design viewpoint, there are ways that the problem can be handled. The objective is that the bacteria in the raw sewage will, in their need for oxygen, seek removal of the oxygen in the nitrates and nitrites of the returned activated sludge. The net result is a reduction in the overall nitrates and nitrites as nitrogen is released as a gas and hence reduces the nitrogen that emerges in the effluent of the plant.

b. Phosphates. The phosphates of the effluent can be reduced by injection of either a ferric or alum salt into the raw sewage as it enters the oxidation ditch plant. If the water is to be recycled for reuse, alum will probably give more color reduction and may in many cases be the most suitable salt to use (if the pH of the system is withih an acceptable range). In some other types of treatment systems, the alum or ferric salt is added after the aeration stage, but in the oxidation ditch it should be added with the raw sewage after the denitrification and before the sewage enters the oxidation ditch. One of the reasons why the oxidation ditch plant works so well by chemical feeding just upstream from the rotor is that the rotor does the necessary complete mixing of the chemical with the waste. Then as the mixed material moves further downstream from the rotor, flocculation actually occurs forming excellent floc that, once passed on to the settling tank, settles very readily for removal from the final tank. No other process has this flocculation zone in the aeration reactor.

With a few design changes in the plant, it is possible to end up with an economical water recycling plant with a minimum of additional apparatus. In some design cases, depending on the waste, the loading rate may be varied substantially. For normal domestic sewage, the volume of the ditch is sized based on 74 ft³/lb BOD applied per day. This is equal to a daily loading of 13.5 lb of BOD per 1,000 ft³ of ditch volume. Loadings from as low as 9 lb for weak wastes to 40 lb for industrial wastes are not uncommon. In very exceptional cases, loadings in excess of 100 lb may be used. The fine points of these extremes exceed the intent and scope of this presentation.

Short, wide, stubby ditches have the possible problem of harmonic-wave action or reverse-flow eddies at the bends. These problems can be corrected by the use of horizontal baffles and/or by return flow baffles. It has also been found that excessive velocities in the ditch caused by large lengths of rotor per ditch volume can cause streaming velocities and poor overall mixture of the ditch contents. This again can be corrected by a horizontal baffle. Ditch velocities above $2\frac{1}{2}$ ft/s are uneconomical to operate from the viewpoint of horsepower consumption. Ditches using $27\frac{1}{2}$-in.-diameter rotors that are constructed 6 ft deep should have baffles to achieve complete mixing. All the considerations set forth in this paragraph should be considered with particular care when chemicals are to be added or water recycling systems designed.

c. Altitude Correction. The equation for altitude correction is as follows (refer back to Eq. (9-1)):

$$C_{sw} = \frac{A_{EL} \times C_s}{A_0} \qquad (A_{EL} \text{ and } A_0 \text{ must be in same units, such as inches.})$$

$$(9\text{-}40)$$

where A_{EL} = barometric pressure at plant elevation,

$\quad A_0$ = barometric pressure at sea level,

$\quad C_s$ = saturation value of oxygen in water at operating temperature.

PROBLEMS

1. Design an oxidation ditch waste treatment plant for each of the following domestic wastewater flow conditions.

Plant	Flow (MGD)
AAA	0.05
AA	0.10
A	0.50
B	1.00
C	10.00
D	100.00

In the above designs, use any size rotor or combination of rotors your engineering judgment deems best. Multiple ditch construction is permissible in any of the problems. Use any scheme indicated in the chapter in order to produce the best solution for your client. Show all assumptions made. Design the plant up to the point of sludge disposal. In sizing clarifiers, it is permissible to use manufacturers' literature and any type of unit on the market as long as you clearly show the source and calculations. You can use different makes of equipment on each plant if desired.

2. Design a sludge drying bed (or beds) for each of the above plants. Use material in Chapter 7 to arrive at a suitable loading rate, etc. Assume that the sludge drying beds are located at Greenville, Ohio.

3. Design a sludge-holding lagoon for each of the above plants. Assume that the climatic conditions are similar to those in Dallas, Texas.

4. Design a sludge-holding tank (or tanks) for each of the plants indicated in Problem 1. Size on basis of 20 days' holding time.

5. Assume that you have a trailer tank truck (or trucks) capable of a net payload of 35,000 lb. How often would this truck (or trucks) have to make a trip carrying liquid sludge from the sludge-holding tanks in each of the plants in Problem 1?

6. Determine the daily electrical consumption for operating the rotors in each of the above plants.

7. Using the oxidation ditch principle purely to supply aeration and re-aeration, size the plants to operate by contact stabilization instead of extended aeration, for plants B, C, and D in Problem 1. This will involve two separate ditches as a minimum for each plant.

8. Design an oxidation ditch for Greenville, Ohio. Interest rate on bonds is 3%. Show all assumptions. Use sludge holding tanks.

9. Design an oxidation ditch for Union, S.C. The interest rate is 3%. Include sludge drying beds.

10. Design an oxidation ditch for Poplarville, Mississippi. Include sludge lagoons. The interest rate is 3%.

10 *Mechanical Aeration*

Chapter 9 was primarily concerned with the use of horizontal types of brush aerators frequently called cage or paddle rotors. This chapter focuses on a more general type of mechanical aerators. If the basic principles of the use of mechanical aerators are fully understood, aerators may more often be employed separately or in conjunction with other modes of treatment.

A substantial portion of the practical knowledge of mechanical aeration has been developed by equipment manufacturers. Substantial assistance for this chapter was received from Aqua-Aerobic Systems, Inc., Peabody Welles Products Corporation, Infilco Division of Westinghouse Corporation, and Lakeside Equipment Corporation.

The material in Sections 10-1 through 10-12 has either been taken directly from, or follows closely, excerpts of material provided by Aqua-Aerobic Systems, Inc. Supplemental information on design are contained in references 1 through 17.

AERATED LAGOONS

10-1. Aerated Lagoon Requirements. Proper design of an aerated lagoon requires consideration of:

1. BOD removal characteristics.
2. Biological oxygen requirements.

3. Oxygen transfer relationships.
4. Mixing requirements.
5. Basin geometry and aerator placement.
6. Sludge production.
7. Clarification and polishing.

10-2. BOD Removal Characteristics. Considering the case of a single aeration cell system, the mass balance based upon BOD (in lb) for a lagoon operating at steady-state conditions can be described as follows:

$$\text{lb BOD removed/day} = \text{(lb BOD in)/day} - \text{(lb BOD out)/day} \qquad (10\text{-}1)$$

Assuming that loss or gain of water in evaporation, percolation, and rainfall is small and can be neglected, the mass balance can be written mathematically as:

$$rV = C_i Q - C_e Q \qquad (10\text{-}2)$$

where r is the BOD removal rate, V is the lagoon volume, Q is the flow rate of sewage, C_i is the concentration of BOD in the influent, and C_e is the concentration of BOD in the effluent. As noted earlier in the book, unless otherwise stated, all BOD is the five-day, 20°C BOD.

If the liquid in the lagoon is completely mixed, the concentration of the BOD in the lagoon is equal to the concentration of BOD in the effluent (C_e). The removal of BOD is considered to be a first-order differential equation, i.e., the rate of removal is proportional to the amount of BOD present.

$$\frac{dC}{dt} = K_e C \qquad (10\text{-}3)$$

where K_e is the BOD removal rate coefficient.

The rate of removal, r, at equilibrium is equal to $K_e C_e$; therefore

$$r = K_e C_e = \frac{Q}{V}(C_i - C_e) \qquad (10\text{-}4)$$

or

$$\frac{K_e C_e V}{Q} = C_i - C_e \qquad (10\text{-}5)$$

Note that $V/Q = t$ (where t = retention time); the equation becomes

$$K_e C_e t = C_i - C_e \qquad (10\text{-}6)$$

or

$$K_e t = \frac{C_i - C_e}{C_e} = \frac{C_i}{C_e} - 1 \qquad (10\text{-}7)$$

or

$$\frac{C_e}{C_i} = \frac{1}{1 + K_e t} \qquad (10\text{-}8)$$

Aerated lagoons represent a system of treatment that is intermediate between oxidation ponds and activated-sludge systems. Generally, there is sufficient power supplied to completely mix the liquid of the lagoon, i.e., the concentration of dissolved oxygen or any other substance in solution is the same throughout the pond, but the power input is not great enough to insure the complete suspension of solids. C_e/C_i is the fraction of BOD remaining, so the percent BOD removal is:

$$\% \text{ removal} = 100 - \frac{100}{1 + K_e t} \qquad (10\text{-}9)$$

Since the designer generally knows the percentage of removal desired and wishes to solve for the retention time necessary to effect this removal, the equation is rearranged:

$$t = \frac{\% \text{ removed}}{(100 - \% \text{ removed})K_e} \qquad (10\text{-}10)$$

Thus, the value of K_e and the desired percentage of BOD removal fixes the retention time. For domestic sewage, the value of K_e falls between 0.50 and 1.0. K_e values up to 3.0 have been reported for industrial wastes. A value of 0.5 is generally accepted as conservative design for cold climates and predicts a 75% BOD removal in 6 days, with a 90% removal in 18 days. For warmer climates, a value of 1.0 for K_e predicts a 75% removal in 3 days, with a 90% removal in 9 days. For design accuracy, K_e should be determined from laboratory or pilot-plant data. K_e varies with temperature as indicated in the previous paragraph. It is normally reported at 20°C, but to convert to any other temperature, the following equation is used:

$$K_T = K_{20}\theta^{(T-20)} \qquad (10\text{-}11)$$

where θ = temperature correction factor = 1.075 and T = temperature in °C.

The absence of nitrogen or phosphorus, although present in sufficient quantities in domestic sewage, can cause a depressed K_e value. Generally, nitrogen should be present in quantities of 5 lb (as ammonia nitrogen) per 100 lb of BOD, and phosphorous should be present at the rate of 1 lb (as phosphorous) per 100 lb of BOD.

10-3. Multiple-Cell BOD Removal Characteristics. The same basic equations may be used to compute the BOD removal characteristics for a multiple-cell system. If K_e remained constant, obviously a three-cell system would give far greater efficiency of treatment than would a single-cell system. Unfortunately, K_e does change, but not linearly; so there is usually some advantage in using a multiple-cell system. Some designers decrease the value of K_e by 20% for each additional cell.

10-4. Biological Oxygen Requirements. In a single-cell system, the oxygen required by the microbiology to destroy a pound of BOD_5 is generally reported as a pound-to-pound ratio, e.g., 1.3 lb O_2 required/lb BOD_5 destroyed. This is done as a convenience; however the actual requirement for oxygen is not based upon a two-day BOD, or a five-day BOD, but upon the total BOD. The actual oxygen requirement is expressed as:

$$O_2 \text{ required/hr} = (L)(\text{BOD}_5 \text{ removed/hr}) \qquad (10\text{-}12)$$

where $L = \text{BOD}_L/\text{BOD}_5$.

Most wastes have an L value that falls between 1.1 and 1.5. Domestic sewage generally has an L value of 1.30. Thus, the oxygen required per hour (O_{2r}) is equal to the BOD_5 applied per hour times the % removal expected, P.

$$O_{2r} = LPQ(\text{BOD}_5)(8.34 \text{ lb/gal})(10)^{-6} \qquad (10\text{-}13)$$

where Q is the flow rate of waste in gal/hr, BOD_5 is the five-day BOD, ppm or mg/l, and 8.34 lb/gal is the weight of water.

The aeration requirements for a multiple-cell system are determined by the same equation; however, since the rate of BOD removal is considered to be a first-order differential equation, the first cell will account for the bulk of the BOD removal, the second cell for a lesser portion, the third cell for even less, etc.

The first-cell retention time is designed to give a desired BOD removal percentage. This will govern the oxygen requirements, and the total installed HP will be apportioned accordingly.

10-5. Oxygen Transfer Rate. The rate at which oxygen is transferred to the wastewater contained in the lagoon depends on the temperature, the dissolved oxygen concentration in the lagoon, and the amount of impurities contained in the lagoon which would alter the transfer rate from the rate that could be expected under the same conditions if the water were pure. This factor is the ratio of $K_L a$ for the wastewater as compared to the $K_L a$ value for pure water and is referred to as the *alpha factor*. For domestic sewage, alpha values are commonly in the range of 0.80 to 0.95. Industrial waste may exhibit alpha values from 0.5 to 1.3. The field transfer rate that can be expected can be calculated via the following equation:

$$FTR = \frac{(CWTR)[(C_{DC})B - C_r]1.024^{T-20}\alpha}{C_{sc}} \qquad (10\text{-}14)$$

where FTR is the field transfer rate (lb/HP/hr), $CWTR$ is the clean-water transfer rate (lb/HP/hr), supplied by the aeration equipment manufacturer, C_{DC} is the saturation concentration of O_2 at design temperature and altitude in ppm or mg/l. C_r is the residual concentration of dissolved oxygen to be maintained in aeration basins (usually 1.0 to 1.5, and it should never exceed 2.0) expressed in ppm or mg/l. C_{sc} is the saturation concentration of O_2 at standard conditions of 760-mm Hg and $20°C = 9.17 = $ constant (ppm or mg/l). T is the design temperature in °C.

$$B = \frac{\text{saturation conc. of } O_2 \text{ in wastewater}}{\text{saturation conc. of } O_2 \text{ in clean water}} \qquad (10\text{-}15)$$

$$\alpha = \frac{\text{rate of transfer of } O_2 \text{ in wastewater}}{\text{rate of transfer of } O_2 \text{ in clean water}} \qquad (10\text{-}16)$$

Once the *FTR* and the O_{2_r} have been determined, the amount of HP required to fulfill the oxygen requirements can be calculated:

$$\text{HP required} = \frac{\text{oxygen required (lb/hr)}}{\text{field transfer rate (lb } O_2/\text{HP/hr)}} \qquad (10\text{-}17)$$

By combining the previous equations, we obtain

$$\text{HP required} = \frac{(O_2 \text{ required, lb/hr})(9.17)(1.024)^{(20-T)}}{(CWTR)[(C_{DC})(B) - C_r]\alpha} \qquad (10\text{-}18)$$

10-6. Mixing Requirements. Although it is not necessary to maintain scouring velocities and the suspension of all solids in an aerated lagoon, it is imperative to provide sufficient power to accomplish a complete mixing of the liquid and to maintain a uniform dispersion of dissolved oxygen throughout the lagoon. *In dealing with dilute wastes and wastes of weak strength, especially in lagoons having long retention times, it is quite common for the horsepower requirements for mixing to exceed the horsepower required for aeration.* Normally, 6 to 10 HP per million gallons will accomplish this goal. To maintain all solids in suspension, 60 to 100 HP per million gallons is required.

10-7. Basin Geometry and Aerator Placement. Since aerated lagoons do have long retention times, it is not necessary to have the inlet pipe located under or adjacent to an aerator. The aerators should be uniformly spaced throughout the lagoon. In irregular shaped lagoons, the same approximate surface area should be apportioned to each aeration unit.

In the design of a new lagoon, the lagoon average area should be apportioned into a number of squares coinciding with the number of aerators anticipated. One aerator should then be placed in the center of each square.

10-8. Sludge Production. Since a long retention time is inherent in the design concept of the aerated lagoon, nearly all organic matter is destroyed. There will be a gradual buildup of inorganic and low organic solids on the bottom of the lagoon. The organic will build up until the mass of organic material on the bottom has a total decay rate equal to the daily deposition rate of organic material. Then a steady-state equilibrium situation is reached which can be expressed by the following equation:

$$Ma = d \qquad (10\text{-}19)$$

where M is the total organic mass, a is the anaerobic decay rate, and d is the rate of deposition (lb)(1/day = lb/day).

This usually occurs at a depth of $\frac{1}{2}$ to 3 in. of solids. The decay process on the lagoon bottom is anaerobic, and occasionally some particles gasify and rise to surface where they are degraded aerobically. Due to the inorganic buildup, a lagoon may require draining and dredging every 10 to 20 years.

10-9. Horsepower Required for Aeration. Equation (10-18) can be simplified to the form

$$\text{HP required} = \frac{QLPL_a(1.024)^{20-T}(3.187)(10)^{-6}}{(CWTR)[(C_{DC})(\beta) - C_r]\alpha} \tag{10-20}$$

where Q is the flow in gal/day, L is the ratio of ultimate BOD to five-day BOD, P is the percentage removal desired, L_a is the five-day BOD in ppm or mg/l, and T is the temperature in °C, $CWTR$ is the clean-water transfer rate of the aerator which is supplied by the equipment manufacturer; it ranges from 2.5 lb/HP/hr to 3.5 lb/HP/hr. C_{DC} is the saturation concentration of O_2 at the design temperature and altitude in ppm, and C_r is the D.O. residual to be maintained in the aeration basin, usually 1.0 to 2.0 in ppm or mg/l.

$$\beta = \frac{\text{saturation conc. of } O_2 \text{ in wastewater}}{\text{saturation conc. of } O_2 \text{ in clean water}} \tag{10-21}$$

$$\alpha = \frac{\text{rate of transfer of } O_2 \text{ into wastewater}}{\text{rate of transfer of } O_2 \text{ into pure water}} \tag{10-22}$$

10-10. Mixing Requirements for Liquid Mix. The minimum value is 7.5 HP/one million gallons.

$$(\text{HP required})_m = V \times 7.5 \times 10^{-6} \tag{10-23}$$

where V is the volume of the aeration basin in gallons.

10-11. For Solids Suspension. If V is the volume of aeration basin in gallons, then

$$(\text{HP required})_m = V \times 75 \times 10^{-6} \tag{10-24}$$

10-12. Aerated Lagoon Design Example. Assume that a large subdivision is to be built containing 1,400 homes. All homes will be equipped with garbage shredders. The subdivision is to be isolated—a "country estate" type. Design a waste treatment facility to produce 90% BOD removal during the summer and 85% in winter. Land costs are low, and there are five acres available on which to develop a system.

Assume 3.6 people/house, 100 gal/capita/day, 0.22-lb BOD/capita/day, $\alpha = 0.90$, $\beta = 0.95$, $K_e = 0.65$ at 10°C, minimum lagoon water temperature $= 1$°C, maximum lagoon water temperature $= 30$°C, and that this will be a single-cell aerated lagoon.

    a. *Detention time* required to accomplish 85% BOD removal:

$$t = \frac{\% \text{ BOD removed}}{(100 - \% \text{ BOD removed})K_e} = \frac{85}{15K_e} \tag{10-25}$$

Winter Conditions: Under winter conditions, 85% BOD removal is required. The K_e value is given for 10°C, but it must be corrected to the 1°C temperature, as this is an extreme condition. Solving for K_e at 20°C, we obtain

$$K_e t = K_{e(20)}1.075^{T-20} \tag{10-26}$$

$$K_{e(20)} = \frac{K_{e(10)}}{(1.075)^{T-20}} = K_{e(10)}1.075^{20-T} = 1.34 \tag{10-27}$$

Solving for $K_{e(1°)}$, we obtain

$$K_{e(1°)} = 1.34 \times 1.075^{-19} = 0.34 \tag{10-28}$$

$$t = \frac{85}{15 \times 0.34} = 16.7 \text{ days} \tag{10-29}$$

Thus 16.7 days of retention time is required for 85% BOD removal for winter conditions.

Summer Conditions: Under summer conditions, 90% BOD removal is required. The maximum lagoon temperature in summer is 30°C.

$$K_{e(30)} = K_{e(20)}(1.075)^{T-20} = 1.34 \times 1.075^{10} = 2.76 \tag{10-30}$$

$$t = \frac{90}{10 \times 2.76} = 3.26 \text{ days}$$

Winter conditions control, therefore use a 16.7 day retention time.

   b. *Biological Oxygen Requirements.*

$$O_{2r} = (BOD_s)Q(8.34)LP \times 10^{-6} \tag{10-31}$$

Daily BOD_s load
$$= (3.6 \text{ people/house})(1,400 \text{ homes}) (0.22 \text{ lb BOD/capita/day})$$
$$= 1,111 \text{ lb BOD}_s\text{/day} \tag{10-32}$$

$$(BOD_s)(Q)(8.34)(10)^{-6} = 1,111 \text{ lb BOD}_s\text{/day} \tag{10-33}$$

$$O_{2r} = \frac{(1,111 \text{ lb BOD}_s\text{/day})(L)(P)}{24} = \frac{1,111 \times 1.3 \times 0.90}{24}$$

$$= 54.0 \text{ lb/hr} \tag{10-34}$$

   c. *Oxygen Transfer Rate.*

$$FTR = \frac{(3.0)[(7.6)(0.95) - (1.0)](1.024)^{10}(0.90)}{9.2}$$

$$= 2.33 \text{ lb/HP/hr at 30°C} \tag{10-35}$$

$$FTR = \frac{(3.0)[(14.4)(0.95) - (1.0)](1.024)^{-9}(0.90)}{9.2}$$

$$= 2.37 \text{ lb/HP/hr} \tag{10-36}$$

2.33 lb/hr < 2.37 lb/HP/hr, summer conditions control, use a field transfer rate of 2.33 lb/HP/hr.

   d. *HP Required for Aeration.*

$$HP = \frac{O_{2r}}{FTR} = \frac{54.0 \text{ lb/hr}}{2.33 \text{ lb/HP/hr}} = 23.2 \text{ HP} \tag{10-37}$$

e. HP Required for Mixing.

$$HP = (7.5)(10)^{-6}(V) \tag{10-38}$$

$$\begin{aligned}V &= (16.7 \text{ days})(100 \text{ gpcd})(1{,}400 \text{ houses})(3.6 \text{ people/house})\\ &= 8{,}430{,}000 \text{ gal}\end{aligned} \tag{10-39}$$

$$\begin{aligned}HP &= (7.5)(10)^{-6}(8{,}430{,}000 \text{ gal}) = 7.5 \times 8.43\\ &= 63.4 \text{ HP}\end{aligned} \tag{10-40}$$

The mixing HP controls in this instance. Use 60 HP, comprised of three each 20-HP aerators.

f. Basin Geometry and Aerator Placement.

$$\text{Basin volume} = 8{,}430{,}000 \text{ gal} = 1{,}128{,}000 \text{ ft}^3$$

$$\text{Average depth} = \bar{A} = \frac{1{,}128{,}000 \text{ ft}^3}{10 \text{ ft}} = 112{,}800 \text{ ft}^2 \tag{10-41}$$

Compute the basin dimensions: Since three 20-HP units are to be used, use a basin with a length equal to 3 times the width ($r = 0.333$).

$$L_T = \frac{[4r\bar{A} - 6r(SD)^2]^{1/2} + SD(1 + r)}{2r} \tag{10-42}$$

where L_T is the length of the basin at water surface, r is the width at water surface/length at water surface (0.333), and S is the side slope (assume 3/1). For D = depth, use 10 ft.

$$\begin{aligned}L_T &= \frac{\begin{aligned}[(4)(0.333)(122{,}800) &- 6(0.333)(3 \times 10)^2\\ + 2(3 \times 10)^2]^{1/2} &+ (3)(10)(1 + 0.33)\end{aligned}}{(2)(0.333)}\\ &= 642\end{aligned} \tag{10-43}$$

$$W_T = L_T r = (642)(0.333) = 214 \text{ ft} \tag{10-44}$$

$$L_B = L_T = 2SD = 642 - 2(3)(10) = 154 \text{ ft} \tag{10-45}$$

Assume a 2-ft freeboard, the basin dimensions are 648 ft × 220 ft at the top and 582 ft × 154 ft at the bottom; the overall depth is 12 ft and H_W depth is 10 ft.

Locate aerators on the centerline of the basin: first aerator 107 ft from shore along the longitudinal axis, then 214 ft to the next unit, 214 ft to the next unit, and 107 ft to shore.

10-13. Mechanical Aeration Equipment. Mechanical aerators are available in various forms. Some are mounted on permanent platforms and access can be obtained by walkways. Other units are on individual platforms in a cluster of poles. Some of the platforms are floating and move up and down with water level, with the poles acting as a guide. There are units rigidly attached to poles and units that float on the lagoon surface and are held in place by mooring lines. Some of the manufacturers can provide equipment for more than one type of mounting.

Figure 10-1 shows a platform-mounted aerator with a horizontal drive. Figure 10-2 shows a pole-supported mechanical aerator with a vertical drive. Figure 10-3 shows the turbulence pattern developed by one type of vertical aerator unit, and a waste treatment plant using rigid-platform vertical aerators is shown in Fig. 10-4. Figure 10-5 shows an aerial view of 14 raft-mounted aerators used to aerate a paper mill waste pond. Figure 10-6 shows one type of floating aerator, and Fig. 10-7 shows floating aeration units treating refinery waste. Figure 10-8 shows an aerial view of a pond with 43 floating aerators in use.

Figure 10.1. Platform Mechanical Aerator with Horizontal Drive. (Courtesy Infilco Division, Westinghouse Electric Corp.)

Figure 10-2. Pile Supported Mechanical Aerator with Vertical Drive. (Courtesy Infilco Division, Westinghouse Electric Corp.)

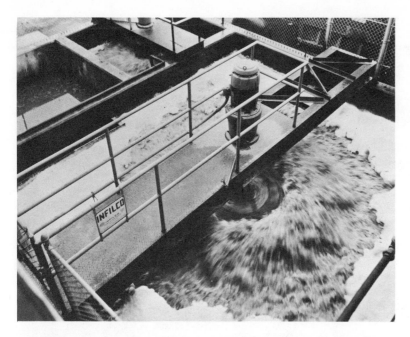

Figure 10-3. Turbulence Pattern of Vertical Aerator Unit. (Courtesy Infilco Division, Westinghouse Electric Corp.)

Figure 10-4. Waste Treatment Plant Using Rigid Platform Vertical Aerators. (Courtesy Infilco Division, Westinghouse Electric Corp.)

Figure 10-5. Aerial View of 14 Raft Mounted 60 HP Aerators Employed for Treatment of Waste at a Large Paper Mill. (Courtesy Infilco Division, Westinghouse Electric Corp.)

Figure 10-6. One Type of Floating Aerator. (Courtesy Peabody Welles, Inc.)

Figure 10-7. Floating Aeration Units Treating Refinery Waste. (Courtesy Peabody Welles, Inc.)

Figure 10-8. Pond with 43 Floating Aerators. (Courtesy Peabody Welles, Inc.)

In the sizing of mechanical aeration equipment, two of the factors involved are often misunderstood and misused. These are the K factor and the alpha factor. Eckenfelder has made substantial contribution on the K factor, and references at the end of the chapter bearing his name should be consulted for additional information. The alpha factor is defined as the ratio of the rate of oxygen transfer in wastewater to the rate of oxygen transfer in clean water.

10-14. K *Factor*. The K factor, in the terminology of Eckenfelder (1) is the rate of BOD removal in an aerated lagoon. Table 10-1 indicates the progressive daily decay of a material which is being destroyed at a constant rate

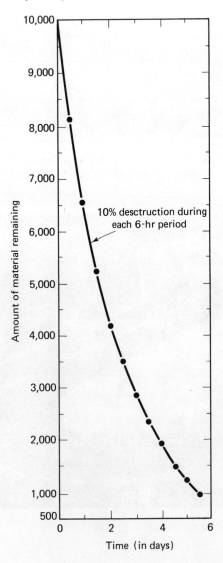

Figure 10-9
Material Remaining vs. Time in Days. (Courtesy Peabody Welles, Inc.)

(10% every 6 hours, in this case). Figure 10-9 shows a curve obtained by plotting the amount of material remaining versus time. Figure 10-10 indicates the result when the *logarithm* of the amount of material remaining is plotted against time. The curve in Fig. 10-9 and the straight line in Fig. 10-10 are characteristic of a *first-order reaction*.

A first-order reaction is a reaction in which the amount of material destroyed in a given time period, *t*, is directly proportional to the amount of material available at the beginning of the time period. Radioactive decay is the perfect example of a first-order reaction. The actual shape of the curve obtained by plotting a first-order reaction depends on the rate at which the material is being destroyed or removed. This is illustrated by the two curves in Fig. 10-11, which indicates removal rates of 10% every six hours and 6% every six hours, respectively.

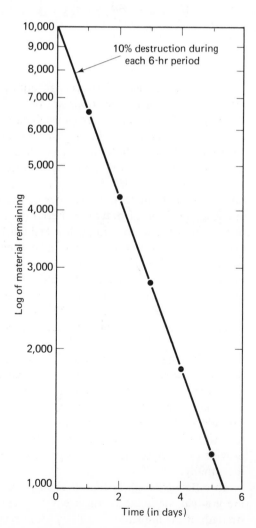

Figure 10-10
Logarithm of Material Remaining
vs. Time. (Courtesy Peabody Welles,
Inc.)

TABLE 10-1. Progressive Daily Decay

Time (days)	Amount of Material Remaining (lbs. BOD)
0	10,000
0.5	8,100
1	6,561
1.5	5,314
2.0	4,305
2.5	3,487
3	2,824
3.5	2,287
4	1,851
4.5	1,498
5	1,213
5.5	982

(Courtesy of Peabody Welles, Inc.)

Both curves indicate first-order reactions, but their shapes differ because the rates at which the reactions are occurring are different. The difference between the first-order curves can be characterized conveniently simply by plotting the logarithm of the amount of material remaining versus time, and determining the slopes of the straight lines thus formed. The slope of each line is a direct indication of the reaction rate. The slope is often designated as the K value, with K referred to as the (first-order) *reaction rate*. Sawyer (2) has pointed out that the kinetics of bacterial degradation are such that the decomposition of organic material (i.e., the pollutant matter in sewage or wastewater) proceeds in such a fashion as to be similar to a first-order reaction. He notes that the deviation between the curve produced by plotting actual organic material remaining versus time differs from a first-order reaction curve because the bacterial population, as well as the amount of organic material present, will vary. That is, there are in fact two factors which can vary, as opposed to the situation in a unimolecular (i.e., completely first-order) reaction where there is only one variable, the amount of material reacting.

The fact that there are two variable factors involved in bacterial degradation does not necessarily preclude the system kinetics from being defined in a manner analogous to a first-order reaction. The reason for this is that, in the case of bacterial stabilization, the bacterial population is a dependent variable. Over a wide range of substrate concentration, the bacterial population will adjust itself in proportion to the amount of substrate available, provided the concentration of organic material available does not fluctuate too rapidly.

In a flow-through system, if changes in influent substrate concentration are gradual, or if the physical environment damps out fluctuations in the strength of the influent waste flow, the effect of the variation of the bacterial

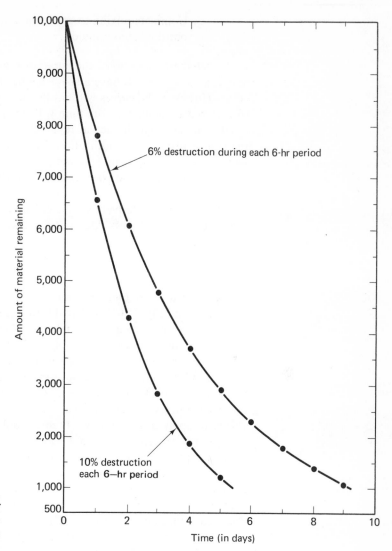

Figure 10-11
Effect of Varying the Rate of
Reaction. (Courtesy Peabody
Welles, Inc.)

population will be minimized. In theory, a point could be reached where
the bacterial population would reflect the substrate concentration so faith-
fully as to in effect represent a system in which there is only one significant
variable—the substrate concentration. This is a somewhat unusual condition,
however, and in practice cannot be realistically achieved. The aerated lagoon
is designed to approach this condition as nearly as practicable by incorpo-
rating the following two features:

1. A relatively long detention time, so that considerable purified
   water is retained to dilute the incoming waste flow and prevent
   rapid, drastic fluctuation of the concentration of the lagoon
   organic material.

2. Complete mixing to maintain a homogeneous situation through-out the lagoon liquid.

Under these conditions, the curve of the actual removal of the organic matter versus time will have a shape similar to that of a first-order reaction curve. However, the curves will not be exactly the same, and a semilog plot of material remaining versus time will not produce a straight line. This is due to the fact that the actual removal rate is decreasing gradually during the course of the waste utilization. Table 10-2, Fig. 10-12, and Fig. 10-13 indicate typical values and reaction rate relationships for a municipal waste-water with no industrial wastes. Note that the semilog plot in Fig. 10-13 is not a straight line.

Fortunately, a relatively simple equation exists which can be applied to this type of curve. Eckenfelder and O'Connor (1) have pointed out that the formula

$$E = \frac{K_t}{1 + K_t} \tag{10-46}$$

can be applied to the biological kinetics of an aerated lagoon, as typified by Table 10-2 and Figs. 10-12 and 10-13. In this equation, E is the BOD removal efficiency, t is the time in days, and K is a BOD removal constant.

For certain types and classes of wastewater, the applicable K value may be known, or it can be estimated with a high degree of confidence on the basis of experience. However, since rational aerated lagoon design is a rela-tively new concept, many wastes are encountered for which K values have not been conclusively determined. Also, it is not uncommon for different K values to apply to the same waste under different circumstances. Municipal sewage at town A may have a K value of 0.4, whereas at another location with an apparently similar waste, a K value of 0.6 might be used successfully.

For this reason, actual testing of the waste involved using rapid, straight-forward, and relatively inexpensive procedures is by far the best means of selecting the K value to be used in a particular design. This is not always done, sometimes due to a false idea of economy. Preliminary pilot studies are normally not difficult to perform. They can be done in a fairly short period of time (on the order of 4–6 weeks) and can produce savings far in excess of their cost. As an example, if 90% BOD reduction is required for a daily waste flow of 5 MGD, use of a K value of 0.50 rather than the correct value of 0.55 will produce a design requiring the excavation and movement of an extra 43,000 cubic yards of earth. It will also entail approximately 2.5 additional acres of lagoon surface area. Obviously, considerable expense can be expended upon pilot-plant studies in order to avoid such incremental construction and land costs.

10-15. *Alpha Factor.* This is somewhat less complex and easier to explain than the K-factor concept. Simply stated, the alpha factor which applies to a particular wastewater is the ratio of the rate of oxygen transfer in the

TABLE 10-2. Destruction of Degradable Organic Material in an Aerated Lagoon

Detention Time (days)	Amount of BOD* Remaining (lb)
0	10,000
1	6,450
2	4,760
3	3,770
4	3,120
5	2,670
6	2,330

*Ultimate BOD.
(Courtesy of Peabody Welles, Inc.)

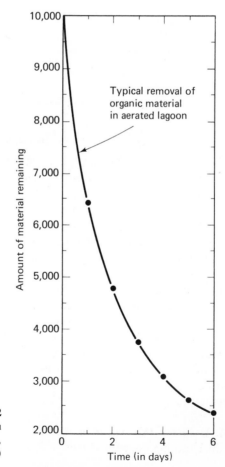

Typical removal of organic material in aerated lagoon

Figure 10-12
Material Remaining vs. Time in Days. (Courtesy Peabody Welles, Inc.)

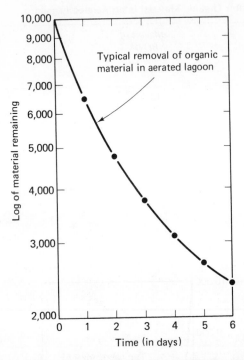

Figure 10-13
Log of Material Remaining vs. Time.
(Courtesy Peabody Welles, Inc.)

wastewater to the rate of oxygen transfer in clean water.

$$\alpha = \frac{\text{rate of oxygen transfer in waste}}{\text{rate of oxygen transfer in water}} \tag{10-47}$$

The discrepancy between the two transfer rates is due to the existence of substances in wastewater (surfactants and other organics) which affect the transfer rate. (*Note*: the rate is normally affected adversely, but *not always*.)

Ford (16) describes the laboratory procedures required to approximate the alpha value as follows: "Samples of distilled water and wastewater can be de-aerated with nitrogen and individually re-aerated by mechanical agitation or diffused aeration. The re-aeration means used in the laboratory should represent the anticipated field system as closely as possible. The increase of dissolved oxygen in the water and wastewater samples can be measured using an oxygen probe and continuous recorder." He notes that the conditions that will exist in actual operation are difficult to reproduce in the laboratory, and he emphasizes that the alpha value obtained in the lab is an approximation of the alpha value to be expected in the field, rather than an exact determination of it. Alpha is used, of course, in Eq. (10-14) which is used to convert from the oxygen transfer rate under standard conditions to the actual transfer rate which will exist in the actual lagoon at temperature t.

10-16. K-Factor Determination. For determining K, it is probably best to start by choosing detention times of 2, 5, 7, and 12 days. Vessels should be set up and should be fed samples of the waste on the basis of these detention times. Aeration should be continuous. COD determinations should be made daily on both the influent and effluent. When the effluent values of each unit have stabilized, i.e., are approximately the same every day, then influent and effluent BODs and CODs obtained after this time will indicate the K rate that applies. Feed should preferably be through low-volume continuous-type pumps, with the feed carefully controlled to give the detention times selected. The fed waste ought to be from one central sample, preferably stored at just about 0°C, warmed to room temperature before feeding, and with the same sample fed to each unit. It is important to see that the test vessels approximate one mixed liquor dissolved oxygen and mixing power level (i.e., 8 HP/mg) to be expected in one aerated lagoon.

10-17. Mechanical Versus Diffused-Air Systems. Diffused-air systems have traditionally been rated and sized on the basis of the amount of *air pumped*, rather than the amount of *oxygen transferred*.

The comparative performance characteristics of floating mechanical aerators and diffused-air systems appear to be with respect to *oxygenation*. Each HP of floating mechanical aerator capacity is equivalent to roughly 34 ft³/min of air from a diffused-air system. With respect to *tank mixing*, each HP of floating mechanical aerator capacity is equivalent to about 48 ft³/min of air supplied by a diffused-air system.

EXTENDED AERATION

10-18. Extended Aeration. Mechanical aerators can be arranged in basins to form regular custom-designed plants using any of the activated-sludge principles. The extended aeration process is probably the most popular and will lend itself better to most applications.

10-19. Design Procedure. The design of an extended aeration system for the stabilization of wastewater involves the use of two basic relationships. These relationships are summarized mathematically by the following two equations. Equation (10-48) defines the degree of BOD removal, and Eq. (10-49) provides a means of predicting the oxygen requirements.

$$L_r = \frac{V \times M \times f \times b \times V_a}{120000 \times a} \tag{10-48}$$

and

$$NOR = \frac{a' \times L_r}{24} + \frac{b' \times M \times V_a}{24} \tag{10-49}$$

The meaning and significance of the terms are explained below, and some of them are illustrated in Figs. 10-14, 10-15, and 10-16.

Tank volume is represented by the V term in Eq. 10-48. It is the volume

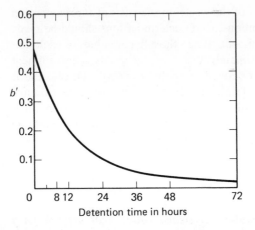

Figure 10-14
Estimation of the b' Factor for Design of Extended Aeration Systems. (Courtesy Peabody Welles, Inc.)

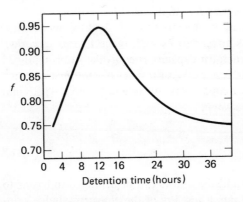

Figure 10-15
Estimation of f-Factor Values for Design of Extended Aeration Systems. (Courtesy Peabody Welles, Inc.)

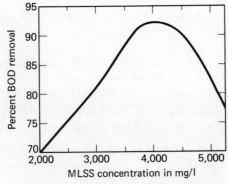

Figure 10-16
Extended Aeration BOD Removal vs. MLSS. (Courtesy Peabody Welles, Inc.)

of the aeration tank, expressed in gallons. Extended aeration systems are very often designed for a 24-hour detention time; in these cases, $V = Q$, where Q is the wastewater flow in gal/day.

L_r indicates the degree of BOD removal which can be expressed in terms of pounds of BOD_5 per day.

M indicates the total amount of mixed liquor suspended solids in the aeration tank. It includes nonvolatile solids and nondegradable volatile solids, as well as the biodegradable cell mass. When used with Eq. (10-48), it is expressed in terms of *milligrams per liter;* when used in Eq. (10-49), it is expressed in terms of *pounds of solids.*

V_a is the ratio of the weight of the volatile mixed liquor suspended solids to the weight of the total mixed liquor suspended solids. That is, it is the percent volatile, expressed as a decimal. For domestic sewage, the V_a value is in the 0.50 to 0.85 range, with 0.7 being an acceptable general value.

f is used to represent the biodegradable portion of the mixed liquor volatile solids. For domestic sewage, the f value is generally in the 0.75 to 0.95 range, depending upon the duration of the aeration period. An f value of 0.90 is normally applicable when the aeration tank detention time is 16–24 hours, and smaller f values are generally associated with either very short or rather long detention times.

b is the endogenous respiration factor. It indicates the fraction of the mixed liquor volatile suspended solids (MLVSS) which is destroyed per day as a result of endogenous respiration. It is not constant, and it often decreases very rapidly when the aeration detention time exceeds 5–7 days. For an extended aeration system operating with a theoretical detention time of 24 hours, the applicable b value should fall in the 0.08–0.20 range. A value of 0.145 is a satisfactory general value to use; selecting a value of $b = 0.10$ would produce a fairly conservative and therefore "safer" design.

a is the fraction of the BOD$_5$ removed per day (in pounds) which is converted to cell mass. The separate a value normally ranges from 0.5 to 0.75, although values as high as 0.93 have been reported (for brewery wastes). For a heterogenous (i.e., mixed and particle-bearing) waste such as domestic sewage an a value of 0.65 is generally applicable, while for a completely soluble, low-carbohydrate waste, an a value of 0.55 would probably be correct.

b' indicates the amount of oxygen that is required for endogenous respiration. It is expressed in terms of pounds of oxygen per day per pound of MLVSS, that is,

$$b' = \frac{\text{lb O}_2/\text{day}}{\text{lb of volatile suspended solids}} \qquad (10\text{-}50)$$

The b' factor varies according to the detention time in the aeration tank. For relatively short (i.e., 6–8 hours) detention times the b' factor should fall somewhere in the 0.24–0.33 range; for systems with long detention times (30–36 hours), the b' factor will more probably have a value somewhere between 0.10 and 0.04.

a' represents the amount of oxygen required to provide the necessary energy for synthesis of cell materials from the organics originally present, that is,

$$a' = \frac{\text{lb O}_2 \text{ required}}{\text{lb BOD}_5 \text{ removed}} \qquad (10\text{-}51)$$

with a' being expressed on a per-day basis. The a' value can vary from 0.3 to 0.8 according to the nature of the pollutant organics; it is usually 0.45–0.55 for domestic wastes.

Once the appropriate factors and values have been determined or assumed, Eq. (10-48) and Eq. (10-49) are used to determine the aeration tank volume and the aeration equipment required. The following extended aeration design example illustrates how this is done.

There is a second and equally important component of the complete extended aeration system. This component is, of course, the *settling chamber*, which is also known as the *final tank* or the *secondary clarifier*. Use of these terms in conjunction with an extended aeration system is a carryover from conventional activated-sludge plant design.

The settling tank provides a quiescent holding period during which the flocculated bacteria settle out of the carrier water. Since essentially all of the organics originally present in the wastewater have been assimilated by the bacteria by this time, removal of the particulate material via settling results in a clarified and nuisance-free effluent water. The purified carrier water is therefore suitable for discharge from the system.

Clarifiers are discussed elsewhere in the book and will not be repeated here. The surface settling rate should not exceed 600 gal/ft²/day. The detention time should be 3 hours. The system is illustrated in Fig. 10-17.

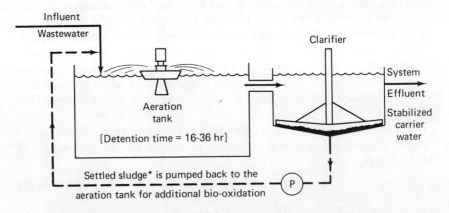

Figure 10-17. Extended Aeration System. (Courtesy Peabody Welles, Inc.)

10-20. *Characteristics of Extended Aeration Treatment*

1. Use the same approach and the same basic principles as for aerated lagoon.

2. It is a *two-stop process* (aeration and settling), as opposed to an aerated lagoon, which is a one-step process (aeration alone).

3. Effectiveness depends upon good flocculation and settling of biosolids.

4. Optimum tank volume depends upon biodegradability of the particular organic materials involved.

5. The *liquid* detention time is much less than the detention time in an aerated lagoon.

6. The aeration chamber solids content is much greater than the suspended solids concentration in an aerated lagoon.

7. *All* aeration tank solids are kept in suspension:

$$P/V = 75 - 100 \text{ HP/million gallons} \qquad (10\text{-}52)$$

8. Dual oxygen requirement—bio-oxidation and sludge respiration.

9. Requires significantly more operation and maintenance than an aerated lagoon.

10. More susceptible than aerated lagoon to biological upset due to shock loading, "slugs," etc.

11. Periodic separate disposal of accumulated sludge solids is sometimes required.

12. This is a custom-designed plant, not a package plant.

10-21. Extended Aeration Design Example

a. *Objective.* It is desired to design an extended aeration system to serve a 400-house subdivision.

b. *Data.* The engineers have provided the following information and data.

(1) The design population is 1,480 persons. This is based upon a 3.7 person per dwelling estimate.

(2) The design flow is estimated at 115 gallons per capita, giving a total daily flow of 170,000 gallons per day, i.e., 0.17 million gallons/day.

(3) The five-day BOD is estimated at 200 mg/l (or 200 ppm, since ppm is essentially the same as mg/l). Therefore the daily BOD loading in lb is

$$\begin{aligned} \text{BOD}_5 \text{ (lb/day)} &= \text{flow (MGD)} \times \text{BOD}_5 \text{ (ppm)} \times 8.345 \\ &= 0.17 \times 200 \times 8.345 = 284 \text{ lb/day of BOD}_5 \end{aligned} \qquad (10\text{-}53)$$

(Note that this is equivalent to 0.192 lb of BOD_5 per capita. The BOD_5 loading in the past was often estimated at 0.17 lb/capita; however, there is a current trend toward the use of values as high as 0.22. The BOD loading in this case therefore appears reasonable.)

(4) There are to be no nondomestic wastes in the system. Therefore a *bottle BOD K* of 0.10 is inferred, since this is the most commonly reported

K value for domestic wastewater. At $K = 0.10$, approximately 68% of the total oxygen demand (as determined by the BOD test) is exerted during the first five days; therefore, $BOD_L = BOD_5/0.68$ and

$$L = \frac{BOD_L}{BOD_s} = \frac{1}{0.68} = 1.47 \qquad (10\text{-}54)$$

(*Note*: The significance and importance of determining the correct and applicable bottle BOD K value is often overlooked. When one considers that any rational design must involve the total oxygen demand rather than the five-day demand, the desirability of at least confirming the assumed K value is obvious. As an example, an organic material with an applicable K (bottle BOD) rate of 0.1 which has a BOD_5 value of 200 ppm will exert a *total oxygen demand* of 294 mg/l for complete biological stabilization. On the other hand, an organic which has a K value of 0.15 and a BOD_5 of 200 ppm will exert a total oxygen demand of only 243 mg/l for complete biodestruction. The K and L values can vary over a rather wide range, particularly when industrial wastes are involved. It is therefore extremely desirable, and at times absolutely imperative, to accurately determine the correct values.)

(5) The desired BOD removal is 85%.

c. *Design Parameters.* No additional information is available, since the consultant feels that the various factors and values generally associated with domestic-municipal sewage will apply to the wastewater in question. Therefore, the next step is to select the remaining factors and values on the basis of prior experience and the reported findings of research workers.

At 15°C

$$t_{-20} = 1.075^{(15-20)} = 0.64797 \qquad (10\text{-}55)$$

therefore

$$K_{15} = K_{20} \times 0.69656 = 0.383 \qquad (10\text{-}56)$$

and similarly,

$$K_{24} = K_{20} \times 1.33546 = 0.7345 \qquad (10\text{-}57)$$

The required lagoon detention time will be

$$T = \frac{0.85}{0.15 \times K_{14}} = 14.8 \text{ days} \qquad (10\text{-}58)$$

The maximum removal efficiency at $T = 14.8$ days is

$$E = \frac{K_{24}T}{1 + K_{24}T} = \frac{0.7345 \times 14.8}{0.7345 \times 14.8 + 1} = 91.6 \qquad (10\text{-}59)$$

The lagoon volume required is

$$14.8 \text{ days} \times 0.17 \text{ MGD} = 2.52 \text{ million gallons} \qquad (10\text{-}60)$$

The maximum *NOR* value will occur at the time of maximum removal; thus

$$P = 0.916 \quad \text{and} \quad D = \frac{292}{24}$$

and

$$NOR = 1.47 \times \frac{292}{24} \times 0.916 = 16.4 \text{ lb/O}_2\text{/hr} \qquad (10\text{-}61)$$

For a 10-HP Aqua-Lator unit with a standard oxygen transfer rate (R_a) of 3.0 lb/HP-hr, the actual transfer under the critical (i.e., summer) condition will be

$$R = 3.0\frac{(8.5 - 1.5)}{9.17} \times 0.9 \times 1.102 \qquad (10\text{-}62)$$
$$= 2.27 \text{ lb of oxygen per HP/hr}$$

d. Design Calculations. Equation (10-47) is first used to determine the aeration tank volume. Rearranging the equation, we have

$$V = \frac{L_r \times 120,000 \times a}{M \times f \times b \times V_a} \qquad (10\text{-}63)$$
$$= \frac{250 \times 120,000 \times 0.65}{3,500 \times 0.80 \times 0.10 \times 0.65} = 107,143 \text{ gal} \qquad (10\text{-}64)$$

This means that the detention time will be

$$\frac{107,143 \text{ gal}}{170,000 \text{ GPD}} = 0.63 \text{ days} = 15 \text{ hr} \qquad (10\text{-}65)$$

which means that the BOD loading is 20.4 lb of BOD_5 applied per day per 1,000 ft³ of aeration capacity. This is an acceptable figure, since permissible loadings in treating domestic-municipal sewage have been reported to be in the range between 12.5 to 26 lb of BOD_5 per 1000 ft³ per day. For soluble wastes, loadings of up to 40 lb BOD/1000 ft³ have been suggested.

Equation (10-48) is next used to determine the aeration equipment requirements. Remembering that the applicable L_r here is 264 lb we have

$$NOR = \frac{0.50 \times 264}{24} + \frac{0.15 \times 3129 \times 0.65}{24} \qquad (10\text{-}66)$$
$$= 18.2 \text{ lb O}_2\text{/hr}$$

(Note that M is expressed in pounds of solids (under air). This figure is arrived at by considering the tank volume and the MLSS concentration; thus

$$0.107143 \times 8.345 \times 3500 = 3129 \text{ lb} \qquad (10\text{-}67)$$

Assuming temporarily that the *actual* oxygen transfer rate (i.e., after adjusting to operating conditions) will be 2 lb/HP-hr, one 10-HP Aqua-Lator unit appears to be required. The R_a value for the 10-HP Aqua-Lator unit is 3.0; therefore, using a 10-HP Aqua-Lator unit and assuming a minimum D.O. concentration of 1.5 mg/l, we have

$$R_s = 3.0 \frac{(9.2 - 1.5)}{9.17} \times 0.9 \times 1.0 \qquad (10\text{-}68)$$
$$= 2.27 \text{ lb. of oxygen transferred per nameplate horsepower per hour}$$
$$\text{of } \textit{power consumed} \text{ under actual operating conditions}$$

One 10-HP Aqua-Lator unit will therefore provide

$$10 \times 2.27 = 22.7 \text{ lb } O_2/\text{hr} \tag{10-69}$$

This is equal to the *NOR* value. The extended aeration system should therefore have a 107,143-gallon aeration tank utilizing one 10-HP Aqua-Lator unit. Complete suspension of solids will be achieved in this system.

PROBLEMS

Listed below are five sewage treatment plants located at 42°N, sea-level elevation, near the east coast of the United States. When the necessary data cannot be derived from the facts given, make the necessary assumptions in order to complete the problem. Explain why and on what basis each assumption is made.

Plant	Flow MGD
AA	0.1
A	0.5
B	1.0
C	10.0
D	100.0

1. Assume a domestic waste from a community using garbage grinders. A treatment of 90% BOD removal is required during the summer and 85% during the winter. Land cost is low and adequate land is available. A minimum lagoon temperature of 1°C and a maximum lagoon water temperature of 30°C have already been determined. Design a single-cell aerated lagoon for each plant size indicated above.

2. Design an extended aeration plant in each case instead of an aerated lagoon for the cases indicated in Problem 1.

REFERENCES

1. ECKENFELDER, W. W., and O'CONNOR, D. J., *Biological Waste Treatment*. Pergamon Press, Oxford, 1961.

2. SAWYER, C. N., *Chemistry for Sanitary Engineers*, McGraw-Hill, New York, 1960.

3. ECKENFELDER, W. W., *Water Quality Engineering for Practicing Engineers*. Barnes & Noble, Inc., New York, 1970.

4. ECKENFELDER, W. W., "Design and Performance of Aerated Lagoons for Pulp and Paper Waste Treatment." Industrial Waste Conference, 1961, Engineering Extension Series No. 109, p. 115, Purdue University, Lafayette, Indiana.

5. O'CONNOR, D. J., and ECKENFELDER, W. W., "Treatment of Organic Wastes in Aerated Lagoon." *Water Poll. Control Jour.* (April, 1960).

6. MANCY, K. H., and OKUN, D., "The Effects of Surface Active Agents on Aeration." *Jour. WPCF*, 37 (1965).

7. DOBBINS, W. E., "Mechanism of Gas Absorption by Turbulent Liquids," in *Advances in Water Pollution Research*, Vol. 2. Pergamon Press, Oxford, 1964, p. 61.

8. LEWIS, W. K., and WHITMAN, W. G., "Principles of Gas Absorption." *Ind. Eng. Chem.*, 16, 1215 (1924).

9. KNOP, E. H. E., *Versuche mit Verschiedenen Beluftungssystemen im technischen Massstab, Emschergenossenschaft*. Essen, West Germany, 1964.

10. CONWAY, R. A., and KUMKE, G. W., "Field Techniques for Evaluating Aerators." *Proc. ASCE*, 92, SA2, Paper 4757, 4, 21 (1966).

11. BENJES, H. H., and McKINNEY, ROSS E., "Specifying and Evaluating Aeration Equipment." *Proc. ASCE*, 93, SA6, 12, 55 (1967).

12. PFEFFER, JOHN T., HART, FRED C., and SCHMID, LAWRENCE A., "Field Evaluation of Aerators in Activated Sludge Systems," Vol. 1. Industrial Waste Conference, 1968, Engineering Extension Series No. 132, p. 183, Purdue University, Lafayette, Indiana.

13. BEYCHOK, MILTON R., *Aqueous Wastes from Petroleum and Petrochemical Plants*, 1967.

14. *Biological Treatment of Petroleum Refinery Wastes*. American Petroleum Institute, New York.

15. PURSELL, W. L., and MILLER, R. B., "Waste Treatment of Skelly Oil Company's El Dorado, Kansas Refinery." Industrial Waste Conference, 1961, Engineering Extension Series No. 109, 292, Purdue University, Lafayette, Indiana.

16. FORD, DAVIS, and GLOYNA, EARNEST F., "Development of Biological Treatment Data for Chemical Wastes." Twenty-Second Industrial Conference, Purdue University, May 1967.

17. ECKENFELDER, W. W., and FORD, D. L., *Advances in Water Quality Improvement*, Vol. 1. University of Texas Press, Austin, Texas, 1967.

11 *Advanced Waste Treatment*

Advanced waste treatment may be physical, biological, chemical, or a combination of these processes. Tertiary implies a single treatment process subsequent to secondary treatment; hence advanced waste treatment is a more extensive term. Some of the advanced waste treatment processes can accept raw sewage, others are capable of accepting sewage after primary treatment, and still others are designed to reduce the concentration of undesirable components from secondary waste treatment processes. It seems reasonably well established that biological secondary treatment following primary treatment is and will continue to be the most economical and most satisfactory means of processing wastewater up to the quality that these processes are capable of producing. Thus, unless otherwise indicated, this chapter assumes that the waste to be further treated is a domestic wastewater effluent from a secondary treatment process.

Even when additional treatment is not presently required, the engineer should give thought in secondary treatment plant designs as to how advanced waste treatment provisions can be provided in the future. Advanced waste treatment may in many cases be handled in a comparatively simple manner. Spray irrigation of the waste, or a simple arrangement for a microstrainer, or diatomaceous earth filtration, or filtration by a dual-media filter may be all that is necessary to provide for the treatment required. In most of the advanced waste treatment processes, the presence of solids is undesirable and one of the above schemes, or modifications thereof, should be given consideration ahead of the more exotic units.

11-1. Ultimate Disposal. Ejecting wastes into space for conveyance into the sun would dispose of the waste permanently but it is not economically feasible for any waste product within the scope of this text and may very likely never be feasible for any waste. This leaves us the atmosphere, freshwater sources, the oceans, land surface, land burial, underground injection, and reuse as potential disposal methods. For incineration and atmospheric disposal, it appears that a chain reaction of problems may be initiated that in the end will be greater than the benefits from the original disposal system. It is the author's opinion based on all available information that ultimate disposal methods not involving the atmosphere should be considered.

When we begin to consider future population growth, pollution generated per capita, and the results of pollution that exist today, several additional conclusions begin to emerge. Dilution by our freshwaters and the oceans cannot meet the increase in wastes that is coming. It is therefore essential that we reduce, not increase, the waste quantities imposed upon these disposal sources.

All treatment plants show variation in BOD and settleable solids removal when viewed over a long time. It can be expected that there will be a steady trend toward more and more automation of operating control, and design should give consideration to systems that will be amenable to future automation. The first line of attack for ultimate disposal should be reuse. The end product sold by the treatment plant will in most cases probably be sold at a distinct loss; however, any return realized will assist in reducing the cost of the treatment.

11-2. Spray Irrigation. Sewage irrigation was discussed briefly in Chapter 4 and references were provided on a number of locations where it has been practiced. This section offers more specific suggestions on how to design spray irrigation systems.

The successful use of spray irrigation requires a careful investigation of the proposed disposal site. A variety of factors affect spray field operation. These include the slope of the surface, application rate, surface soil permeability, subsurface soil characteristics, type of cover vegetation, period of time of application, total quantity of waste applied during one time period, the chemical composition of the soil, and the characteristics of the waste applied. A number of spray irrigation projects have been designed to handle canning, food processing, and other industrial wastes. Most of these systems have been designed to take a waste after primary treatment.

Insofar as the vegetation itself is concerned, there is a limit on the amount of water that can be applied to most plants and still result in beneficial growth. Figure 11-1 is an example of what happens to the growth of one type of grass when increased amounts of water are applied. If the figure had been extended, it would have been found that at an application rate of 100 in. of water annually, the same production would result as with 5 in. annual irrigation. Forests having good hydrologic floor covering can receive large quantities of spray irrigation. It is advisable to consult an agricultural specialist either

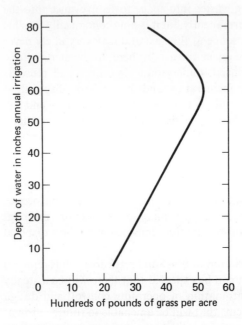

Figure 11-1
Grass Production Variation with
Water Supplied.

from the county extension agents office, or from the state agricultural college, on the soils of the tract of land to be sprayed and the vegetative cover existing on it. It may be that the soil is unsuited for spray irrigation, or that the existing cover must be replaced, or some form of chemical treatment such as liming may be required periodically. The agricultural specialist can assist you in arriving at a suitable application rate that the vegetation can take without being killed.

a. Spray Irrigation Treatment Efficiency. A spray irrigation system is the final product of the design engineer, and the results will depend on the skill put into the design. Two tables have been created to indicate a form of guide. Data are not available on a national basis, and these tables must be used with substantial discretion. Results exceeding the values in the "excellent" columns have been achieved or exceeded in practice in some sections of the United States. Table 11-1 indicates the percentages of reduction on the concentration basis which is a rather poor method to use in association with the system. In actual practice, only a portion of the waste will become runoff from the watershed. Therefore, Table 11-2 indicating the mass removal basis is more appropriate. The figures apply to the waste as applied; thus a secondary effluent properly applied would be significantly reduced further, and microscreening or other filtration is not necessary.

It is recommended that the effluent be heavily chlorinated, retained in a contact chamber for a period of at least two hours, preferably longer, and then sprayed.

b. Spray Irrigation Design. Figure 11-1 was used to make the designer

TABLE 11-1. Treatment Performance on a Concentration Basis

Pollution Parameter	Excellent % Reduction	Good % Reduction	Poor % Reduction
BOD	75	65	55
COD	80	70	60
Suspended solids	85	80	75
Total N	70	60	50
Phosphates	45	40	35
TVS	30+	20	10
TOC	80	70	60

TABLE 11-2. Treatment Performance on a Mass Removal Basis

Pollution Parameter	Excellent % Reduction	Good % Reduction	Poor % Reduction
BOD	90	80	75
COD	95	85	75
Suspended solids	95	90	80
Total N	90	85	75
Phosphates	85	75	70

aware of optimum application rates if crops were being grown for profit in conjunction with the irrigation system. A soil-water analysis should be made as a part of the work discussed with the agricultural specialist. This would indicate how heavy a concentration of certain dissolved solids exists in the soil and provide a basis for the increase rate of concentration of these solids as the system is used. High concentration of chlorides and methyl orange alkalinity, as an example, may discourage the use of the spray irrigation system in a particular locality.

The area to be sprayed should be divided into five sections and the piping and valves arranged in a manner that one section is dosed heavily one day at a time in sequence. Thus, each section has four days without waste being applied. During this idle time, bacterial and other microbial action takes place that aids in the treatment process and also keeps down the buildup of undesirable dissolved solids in the soil.

The spray system should be of the medium-pressure type in order to operate most economically by keeping pumping costs down. From an application standpoint, a fog-type spray that does not strike the surface with significant force is desirable. Spray nozzles must be of a type that will minimize clogging from the solids in the waste.

Around a roadside rest area, in portions of a recreation area, or in any location where sprays must be operated in proximity to the public, it may be necessary to restrict the spray distance. The soils and vegetation may also restrict the application rate. If a sufficiently treated effluent is used, particularly one that has passed through a sand filter, the Rain-bird 35PJ-TNT

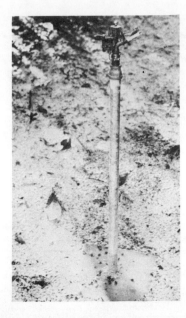

Figure 11-2
Typical Mounting of Sprinkler
Head.

special sprinkler with tamper-proof friction collar might be used. Figure 11-2 indicates the mounting of such a head and Table 11-3 shows the performance of such a head using the 7/32-in. nozzle. The highest point of the stream is 9 ft above the nozzle.

If the soil will readily take a higher application rate, considering the vegetation cover, then a Rain Bird 65D-DA-TNT can be used in tight spray-pattern areas. These sprinklers can be used for either full-circle operation or part-circle operation. When control of the spray distance is not critical, the 65D-TNT may be found suitable. The performance is indicated in Table 11-4. Figure 11-3 incidates a half-circle sprinkler head operating with sewage effluent.

In a larger plant it is desirable to use larger heads if the soils and vegetation will permit. The 70EW is shown operating in Fig. 11-4 and its perfor-

TABLE 11-3. Performance of 35PJ-TNT Special Head

PSI	Radius, Ft	GPM
30	48	7.58
35	49	8.26
40	50	8.87
45	51	9.41
50	52	9.88
55	53	10.30
60	54	10.60

Chart shows nozzles at maximum throw. Highest point of stream is 9 ft above nozzle, at normal operating temperature. Best distribution is obtained at pressures above 45 psi. (Courtesy of Rain Bird.)

TABLE 11-4. Performance of 65D-TNT Sprinklers

PSI Nozzle Pressure	Nozzle 7/32" Radius (Ft)	Discharge GPM	†Nozzle 1/4" Radius (Ft)	Discharge GPM	Nozzle 9/32" Radius (Ft)	Discharge GPM	Nozzle 5/16" Radius (Ft)	Discharge GPM	Nozzle 11/32" Radius (Ft)	Discharge GPM	Nozzle 3/8" Radius (Ft)	Discharge GPM
40	51	8.8	54	11.5	57	14.6	59	17.7	60	21.1	62	24.4
45	53	9.4	55	12.2	58	15.5	61	18.9	63	22.5	65	26.0
50	55	9.9	57	12.9	60	16.3	63	20.0	65	23.8	67	27.5
55	56	10.4	57	13.6	62	17.2	65	21.0	67	25.0	70	29.1
60	58	10.9	58	14.2	65	18.0	67	22.0	69	26.2	73	30.6
65	59	11.4	62	14.8	67	18.8	69	23.0	70	27.4	74	32.0
70	60	11.8	63	15.4	70	19.5	71	23.9	72	28.5	75	33.2
75	62	12.2	64	16.0	71	20.3	73	24.8	74	29.6	76	34.5
80	63	12.6	65	16.5	72	20.9	75	25.7	76	30.6	77	35.7

Highest point of stream is 10 ft above nozzle.
Shaded areas in chart indicate working pressures not recommended for best distribution.
(Courtesy of Rain Bird.)

Figure 11-3. Half-Circle Sprinkler Head Spraying Sewage Effluent.

mance is shown in Table 11-5 for the larger nozzle sizes suggested for sewage effluent service.

Stump holes or other depressions that will result in ponding should be filled or fixed so that they will drain. Otherwise mosquito breeding may be stimulated and become a problem. Locate the spray heads to secure a reasonably uniform distribution. Use a spacing of 60% of sprinkler spray pattern, up to 8–10 mi/hr winds, and 50% and less of diameter in winds above 10 mi/hr.

Waste taken directly from primary treatment plants should have sprin-

Figure 11-4. Full-Circle Sprinkler Heads Operating in Spray Field.

TABLE 11-5. Performance of 70EW Sprinklers

PSI	Nozzle 5/16″ Dia.	GPM	Nozzle 11/32″ Dia.	GPM	Nozzle 3/8″ Dia.	GPM
40	132	17.7	136	21.1	145	24.4
45	136	18.9	142	22.5	151	26.0
50	140	20.0	148	23.8	157	27.5
55	144	21.0	152	25.0	161	29.1
60	148	22.0	156	26.2	165	30.6
65	152	23.0	160	27.4	167	32.0
70	155	23.9	164	28.5	173	33.2
75	158	24.8	168	29.6	177	34.5
80	161	25.7	172	30.6	181	35.7

Shaded areas in chart indicate working pressure not recommended for best distribution. (Courtesy of Rain Bird.)

klers with nozzles that have at least a $\frac{1}{2}$-in. orifice. A Rain Bird 80EW-TNT with $\frac{5}{8}$-in. nozzle would deliver 87.8 gal/min over a diameter of 208 ft at 70 psi. Sprays from primary treatment should be at least 150 ft from the protective fence line to avoid liquid spray contact with the public.

It is possible to produce an effluent (1) that may be sprayed with very little hazard to the public even if children dash through the spray. The effluent from a well-designed oxidation ditch, properly operated, passed through a 60-in. deep sand filter, and chlorinated at 2–3 ppm free residual for not less than 2 hours' contact time should be more harmless to the public than waters approved by most states for public bathing.

Maximum precipitation rates to use on level ground depend upon the type of soil. Light, sandy soils vary from 0.75 to 0.5 in./hr, medium-textured soils 0.5 to 0.25 in./hr, and heavy-textured soils 0.25 to 0.10 in./hr. The allowable rates increase with adequate cover and decrease with land slopes and time. The turf requires one inch of water per week for healthy growth; however, it can of course receive a much larger quantity of water. The exact amount depends upon the species of grass used. You can apply a maximum of 6,787 gal/hr/acre or approximately 6.45 gal/ft²/hr. It is suggested that a design figure of 153,600 gal/acre/24 hr excluding evaporation losses would be a good working figure on turf.

The primary advantage of spray irrigation is that it can be used on any slope or configuration of terrain. It can also be used inside a forest.

11-3. Ridge-and-Furrow Liquid Waste Disposal. There is no reason that ridge-and-furrow as well as spray irrigation cannot be used as a means of wastewater effluent disposal. There is evidence in the literature that it is an acceptable means of disposal (2, 3, 4, 5, 6).

Bendixen (2) concluded that a ridge-and-furrow method of applying liquid waste to land offers the advantage over spray or flood application of longer life before remedial measures are required. He also concluded that no major differences in waste treatment exist from using the three common methods of irrigation and that a vegetated land disposal system provides a high degree of nutrient removal.

Another mode of disposal is to use a seepage lagoon such as that shown in Fig. 11-5. This works particularly well in deep sand. It is necessary to have two lagoons. Use one lagoon at a time. When the seepage rate of the lagoon begins to slow down, start using the second lagoon. When the first lagoon is empty and dry enough, use a farm tool to disc the bottom coating into the sand, and the lagoon can then be reused at original capacity. It is recommended that properly chlorinated high-quality effluents be used.

11-4. Microstraining. The term *microstraining* is a registered trademark of the Crane Co. It is one form of an extremely fine screening device that can be kept clean by periodic backwashing. There are other fine screens on the market, but this discussion will be centered around the Crane unit.

There are several reports in the literature that provide an indication of what microstraining units can be expected to accomplish in polishing the

Figure 11-5. Seepage Lagoon.

effluent of a secondary treatment plant. In one case, a microstraining unit is reported (7) to have removed 38 % of the five-day BOD and somewhat more than 55 % of the suspended solids from trickling filter effluent. Evans (8) reported on a microstrainer unit at Luton, England, which for the poorest year of operation reduced the BOD from 10.7 mg/l to 7.4 mg/l in the effluent (21.5 % reduction) and SS from 17 mg/l to 8.9 mg/l (47.6 % reduction). Another report (9) indicated an influent BOD of 18.5 mg/l and an effluent of 12.4 mg/l (33 % reduction) and suspended solids of 14.7–16.2 mg/l to 6.6 mg/l (59 % removal).

There are several grades of microstraining system fabrics available on the market. One of the finest mesh has approximately 160,000 apertures per square inch, the aperature size being approximately 23 microns. Table 11-6 indicates data pertaining to using two types of fabrics with several different sizes of microstraining equipment. Figure 11-6 shows a cutaway view of a microstraining unit in operation. Figure 11-7 amplifies a portion of the cutaway view in Fig. 11-6 and contains notations that will assist in understanding the operation. Figure 11-8 is an end view showing the general location of components and differential head during operation. Figure 11-9 indicates the differential head across drum fabric of 3 to 6 in. Loss due to valves, and similar factors is 6 to 12 in. Total loss is 12 to 18 in. which avoids pumping in most cases for flow through the unit. Figure 11-10 shows four standard drum sizes with motor HP and flow capacities in MGD. (Flow capacity depends on fabric and application.) Figure 11-11 indicates the relation of drum sizes to tank sizes in the Crane equipment. In a personal letter from the Cochrane Division of Crane Co., Mr. E. W. J. Diaper discussed wash water:

"Percentage of microstrainer wash water discharged to sewer depends on the flow filtered. The amount pumped for each machine is constant. Experi-

TABLE 11-6. Microstraining Unit Fabric Performance

Size of Unit	Aperture of Fabric (microns)	Max Flow (GPM)	% Solids Removal
5 × 1	23	70	70–80
	35	100	60–70
5 × 3	23	240	70–80
	35	360	60–70
7.5 × 5	23	700	70–80
	35	1,500	60–70
10 × 10	23	1,800	70–80
	35	2,800	60–70
12.5 × 30	23	10,400	70–80

Percentage of solids removal is applicable to an efficiently treated secondary effluent with solids ranging between 20 and 30 ppm. Higher solids loading will result in decreased flow capability. In tertiary sewage treatment, a microstraining unit is generally restricted to an effluent with suspended solids below 100 ppm. (Courtesy of Cochrane Division, Crane Co.)

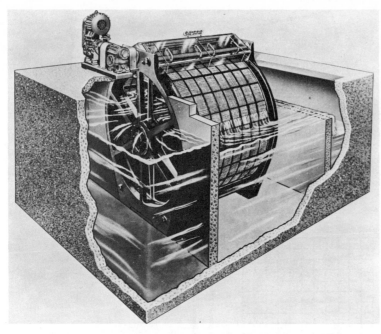

Figure 11-6. Microstraining Unit in Operation. (Courtesy Cochrane Division—Crane Co.)

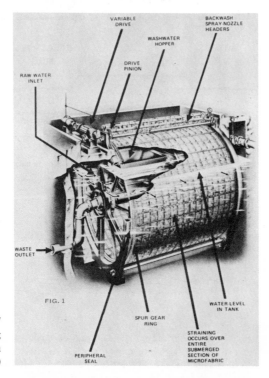

Figure 11-7
Amplified View of Microstraining Unit. (Courtesy Cochrane Division —Crane Co.)

Overflow Weirs
Continuous Wash
WASTE HOPPER
DIFFERENTIAL HEAD
COMBINED MAIN SHAFT & WASTE PIPE
FABRIC

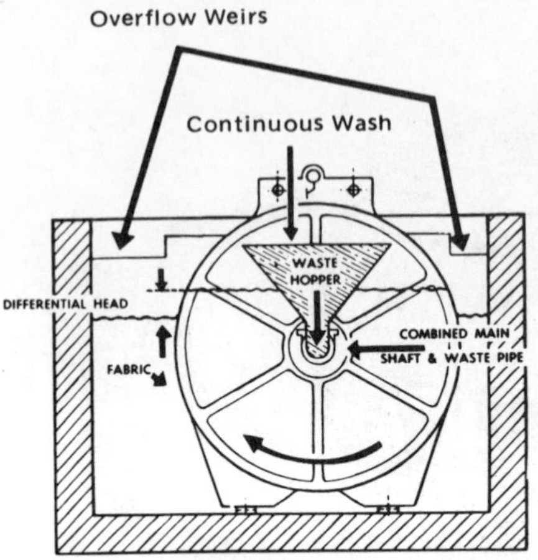

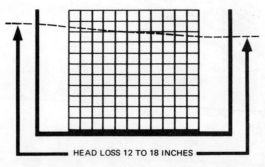

End view showing general location of components and differential head during operation.

Figure 11-8
Component Location and Differential Head During Operation. (Courtesy Cochrane Division—Crane Co.)

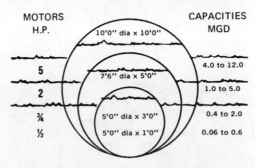

HEAD LOSS 12 TO 18 INCHES

Figure 11-9
Head Loss of Microstraining Unit. (Courtesy Cochrane Division—Crane Co.)

MOTORS
H.P.
CAPACITIES
MGD
10'0" dia x 10'0"
5
4.0 to 12.0
7'6" dia x 5'0"
2
1.0 to 5.0
¾
5'0" dia x 3'0"
0.4 to 2.0
½
5'0" dia x 1'0"
0.06 to 0.6

Figure 11-10
Motor HP and Capacity Chart for Microstraining Units. (Courtesy Cochrane Division—Crane Co.)

DRUM–5'x1"
DRUM–5'x3"
DRUM–7'6"x5'
DRUM–10'x10'
TANK- 5'6"x8'0"x6'0"
TANK- 5'10"x9'0"x14'0"
TANK–7'0"x11'0"x16'0"
TANK–10'2"x14'0"x22'0"

Figure 11-11. Space Requirements for Microstraining Units. (Courtesy Cochrane Division—Crane Co.)

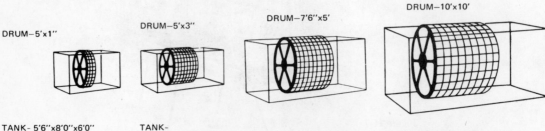

TABLE 11-7. Estimated Wash Water Wasted

Size of Unit	GPM Wasted
5 × 1	3
5 × 3	12
7.5 × 5	16
10 × 10	32

Courtesy of Cochrane Division, Crane Co.

ence shows that only half of the amount pumped is wasted because the remainder runs down the outside of the drum and continues into the supply."

The estimated wash water wasted for four different sizes of machines is indicated in Table 11-7.

11-5. Diatomite Filtration. The information that follows has been extracted and condensed from publications of Johns-Manville Products Corporation (11, 12).

Diatomite is composed of fossil-like skeletons of microscopic water plants called diatoms, ranging in dimensions from under 5 to over 100 microns. Under favorable conditions of light, temperature, and nutrition, such plants grow in great profusion. During the geological past, many deposits of these plant skeletons were built up in different areas of the earth.

Each skeleton is an extremely porous framework of nearly pure silica. Characteristic of diatoms is the diversity of shapes in which they occur. More than 10,000 kinds of diatoms have been identified and catalogued. Not all of these are found in every deposit, for the conditions under which a deposit is formed, namely, whether it is of marine or freshwater origin, of recent or some other geological age, etc., will be reflected in the types of diatoms found. Figure 11-12 shows a photomicrograph of some typical diatoms.

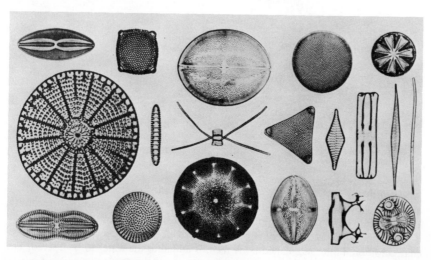

Figure 11-12. Photomicrograph of Some Typical Diatoms. (Courtesy Johns-Manville Products Corp.)

Figure 11-13
Comparison of Coarse
Diatomite and Sand.
(Courtesy Johns-
Manville Products
Corporation)

a. Diatom Size Compared to Sand. The basic difference between a rapid sand filter and a diatomite filter is the difference in the two materials themselves. Figure 11-13 compares photomicrographs of Celite 545 and filter sand at identical magnification. Celite 545 is the coarsest grade of diatomite filter aid used to filter water. The filter sand shown has an effective size of approximately 0.45 mm and a uniformity coefficient of about 1.6.

b. Filtration Cycle. The diatomite filtration cycle consists of the following steps: (a) precoat, (b) filtration and body feed addition, and (c) removal of the filter cake.

c. Precoat. In precoating, filtered liquid is used to deposit a thin layer of filter aid on a filter septum. The amount of precoat should be about 10 lb/100 ft^2 of filter area. This corresponds to a thickness of about $\frac{1}{16}$ in. As can be seen from Fig. 11-14, the filter-aid particles bridge the openings of the septum (filter cloth) and form a microscopically fine sieve, much finer than the septum itself could provide. Consequently, the filter-aid cake is now the filter medium, and the septum serves only as a support for the filter aid. Before the bridges form, however, a slight amount of filter aid may bleed through the septum. For this reason, flow during the precoating operation is either recirculated or run to waste.

The precoat serves two purposes: (a) it forms the actual filter medium and (b) it prevents the filter septum from being fouled by the solid impurities in the water, and thus materially decreases the need for expensive chemical cleaning or replacement of the septa.

The flow rate during the precoating operation should be at least as great as the filtering rate, but no less than 1 gal/ft^2/min.

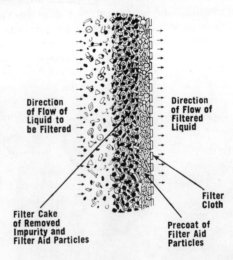

Direction of Flow of Liquid to be Filtered

Direction of Flow of Filtered Liquid

Filter Cake of Removed Impurity and Filter Aid Particles

Filter Cloth

Precoat of Filter Aid Particles

Figure 11-14
Diatomite Filtration. (Courtesy Johns-Manville Products Corporation)

d. Filtration and Body-Feed Addition. The filtering characteristics of a particular water depend upon (a) the amount of turbidity and (b) the nature of turbidity. It is therefore quite possible to plug or slime over the minute openings in a precoat if much fine colloidal turbidity is present. To reduce the tendency to plug the precoat, additional filter aid, called body feed, is added to the water to be filtered. The body feed maintains cake porosity by providing a continually fresh supply of microscopic flow channels so that the plugging tendency of the suspended solids is overcome allowing long, economical filtering cycles.

Adequate body feed is essential for long, economic cycles on raw water. The nature and amount of solids govern the ratio of body feed to turbidity. Low suspended solids usually require higher filter-aid ratios than do waters with higher suspended solids where some of the latter may themselves act as filter aids. As a convenient starting point, a body-feed ratio of two parts of Celite per one part turbidity has been found satisfactory.

e. Removal of Filter Cake. The filtration is continued until the resistance of the filter cake has increased to a point where, depending upon the mode of control, either the pressure dorp, the filtration rate, or a combination of the two has reached the economic limit. The filtration is then stopped, and the filter cake is removed by one of a variety of means, depending on the filter design. It must be stressed that the complete removal of the dirty cake is of paramount importance to successful and economical operation, since it is one of the major factors determining the cycle length and septum life.

A pressure diatomite filter can be operated to any feasible shutoff pressure, but it normally uses an upper limit of 30 to 40 psi. Vacuum unit runs are usually terminated at a terminal pressure of 10 psi. On a constant flow rate system, the rate-of-flow controller applies the terminal head (approximately) to limit flow. Therefore, both systems basically require full power load for the entire cycle. The power savings possible with the vacuum filter, therefore, may partially offset additional precoats and labor.

If a vacuum system is operated at a lower flow rate than a pressure unit, a cycle length can in most cases be tailored to match the pressure unit. The lower installation cost per square foot of filter area, generally possible with the vacuum system, permits the offering of more filtering area and thus lower flow rate. Theory indicates that halving the flow rate may increase the cycle length by a factor of 4.

In a properly designed iron-removal system using the diatomite process, a soluble iron in the effluent from either pressure or vacuum filters would be indicative of a failure in the chemical reaction process rather than a function of vacuum or pressure.

Vacuum filters are easier to operate and easier to inspect for possible failure of the filter cake. There is material in the literature that indicates the possibility of iron appearing in the effluent of a vacuum filter; however, the information cited from Johns-Manville Products Corporation seems to represent the present state of the art. Coogan (13) found that iron removal was good if the iron is completely oxidized to the ferric state by aeration or with the aid of chlorine or potassium permanganate. Manganese produced much shorter runs than iron if the concentrations were the same.

Most of the diatomite filter aid is produced by three sources in the United States: Eagle-Picher Company, Cincinnati, Ohio; Johns-Manville Products Corporation, Denver, Colorado; and Great Lakes Carbon Corporation, Los Angeles. Each manufacturer has a different method for determining filter-aid particle size. Baumann (14) found that externally applied mechanical pressure degraded the filter-aid particles and decreased its permeability. It was reasonable to conclude that hydraulic pressures would produce a similar effect.

There is, however, no clear evidence that hydraulic pressures within the design limits of most filters would do the same. In fact, it is the capability of the diatomite particles to successfully reduce cake compression that makes it an excellent filter medium.

Using a filter aid from a reputable manufacturer and suitable filtering equipment, properly operated and maintained, it should be feasible to secure a 50% reduction of BOD from an influent containing 15 mg/l and a 70% reduction of settleable solids from an influent containing 15 mg/l settleable solids. One type of diatomite filter unit is shown in Fig. 11-15. In other words, a diatomite filter can remove suspended material from water to produce a clarity equal or superior to sand filters.

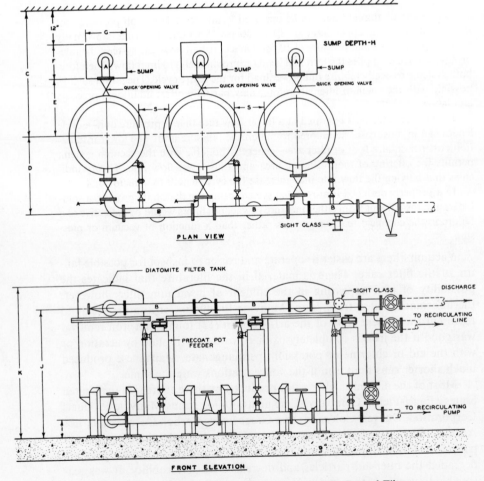

Figure 11-15. Triple Unit Diatomite Filter. (Courtesy General Filter Company)

11-6. Vertical Sand Filters. There are a number of different types of filters using sand and other media that can be used to improve the quality of sewage treatment plant effluent. The intermittent sand filters described in Chapter 7 are one means that can be employed. The size of the waste treatment plant and the quantity of effluent to be polished each day will decide which type would be most suitable.

It is preferable to have the quality of the effluent with as low a load of BOD and suspended solids as practicable from a secondary treatment process before sand filtering. With an influent containing 15 mg/l of BOD and settleable solids, respectively, an efficiency of approximately 50% BOD removal and 70% settleable solids removal can be obtained with *appropriate units* of a reputable manufacturer or custom-designed units made to your specifications.

Small plants may find pressure-type sand filters such as those shown in Fig. 11-16 suitable for their purpose. The minimum recommendation would be a filter bed consisting of 30 in. of filter sand having an effective size of 0.5 to 0.6 mm and a uniformity coefficient of not more than 1.6. The finer the filter media, the less the mg/l of settleable solids in the final effluent, but the lower the rate of filtration feasible for a given square footage of filter and pump pressure.

In practice, the output of the sand filter may range from 5 mg/l to 8 mg/l using sands typically employed. When the filter is backwashed, the sand bed expands, and this is the reason for the space that exists above the sand level in a proprietary unit. During the backwashing, while the filter bed is expanded from 150% to over 200% of its original volume, the grains of sand rub against each other and aid in effecting proper cleaning. An excessive backwash rate would cause the expansion to be too great and the sand grains not to come in contact properly with each other.

There are filters on the market that have artificial porous media flooring that prevents the sand from escaping. If gravel is used, it is usually applied with a coarse layer on the bottom, then a medium layer, and then a layer of fine gravel. The proportions vary, but in some units they range from $1\frac{1}{2}$: 1: 1 to 1: 1: 1. In some installations the volume of wash water used has equaled approximately 2.5% of sewage being treated. A filter run of 8 to 12 hours between backwashings has been typical in some plants. Table 11-8 has been established to provide a minimum-suggested guide in considering pressure-type sand filters.

Vertical-type rapid sand filters will not remove a satisfactory percentage of cercariae of *Schistosoma mansoni* and various other microorganisms. They will polish the effluent. If the operator gets careless and permits his sludge floc to become nitrified, it becomes a pinpoint floc and the sand filter will catch it and remove it. Unless the operator takes the appropriate steps quickly enough to correct the plant performance, he may end up with a mess in the filters.

11-7. Horizontal Sand Filters. There are filters in vendor catalogs called horizontal sand filters which are nothing more than vertical filters built in

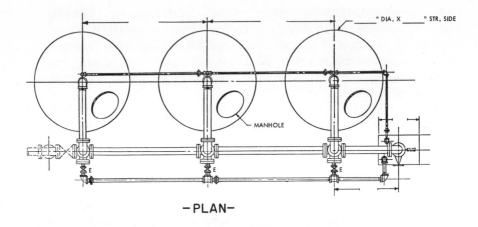

−PLAN−

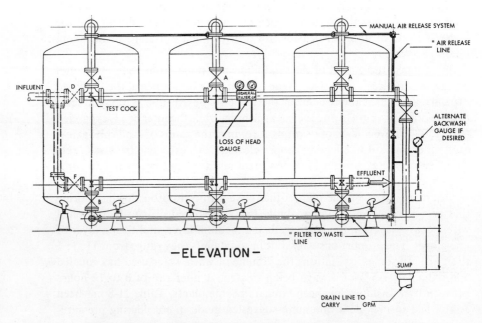

−ELEVATION−

Figure 11-16. Vertical Pressure Filters (Three-Unit Battery). (Courtesy General Filter Company)

a horizontal tank. This section refers to a horizontal sand filter in which the influent enters the sand at one end of the filter and travels horizontally and at a vertical gradient percolating through the filter. The current state of the art on horizontal sand filters more or less requires a field test of a small unit using the locally available sands.

Bernarde and Johnson (15) experimented with a horizontal sand filter to remove cercariae of *Schistosoma mansoni*. Sand particles of 0.35 mm in

TABLE 11-8. Vertical Pressure Filters

Tank Dia. (in.)	Bed Area (ft²)	Capacity Rate/Ft² (2 GPM)	(3 GPM)	Wash Rate	Main Pipe Size (in.)	Tank Steel 100 lb W.P. Heads (in.)	Shell (in.)	Gravel (ft³)	Sand (ft³)	Floor Space (in.)	Operating Weight (lb)
20	2.2	4.3	6.5	26–33	1	3/16	3/16	3.0	4.5	32 × 59	1,800
24	3.1	6.3	9.4	34–47	1½	3/16	3/16	4.0	6.5	36 × 63	2,230
30	4.9	10	15	60–74	2	3/16	3/16	6.5	10.0	42 × 72	3,290
36	7.1	14	21	85–106	2½	1/4	3/16	9.0	14.0	54 × 90	4,810
42	9.6	19	29	115–142	2½	5/16	1/4	13.0	20.0	60 × 96	6,460
48	12.6	25	38	150–188	3	5/16	1/4	16.5	25.0	60 × 102	8,390
54	15.9	32	48	190–240	3	3/8	1/4	21.0	32.0	72 × 114	10,900
60	19.6	39	59	235–300	3	3/8	5/16	26.0	40.0	78 × 120	12,600
66	23.8	48	71	285–360	4	7/16	5/16	32.0	48.0	84 × 126	16,050
72	28.3	57	85	340–425	4	7/16	3/8	37.0	56.0	90 × 132	18,980
78	33.2	66	100	400–500	4	1/2	3/8	44.0	66.0	96 × 138	24,230
84	38.5	77	115	466–580	5	9/16	7/16	52.0	76.0	108 × 156	26,620
90	44.2	88	133	530–660	5	9/16	7/16	60.0	88.0	114 × 162	32,060
96	50.1	100	150	600–760	5	5/8	7/16	67.0	100.0	120 × 168	37,350
102	56.7	113	170	680–850	6	11/16	1/2	76.0	114.0	126 × 174	44,200
108	63.6	127	191	765–960	6	11/16	1/2	85.0	128.0	132 × 180	52,100
114	70.9	142	213	850–1060	6	3/4	9/16	94.0	141.0	138 × 186	61,750
120	78.5	157	235	945–1180	6	3/4	9/16	105.0	158.0	144 × 192	69,400

Notes:
1. All filters listed above to have 5′ 0″ straight side shell.
2. Data are given for a single filter unit. Filters may be constructed with 2, 3, 4, or more units parallel in order to increase plant capacity.
3. Pumping unit usually arranged in multifilter unit such that one filter unit at a time can be in backwashed.

diameter and smaller were found efficient in removing 100% of cercariae added to the influent of a constant head horizontal permeameter.

Figure 11-17 indicates the flow rate obtained in a horizontal filter constructed by the author. It was concluded that such a unit would be most appropriate after a vertical filter that would be simpler to backwash and clean, i.e., in series following a vertical unit.

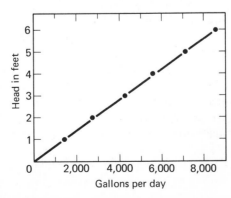

Figure 11-17
Horizontal Filter Flow Rate.

11-8. Mixed-Media Filters. Assume that we select several different filter media such that the finest material has the highest specific gravity, the next finest material has the next highest specific gravity, and so on until the coarsest material is reached with the lowest specific gravity. Assume that we permit the particle sizes to range from 0.15 mm to 1.0 mm in size within a 30-in. filter. When the filter is backwashed, the finest media will be on the bottom and the coarsest media will be on the top. However, the effective equivalent filtering depth may be equal to more than 100 in. of a standard sand filter.

A combination of coal and sand is an example of one type of a dual-media unit. The multiple mixed-media unit has an efficiency of removal and breakthrough capacity that significantly exceeds that of the coal-sand or rapid sand filter. Included in the media being developed are special plastic materials. Since these materials are constantly being improved, it is not practical to try to provide specific figures of removal efficiency for BOD and solids.

11-9. Porous Ceramic. Another type of filter that may be suitable in certain types of applications is a porous ceramic. One unit is sold under the trade name of Poro-Stone. The Poro-Stone filter media is a ceramic structure having a $\frac{5}{8}$-in. thick wall through which the raw water must pass in the removal of solids. The direction of flow is from the outside of the tube to the inside or bore. The removed solids collect on the outside wall of the tube until the backwashing process disposes of the collected solids to waste. The passage of water through the mass of collected solids results in a finer degree of separation than would be normal for the porosity of the tube. Through this normal precoat of solids, it is expected that a separation of 12.5 microns (0.0005 in.) is attained.

A filter aid such as diatomaceous earth can be used with porous ceramic to further improve the filter in terms of solids removal. One source for the porous ceramic unit is R. P. Adams Co., Inc., Buffalo, New York.

11-10. Adsorption. Adsorption by activated carbon for removal of organic compounds has been shown to be highly effective (16). The occurrence in municipal wastewater of organic contaminants resistant to biological degradation presents the major technical challenge in advanced wastewater treatment. For many reuse applications, these materials must be reduced in concentration or completely removed; furthermore, their presence can seriously interfere with other stages of treatment either for inorganic removal or disinfection. Removal of these residual organics therefore becomes a critical aspect of advanced waste treatment. It requires separation techniques that will be suitable for reducing the comparatively large concentration of refractory organic compounds occurring in secondary effluents (50 to 100 ppm COD) and that can effectively scavenge organics at the 3- to 5-ppm concentrations.

Experiments have indicated (16) that:

    1. The rate of adsorption appears to be controlled by the rate of

diffusion of solute within the internal capillary pores of the granular carbon particles.

2. The rate of uptake of ABS by granular carbon increases with decreasing pH of the solution.

3. Rates of adsorption increase with increasing concentration of solute, but adsorption of ABS from water is relatively more rapid for more dilute solutions. This is advantageous for the removal of final traces of material in the renovation of a wastewater.

4. The rate of adsorption per unit weight of material adsorbed decreases considerably as molecular size increases.

5. The structure of the alkyl chain attached to the benzene ring has a small but measurable influence on the rate of adsorption. Molecules having a highly branched structure are removed more slowly than straight-chain or more nearly straight-chain molecules.

6. Relative adsorption rates vary reciprocally as the square of the diameter of the individual carbon particles in the adsorbing surfactants.

7. Temperature effects on adsorption equilibria are generally felt not to be significant over the range of temperature practically encountered in water and wastewaters.

In an evaluation of other commercially available carbons to establish suitability of adsorbents that may be less expensive than the Columbia LC Carbon used in the first experiments, Pittsburgh carbon (a relatively soft carbon) was found to exhibit a rate constant for adsorption of technical grade ABS about five times as large as that for the Columbia LC Carbon (16). This increased capacity is presumably attributable to the larger-pore diameters of the Pittsburgh carbon. Because of their high capacity and high rate of uptake, these large-pore carbons appear more suitable for application to wastewater treatment.

There are over 500,000 synthesized organic chemicals, but only several dozen have been determined safe by the Food and Drug Administration for addition to foods (17). Many of these refractory materials give no indication of their presence and many cannot be completely removed from water supplies or wastewaters by conventional waste treatment methods (18).

In water treatment practice, activated carbon is usually applied, in a water slurry, at a rate from 20 to 40 lb/mil gal, though doses have varied from 2 to 1,800 lb/mil gal (19). Vaughn (20) indicated that a carbon dosage of 240 mg/l has removed 97% of a 20 mg/l synthetic-detergent concentration from raw water.

Activated carbon has a disadvantage of relatively high cost, and there is difficulty regenerating typical water-grade carbons because of their fragility. The hydrogen peroxide, as an example, required to regenerate 1 pound of activated carbon would cost approximately twice as much as the carbon itself. This is a basic reason that chemical regeneration is not economically feasible. Thermal regeneration is feasible. The carbon will lose some of its adsorption ability after each regeneration, but it can be regenerated over a number of cycles.

Consider the case where long-term usage is considered and there are organics present that require removal. First it is necessary to design an effective secondary biological system. The author usually designs for 95% BOD and SS removal before resorting to organic carbon. Depending on the installation, it might be advisable to use a vertical sand filter before entering the units containing activated carbon. It is feasible to use a series of activated carbon filters in order to accumulate the 12-ft of bed depth required. Figure 11-18 show actual filters of the expanded-bed type in service.

It appears that an average retention time of 12 minutes is necessary (empty-bed residence time of 30 minutes with carbon having 40% voids). The minimum bed depth should be 12 ft for most wastes. In a downflow system, the superficial velocity should be between 4 and 8 gal/min/ft². These figures are based on data from Westvaco, Carbon Technical Center at Covington, West Virginia.

While in the past, effort has been more toward downflow columns, it is expected that the future of activated carbon will be more promising in upflow expanded-bed columns. Westvaco, Calgon, and EIMCO have all been active in the activated-carbon column development, and if a plant is to be constructed all of the listed firms should be contacted to determine what they have to offer at the time.

Figure 11-18
Expanded Bed Type Activated Carbon Filter Plant. (Courtesy Calgon Corporation)

Figure 11-19
Granular Carbon Reactivation.
(Courtesy Calgon Corporation)

An integral part of Calgon's granular carbon system is the reactivation process, consisting of carbon storage tanks and a gas-fired multiple-hearth furnace. Exhausted carbon, which no longer has the capacity to remove contaminants, is hydraulically transferred to the furnace where the contaminants are completely burned at temperatures of 1800°F and the carbon is returned to the filter bays ready for reuse. Such a furnace and associated apparatus is shown in Fig. 11-19. The furnace is completely automatic, is equipped with an afterburner and wet scrubber to prevent air pollution, and has an operating capacity of 8,500 lb/day.

11-11. Foam Separation. Foam separation is a process that takes advantage of the foaming properties of a waste stream by passing air bubbles through the stream and causing surface-active organic impurities to concentrate at the bubble surfaces. The process is capable of removing 30 to 40% of the organic contaminants from a conventionally treated waste effluent (16). It also removes 70 to 80% of the synthetic detergent present. The cost is less than half the cost of adsorption techniques.

When a solution containing surface-active solutes is foamed, the foam becomes enriched with these solutes, and the residual in the bulk liquid is depleted. The foam can be subsequently collected and collapsed to produce a solute-rich liquid product. In addition to the concentrating of soluble surfactants in surface films, there is also a concentration in the foam of suspended solids by a froth flotation mechanism.

One method of construction that could be used is shown in Fig. 11-20, but the efficiency of this unit is not as high as the laboratory column-type unit shown in Fig. 11-21. Figure 11-22 shows the air-volume performance, Fig. 11-23 illustrates the air-to-liquid flow-rate ratio, and Fig. 11-24 shows the ABS in the secondary effluent before and after foaming, all with the column-type foamer. It can be summarized (16) that optimum results are obtained when:

1. Gas-to-liquid ratio is 5:1.
2. Detention period is 5 minutes.
3. Minimum liquid depth is 5 to 6 ft.
4. Foam height over liquid is 1.5 to 2.0 ft.
5. Foamate volume is 1 to 2% of waste flow.

11-12. Electrodialysis. Electrodialysis is a process for separating ionized materials from water by the use of cation- and anion-permeable membranes placed in alternate sequence between two electrodes (16). Although not practical for removing inorganic molecules, it is highly effective in removing

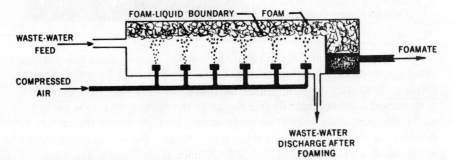

Figure 11-20. Trough-Type Foamer (Reference 16).

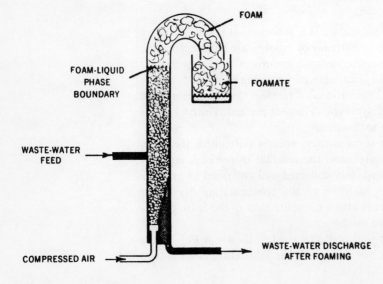

Figure 11-21
Column-Type Foamer (Reference 16).

dissolved inorganic salts and would work well as part of a two-stage inorganic separation system. Exclusive of brine disposal, this system costs approximately $1\frac{1}{2}$ to 2 times as much as the adsorption process.

When an electric potential is impressed across a cell containing mineral-

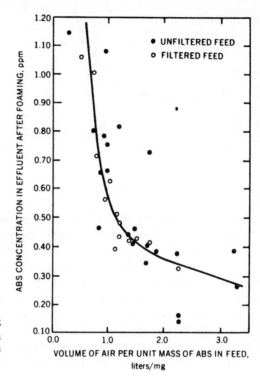

Figure 11-22
Air-Volume Performance on Column-Type Foamer (Reference 16).

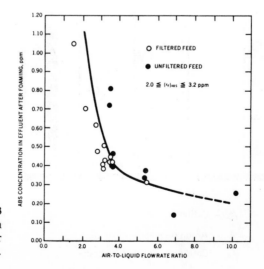

Figure 11-23
Air-to-Liquid Flow-Rate Ration With Column-Type Foamer (Reference 16).

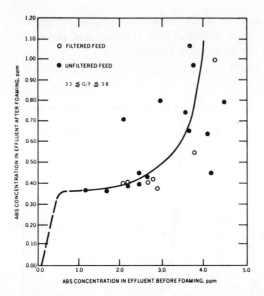

Figure 11-24
ABS in Effluent Before and After Foaming (Reference 16).

ized water, positively charged ions (cations) migrate to the negative electrode, and negatively charged ions (anions) migrate to the positive electrode. If cation- and anion-permeable membranes are placed alternately between the electrodes, ions will concentrate in alternate compartments and become more dilute in the intervening compartments. The basic principle is indicated in Fig. 11-25. If the apparatus is arranged so that concentrated and dilute streams flow continuously, a device useful for large-scale dimineralization of water results (16). Only partial dimineralization by this method is practical, however, because electrical power requirements become excessive if the ion concentration is reduced too far. Fortunately, power consumption remains

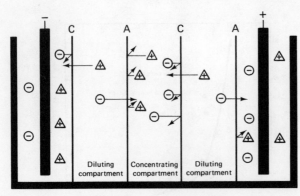

⊖ – Anion

⬛ – Cation

A – Anion permeable membrane

C – Cation permeable membrane

Figure 11-25
Principle of Electrodialysis (Reference 16).

reasonable at the mineral ion concentrations occurring in drinking water. Since the electrodialysis phenomenon is specified for ions, application of the process is limited to removal of soluble ionized contaminants.

The process is commercially feasible for the treatment of brackish waters (16). Soluble organic contaminants should be removed from wastewaters before using this process with domestic wastewaters or wastewaters having similar organic content. Electrodialysis is a technically feasible method of dimineralizing municipal wastewater to the extent required for most use.

Electrodialysis will probably be a proprietary process and detail design is beyond the scope of this text. Ionics, Incorporated, is one firm that has accumulated experience with the process. General Electric has a unit that apparently works on the electrodialysis in their shipboard waste disposal plant program.

11-13. Distillation. Distillation, while particularly suited to feeds containing only nonvolatile contaminants, is adaptable to municipal wastewaters containing certain volatile contaminants that may be carried over into the product (16). Although this situation constitutes a significant drawback to the process, the possibility of using follow-up polishing treatments to reduce trace impurities offers a practical solution. The cost for treatment is about ten times more expensive than adsorption in terms of unit cost. This value can be cut more than half by application of a blending principle in this "total removal" process (16).

For distillation to be economically practicable, the latent heat of evaporation must be reused many times. Three types of systems are commonly used to achieve heat economy. The first type, in Fig. 11-26, is the multiple-effect evaporator, in which each effect is maintained at slightly lower pressure and, hence, a lower temperature than the preceding one. This allows the steam produced in one effect to be used as the heating medium for the next. The result is that, for each pound of steam supplied to the first effect, there is produced a number of pounds of product water approximately equal to the number of effects (16).

The second type, known as the multistage flash still (Fig. 11-27), is constructed so that water flash-evaporates in a number of stages at successively lower temperatures. The heat of condensation is used to preheat the feed. It is necessary to add only a small amount of heat to the hot feed to raise its temperature the few degrees needed so that it is hot enough for flashing in the first stage. The heat economy depends on the number of stages and on the amount of temperature boost given to the preheated feed (16).

Heat reuse is accomplished in the third type, the vapor compression still, Fig. 11-28, by raising the temperature high enough that its own heat of condensation can be used to evaporate more water. Although no heat energy is required after startup, mechanical energy is necessary to compress the water vapor (16). This latter method is probably the least appropriate from an economical viewpoint.

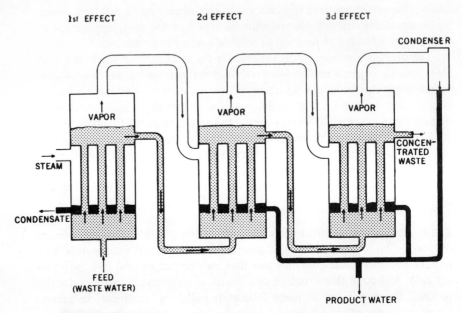

Figure 11-26. Schematic Arrangement of Multiple-Effect Evaporator (Reference 16).

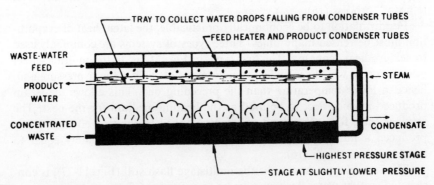

Figure 11-27. Schematic Arrangement of Multistage-Flash Evaporator (Reference 16).

Ammonia is a normal component of municipal waste water that presents a particular challenge in water renovation, especially in the distillation process. Ammonia is the principal contaminant in the first 10% of distillate. Under laboratory conditions, acidification of the feed effectively eliminated ammonia from the product. Similar treatment is likely to be required during full-scale operation.

11-14. Reverse Osmosis. The following explanation by the U.S. Public Health Service (16) is one of the most appropriate:

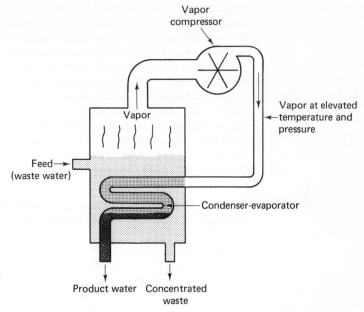

Figure 11-28
Vapor Compression Distillation
(Reference 16)

A natural phenomenon known as osmosis occurs when solutions of two different concentrations are separated by a semipermeable membrane such as cellophane. With such an arrangement, water tends to pass through the membrane from the more dilute side to the more concentrated side and produce concentration equilibrium on both sides of the membrane. The ideal osmotic membrane permits passage of water molecules but prevents passage of dissolved materials. The driving force that impels this flow through the membrane is related to the osmotic pressure of the system, and this pressure is proportional to the difference in concentration between the two solutions. The principle is illustrated in part A of Fig. 11-29. If the liquid on the more concentrated side is allowed to rise in a standpipe as water passes through the membrane, as shown in B of Fig. 11-29, the hydrostatic pressure on the right side of the membrane gradually increases until it finally equals the

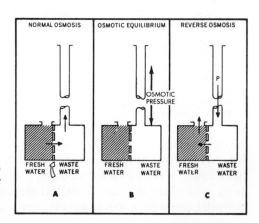

Figure 11-29
Schematic Representation of
Principles Involved in Normal
Osmosis and Reverse Osmosis.

osmotic pressure. At this point, flow through the membrane ceases. If we then intentionally increase the pressure on the more concentrated (right) side of the membrane, as shown in C of Fig. 11-29, the flow of water through the membrane reverses, that is, water moves from the more concentrated compartment to the less concentrated compartment. This method of purifying water is popularly called *reverse osmosis*, though the mechanism of water transfer may not be the same as in normal osmosis.

The properties of a membrane that permit water molecules through but impede the flow of contaminant molecules are not clearly understood. It is believed to be not simply a molecular filtering action, even though individual water molecules are smaller than most of the contaminant molecules of concern. Some scientists believe that water passes through the membrane by the progressive formation of hydrates. Only H_2O molecules can form hydrates, which accounts for the passage of only pure water.

Table 11-9 indicates the results obtained by Aerojet-General Corporation, of Azusa, California, in experimental work on the AWTR program.

TABLE 11-9. Composition of Water Produced by Treating Municipal Wastewater by the Reverse-Osmosis Process (16)

Component	Feed Water (typical) (ppm)	Product Test 1 (ppm)	Product Test 2 (ppm)
TDS	550	15	28
ABS	4.5	0.1	0.1
COD	95	2	6.0
pH	5*	6.3	5.5
Cations			
Na^+	85	4.6	6.1
K^{++}	40	4.9	2.7
NH_4^+	25	0	3.9
Ca^{++}	125	0	2.6
Mg^{++}	50	0.7	0
Fe^{++}	N.D.†	0	0
Anions			
Cl^-	65	20.3	22.1
NO_3^-	2	0	0
HCO_3^-	260	8.1	3.5
$CO_3^=$	0	0	0
$SO_4^=$	200	1.7	3.3
$SiO_3^=$	30	2.9	2.9
$PO_4^=$ (total)	25	0	0

*pH adjusted during test.
†N.D. = not determined.

11-15. Freezing. When impure waters are frozen, the ice crystals formed have long been known to be composed of essentially pure water. This principle has frequently been used for laboratory scale purification of water and solvents of all kinds. In recent years, this process has been recognized as having potential application for large-scale operation as a method of separating commercial quantities of water from undesirable components contained in solution (16). Much research and development work has been done to determine the applicability of freezing as a method of separating contaminants from water on a practical scale.

Three steps are involved in applying freezing to the purification of water (16). First, heat is removed from the feed water to cool it to its freezing point. Additional heat is then removed by the vaporization of a refrigerant directly in contact with the feed, causing fine crystals of ice to freeze out of the solution. The refrigerant may be a low-boiling hydrocarbon such as butane or it may be water itself. When roughly half of the water has been frozen, the ice-water slurry is transferred to another vessel (the washing column) where the unfrozen liquid is drained off and the crystals are washed by contacting them with pure water (16). As the final step, the washed ice is transferred to a third vessel where it is melted to form the purified product.

Research has shown that up to 85 % of the organic contaminants and 90 % of the inorganic contaminants could be removed from wastewater by the freezing process (16). Removal efficiency for ABS showed, however, an erratic pattern under the conditions tested. It is not feasible to have direct contact between a highly volatile organic refrigerant and wastewater. Noncontaminating units have been tried, but the tendency of ice crystals to fuse in the test equipment proved to be a serious drawback to the washing out of impurities, owing to entrapment of contaminant-bearing water in the interstices between fused crystals (16).

The Carrier Research and Development Co. of Syracuse, New York, has been active in research on this approach. Based on all the facts available, the author was not impressed with the field applicability of this process at the present or in the foreseeable future, unless a new breakthrough is achieved on technology of the process.

11-16. Ion Exchange. Ion exchange is a method of removing certain dissolved solids from water. During the removal process, raw water is percolated through a column or bed of ion-exchange material. Ion exchangers are very similar in construction to pressure-type sand filters, except that an ion-exchange material such as zeolite replaces the sand. The ion-exchange material has the capacity to replace mineral ions in the water with ions from inside the zeolite. No chemicals are added during the exchange process. Eventually the ion-exchange capacity of the zeolite is exhausted and it is then necessary to regenerate it by addition of a chemical.

The common household water softeners frequently use a zeolite in which the sodium in the zeolite is replaced by calcium or magnesium.

$$2NaAlSi_2 + Ca^{++} \longrightarrow Ca(AlSi_2O_6)_2 + 2Na^+$$

To regenerate this zeolite, it is necessary to pass a concentrated brine solution of sodium chloride through the zeolite.

$$Ca(AlSi_2O_6)_2 + 2Na^+ \longrightarrow 2NaAlSi_2 + Ca^{++}$$

The brine regeneration solution is passed to waste with a small amount of rinse water before replacing the equipment back in service.

Fortunately, a number of synthetic ion-exchange resins have been created, so it is now possible to remove either anions or cations from water. By combining a bed of hydrogen-containing cation-exchange resins and a bed of hydroxyl-containing ion-exchange resins, it is possible to virtually completely deionize water. Since the hydrogen and hydrozyl ions entering the water from the resin combine to produce additional molecules of water, this technique completely removes the contaminating ions rather than exchanging one mineral ion in the water for another (16). When these resins are exhausted, they must also be regenerated; an acid is used to supply hydrogen ions to the cation resin, and a base is used to supply hydroxyl ions to the anion resin.

The ion-exchange process is expensive, but it works very effectively. Where water reuse is mandatory, it can be used to keep down the buildup of certain undesirable dissolved solids. The water that is put through the ionization process should first have had as much of the dissolved organics removed as possible in order to minimize the danger of fouling the resin. Activated carbon filtration ahead of the ionization process is desirable. Commercially available equipment may be obtained from a number of sources such as General Filter Co., Permutit, and Crane-Cochrane.

11-17. Solvent Extraction (16). For some separation processes, the cost of operation decreases as the concentration of contaminants in the feed decreases. One such process is solvent extraction, a method whereby water can be separated from its contaminants by a solvent capable of "dissolving" the water but not the contaminants.

The solvent extraction process operates on the principle that certain organic solvents that contain strong electronegative atoms with the molecule are partially miscible with water to an extent depending mainly on the temperature and salt content of the water. One such solvent is di-isopropylamine (DIPA). A solubility diagram of DIPA and water at various salt concentrations is given in Fig. 11-30. In practice, a water containing, for example, 1.0% sodium chloride would be contacted with DIPA at 32°C (see Fig. 11-30). At this temperature, a separate phase

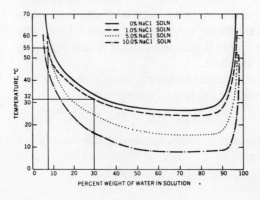

Figure 11-30
Solubility Diagram for
Di-isopropylamine and
Water-NaCl System.

containing 30% water and 70% solvent is formed. The water drawn into solution leaves the bulk of its contaminant load behind. The DIPA-water solution is then separated from the salt solution for recovery of the water. This recovery is accomplished by raising the temperature of the solution. At 55°C, the solubility of water in the solvent is decreased to 8% (see Fig. 11-30), causing water to be released from the solvent. The water thus recovered contains a much lesser concentration of contaminants than it had originally. Solvent remaining with the product water is separated by air-blowing.

A schematic diagram of the solvent extraction process is shown in Fig. 11-31. In a continuous-flow, counter-current system, the raw water and the solvent are intimately mixed in the extraction column under controlled temperature conditions. The extract discharged at the top of the column, representing the usable portion of the liquid stream, is then fed into a phase separation tank where the temperature is increased. At this point the water phase, now relatively pure, becomes the product stream and requires only air-stripping to remove the residual solvent down to low ppm concentrations. Solvent recovered from the product stream, from the phase separation tank, and from the brine or raffinate (discharged at the bottom of the extraction column) stream, is recycled in the system. The brine containing the concentrated contaminants is wasted after recovery of the solvent.

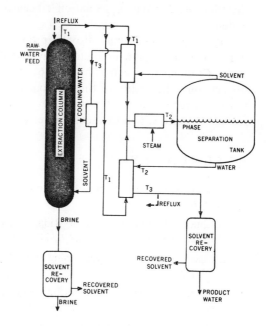

Figure 11-31
Schematic of Continuous-Flow
Solvent Extraction Process.

The cost of solvent extraction appears to have ultimate potential, but the current cost is approximately five times greater than activated-carbon absorption.

11-18. Oxidation. Six oxidation systems offer varying degrees of potential for the direct treatment of wastewaters (16). These systems are:

1. Oxidation by oxidants containing active oxygen (O_3, H_2O_2).
2. Accelerated molecular oxygen oxidation (autooxidation).
3. Catalytic oxidation of adsorbed organics by oxygen.
4. Oxidation by chlorine and its derivatives.
5. Oxidation by oxy acids and their salts, e.g., $KMnO_4$.
6. Electrochemical oxidation.

Three of these systems—the active oxygen, the autooxidation, and the catalytic oxidation of adsorbed organics by oxygen—do not produce secondary pollutants and do not necessarily increase the inorganic salt load in the treated waste water.

Hydrogen peroxide, H_2O_2, decomposes (21) to form the $\cdot OH$ radical. This radical is one of the strongest oxidants known in water (22). Swallow (23) reported approximately a 70% oxidation efficiency. Hydrogen peroxide alone does not oxidize the organic materials in wastewaters within a practical reaction time. The system requires a multivalent metal-salt (e.g., iron salt) catalyst (16). With iron-salt catalyst, a specific pH range of 3 to 4 is required to produce the $\cdot OH$ oxidant efficiently. Ozone could be used to attain about the same percentage of oxidation efficiency, but it would cost more per gallon.

Hydroxyl radicals may also be produced directly in aqueous solution during the adsorption of high-energy radiation from radioactive isotopes or from high-energy electron beams (16). Research does not appear to support the economic feasibility of this approach. None of the foregoing oxidation schemes shows results of immediate practical application on the types of facilities toward which this text is directed.

REFERENCES

1. PARKER, HOMER W., "Spray Irrigation of Roadside Rest Area Effluent." Unpublished paper.

2. BENDIXEN, THOMAS W., et al., "Ridge and Furrow Liquid Waste Disposal in Northern Latitude." *Proc. ASCE*, 94, SA1, 147 (1968).

3. LAVERTY, F. B., et al., "Reclaiming Hyperion Effluent." *Proc. ASCE*, 87, SA6, 1 (1961).

4. ROBECK, G. G., et al., "Factors Influencing the Design and Operation of Soil Systems for Waste Treatment." *Jour. WPCF*, 36, 8, 971 (1964).

5. MCMICHAEL, F. C., and MCKEE, J. E., *Report of Research on Wastewater Reclamation at Whittier Narrows*. W. M. Keck Laboratory of Environmental Health Engineering, Calif. Inst. of Technology, Pasadena, Calif., Sept., 1964.

6. SCHRAUFNAGEL, F. H., "Ridge-and-Furrow Irrigation for Industrial Waste Disposal." *Jour. WPCF*, 34, 11 (1962).

7. KEEFER, C. E., "Operating A Sewage Microstrainer." *W & S. W.*, 101, 8, 346 (1954).

8. EVANS, S. C., "Twelve Months Operation of Sand Filtration and Microstraining Plant at Luton." *Jour. and Proc., Inst. Sew. Purif.*, Part 4, 333 (1952).

9. CASSIDY, J. E., "The Operation of the Hazelwood Lane Works of the Brackness Development Corporation." *Jour. and Proc., Inst. Sew Purif.*, Part 3, 276 (1960).

10. NELSON, O. FRED, "Four Years of Microstraining at Kenosha, Wisconsin." *Water Works and Wastes Eng.*, Reprint July, 1965, from Glenfield & Kennedy, Inc.

11. *The Filtration of Water*. Johns-Manville Celite Filter Aids, Johns-Manville Corp.

12. *Filtering with Diatomite for Efficient Clarity at Lowest Cost*. Johns-Manville Celite Filter Aids.

13. COOGAN, GEORGE J., "Diatomite Filtration for Removal of Iron and Manganese." *Jour. AWWA*, 54, 1507 (1962).

14. BAUMANN, E. R., "A Study of the Effect of Mechanical Compression on the Particle Size and Permeability of Diatomite Filter Aids," section D of final report to Engineer Research and Development Laboratories, Fort Belvour, Virginia. Division of Sanitary Engineering, Department of Civil Engineering, University of Illinois, Urbana, Illinois, 1950.

15. BENARDE, MELVIN A., and JOHNSON, BARRY, "Schistosome Cercariae Removal by Sand Filtration." *Jour. AWWA*, 63, 7, 449 (1971).

16. *Summary Report*, Jan., 1962–June, 1964, Advanced Waste Treatment Research, AWTR-14, U.S. Department of Health, Education, and Welfare, Public Health Service.

17. EITTINGER, M. B., "Proposed Toxicity Screening Procedure for Use in Protective Drinking Water Quality." *Jour. AWWA*, 52, 687, (1960).

18. JOHNSON, R. L., et al., "Evaluation of the Use of Activated Carbons and Chemical Regenerates in Treatment of Waste Water." U.S. Public Health Service, Environmental Health Series, PHS Publ. No. 999-WP-13 (1964).

19. SIGWORTH, E. A., "Activated Carbon—Its Value and Proper Points of Applications." *Jour. AWWA*, 29, 688 (1937).

20. VAUGHN, J. C., and FALKENTHAL, R. E., "Detergents in Water Supplies," *Ind. and Eng. Chem.*, 48, 241 (1956).

21. BARB, W. G., et al., "Reactions of Ferrous and Ferric Ions with Hydrogen Peroxide." *Trans. Faraday Soc.*, 47, 462, 591 (1951).

22. UNI, N., "Inorganic Free Radicals in Solution." *Chem. Rev.* 50, 375 (1952).

23. SWALLOW, A. J., *Radiation Chemistry of Organic Compounds*. Pergamon Press, New York, 1960.

12 Sewage Lift Stations

Sewage pumping stations are used when it is necessary to lift sewage from a given point to a higher elevation. This chapter will consider only centrifugal pumps and pneumatic ejectors for smaller-size pump stations. These small pump stations are often called *sewage lift stations*. It is not uncommon today for life stations to be built as a factory-assembled package, or at least as a partial factory-assembled unit or series of units.

12-1. Applications for Sewage Lift Stations. Sewage lift stations may be used as a matter of economics or to overcome inadequate hydraulic head when it is obvious that no other solution is practical. Examples are:

1. Sewer construction cost increases rapidly with depth. It may be more economical to construct a lift station and raise the sewer closer to the surface.
2. It may be more economical to lift the sewage over a ridge and let it flow by gravity to an existing sewage plant rather than construct a new treatment plant.
3. Lift sewage out of a building whose basement is constructed below the sewer.
4. Elevate sewage to pass through a treatment plant by gravity.
5. Standby station to pump water into a river during flood conditions. Station bypassed under normal river stage.

12-2. Types of Sewage Lift Stations. Sewage lift stations are commonly divided into wet-pit pump stations and dry-pit pump stations. The wet-pit pump station consists of a steel, cast-iron, concrete, or other material forming a single basin at the lowest point of a building or at a point from which it is desirable to elevate sewage. The basin is covered by a steel or cast-iron cover plate on which one or more pumps are suspended. The pump is actually immersed in the sewage and is supported from the baseplate by means of a column pipe. The drive is mounted above the baseplate and transmits power to the pump by means of a drive shaft. A typical duplex wet-pit pump installation is shown in Fig. 12-1.

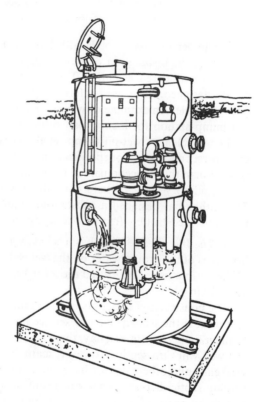

Figure 12-1
Duplex Wet-Pit Sewage Lift Station.
(Courtesy Can-Tex Industries)

A dry-pit pump station consists of a separate wet well in which the sewage is collected and a dry well in which pumps are mounted and an intake pipe from the pump station is connected to the wet pit. This type of pump station is illustrated in Fig. 12-2.

12-3. Wet-Pit Pump Stations. There are thousands of wet-pit pump installations pumping hot and cold waste, drainage, suspended solids, and stringy materials from:

Toilets	Floor drains
Plumbing fixtures	Urinals
Disposal units	Bathtubs
Showers	Sinks
Laundry tubs	Service sinks
Dish washers	Drinking fountains
Clothes washers	Wash fountains
Hose racks	Hose faucets
Boiler pits	Elevator pits
Subways	Underpasses
Basements	Pools
Grade separations	Reservoirs

A number of Boards of Health have regulations either discouraging or preventing the installation of wet-pit pumps for sewage service where reliability is considered essential. The concern is that many installations are poorly designed, have used inferior equipment, and have a reputation for giving trouble. An example of where reliability would be considered essential is a pump station for a suburban residential community.

The problem of wet-pit pump reliability is usually caused by inadequate or incorrect design and/or specifications. A properly designed, carefully documented plan that is accompanied by a detailed specification may be accepted by a regulatory agency where normally a wet-pit station would be rejected. The author has designed a substantial number of wet-pit installations that have provided trouble-free service using the following principles.

a. The Wet Well. If the basin is below groundwater level, the hydrostatic lift forces of the groundwater on the wet well must be considered. The weight of the basin must either exceed these forces, or the wet well must be anchored securely in place.

The construction materials of the basin are commonly concrete, steel, or cast iron. If the seepage flow or service flow is adequate to cause the pumps to function several times a day, there is small probability that the waste in the wet well will turn septic. However, many installations receive little water on weekends and others do not have enough usage to prevent the sewage from turning septic. Septic sewage can result in high sulfide concentrations with associated odors. Gas-tight construction is available to retain the odor in the basin.

Concrete: If the waste effluents are characterized by high sulfide concentrations, then concrete basins may suffer bacterial corrosion. According to Gilchrist (1), this corrosion is caused by sulfur-oxidizing bacteria, *Thiobacillus*, that form in the film of moisture on the walls above the liquid level. This corrosion results in disintegration of the concrete. The causative conditions required are temperature above 18°C, from 1 to 10 ppm of hydrogen sulfide, carbon dioxide, oxygen, NH_3, and condensing dew. Soluble phosphorus compounds and trace elements such as iron must also be present in

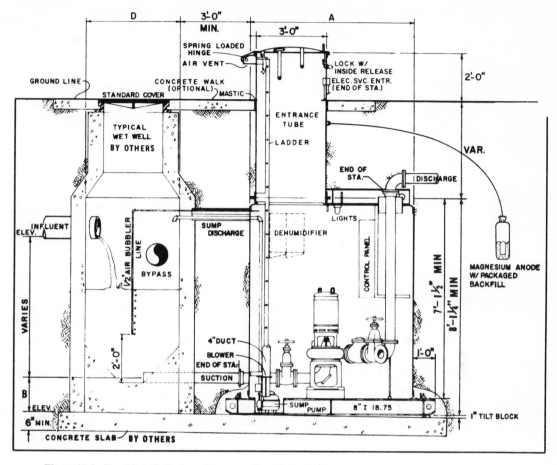

Figure 12-2. Dry-Pit Lift Station. (Courtesy Can-Tex Industries)

the concrete for the use of the bacteria. The corrosion takes place in the film of moisture above the liquid level of the sewage. Aggregate made of limestone or dolomite will substantially inhibit bacterial corrosion (2, 3, 4). Where the lift station operates often enough to prevent the sewage from going septic, the above condition will not occur. The use of a nonclog air-diffusion unit to keep the sewage aerobic would be an adequate preventive measure. The use of an air diffuser increases the initial and operating cost of the installation since a compressed-air source is required.

Coated Concrete: The use of a suitable coating on the concrete will inhibit bacterial action if the coating is free of imperfections. Coal tar is not recommended because it can be chipped free too easily and is often not applied properly by the workmen. If properly applied, some of the asphalt compounds will help to inhibit deterioration. Some of the newer bacteria-resistant coating compounds suitable for use on concrete, such as special epoxy coatings, are

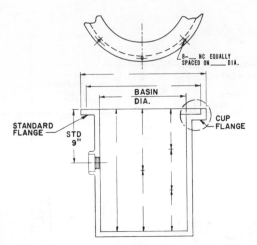

Figure 12-3
Cast-Iron Sump Basin. (Courtesy
Aurora Pump Division)

more appropriate. The surface of the concrete must be well cleaned in order
for any coating to properly adhere. If a coating is used, it should be at least
two coatings thick, one coat applied at right angles to the other.

Vitrified Clay: Vitrified clay pipe with compression-type joints and a concrete bottom is satisfactory, provided a vitrified pipe diameter of the required
size is available. This solution is usually limited to small, single-unit sump
pumps where hydrostatic uplift forces are not a problem.

Cast Iron: A cast-iron sump basin can be obtained in a single casting up
to 60 in. deep. Basins deeper than this are made in two or three pieces.
Figure 12-3 illustrates the dimensions normally required to properly specify
a cast-iron sump basin. Flat-flage and cup-flange basins may be ordered in
24-, 30-, 36-, 42-, 48-, 54-, 60-, and 72-in. diameters. Sump basin lengths may
be selected in increments of 6 in. If an application requires sections of a
specific length, enter these lengths in the spaces provided in Fig. 12-3. It is
necessary to specify whether a flat flange or cup flange is desired. For conventional installations, the flat-flange sump basin is recommended. A flat-
flange basin will most likely be supplied as standard unless an optional design
is specified. The cup flange is one optional design available. It may be specified for applications which require that the basin cover be mounted flush with
the floor. One 4-in. opening 9 in. below the top of the basin is furnished as
standard in most cases, but a total of three openings are available from most
sources at no additional cost.

The minimum pipe tap opening available is 2 in. and the maximum is 6 in.
The number of basin inlets required and their size and location must be
clearly identified on the drawings. Covers must be specified as round or
square, steel or cast iron, simplex or duplex; the sizes (see basin diagram)
must also be specified. A cement flange and a drain plug are optional. If a
basin has to be immersed in water, a drain plug is useful. The basin can be
firmly anchored before the plug is put in place.

Plastic sheets of a minimum of 8 mils thickness should be used to completely encase the exterior of the cast-iron (or steel) sump basin and all cast-iron or steel pipes entering the basin. Entry of water between the basin and the plastic is not critical. It appears that the use of plastic in this manner may significantly reduce galvanic corrosion.

Steel: Steel basins should in all cases be at least $\frac{1}{4}$-in. thick. Concrete may be poured into the space between the steel basin and the edge of the excavation for the sump to assist in anchoring the basin in place. The concrete should be a minimum of 4 in. and preferably 8 in. thick, and it should be well vibrated in order to make a solid wall. It will provides cathodic protection to the basin from the outside but will not protect metal pipes entering the basin unless they are also embedded in concrete.

Steel basins are not as widely used at this time as was the custom some years ago. Cast iron is widely used and fiberglass basins are growing in popularity.

Plastic Basins: Fiberglass basins appear to be satisfactory, provided they are designed to provide an external ring, or rib, molded around the tank to serve as an anchor. Fiberglass tanks are lighter in weight, which saves shipping costs, and they can easily be repaired if necessary.

Cathodic Protection of Basins: Galvanic corrosion is self-generated electrical activity resulting in differences in potentials which develop when metal is placed in an electrolyte. These differences of potential can result from the coupling of dissimilar metals or they can result from variation in conditions which exist upon the surface of a single metal. Coating the anode in a bimetallic couple merely concentrates all of the corrosion activity which would normally occur on the entire anode surface at faults in the coating. Unless the coating is absolutely perfect, leaks will develop more rapidly than if the metal had been left uncoated (5). Covering with plastic sheeting, however, as previously described, may substantially reduce the galvanic action. The technique is still in the research stage, and while the results appear very promising, the test of time and variety of conditions is necessary.

b. Pump Construction. Where economy is concerned, the pump case should be cast from a close-grain cast iron, and the impeller made from an extra hard cast iron. Figure 12-4 shows a recommended type of pump construction with the exception of the lower bearing. The lower bearing is mounted on the shaft immediately above the impeller. The bearing shown is the standard type. There are several features that are not indicated in Fig. 12-4. Note that the motor mounting head is of the elevated type. This places the upper bearing on the shaft above floor level and assists in preventing moisture from getting into the bearing. It is also much easier to work on an elevated-head type of pump from most manufacturers. The metal is drilled and tapped in most places, making it unnecessary to use separate nuts. When nuts are employed, they are easily accessible.

c. Bearing Lubrication. If a standard type of bearing is used for the lower

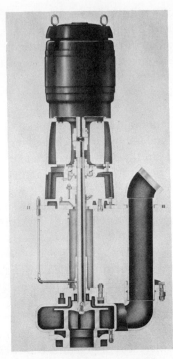

Figure 12-4
Pump Construction. (Courtesy
Aurora Pump Division)

bearing, then force-fed grease lubrication should be applied to all the pump
and line bearings. Grease, forced directly into the bearings through flexible
nylon tubing, will greatly prolong the life of these bearings.

Zerk fittings are usually provided as standard by pump manufacturers.
It is advisable to pay extra and obtain an automatic grease-cup arrangement
as shown in Fig. 12-5. A separate grease line to each bearing is advisable.
If each bearing is provided with a constant supply of grease, the expense and
necessity of replacing bearings is held to a minimum. Grease has been found
more desirable than oil in most applications. A standard bearing is shown
in Fig. 12-6.

d. Special Lower Bearing.   The configuration described above is adequate
if the pump has very little usage. When a pump must operate frequently, or
at 1750 rpm, it is recommended that a bearing construction be used as shown
in Fig. 12-7. Add a grease line with a constant-pressure lubricator and the
bearing will wear in sewage service for many years. There are other special
bearings, such as shown in Fig. 12-8, used to meet certain types of sump
pump applications, but the type shown in Fig. 12-7 is the finest type that has
been encountered in practice.

e. Floats and Rods.   If floats and rods are used, it is advisable to use
stainless steel. Bronze float rods have a tendency to fail at the water level line
on the rod, and bronze floats develop leaks if used in septic sewage.

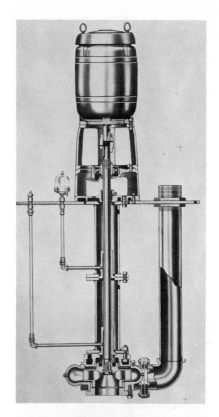

Figure 12-5
Grease Lubrication. (Courtesy
Aurora Pump Division)

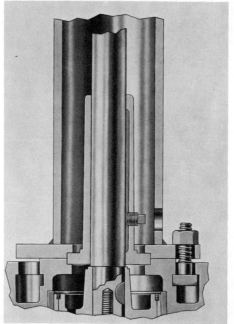

Figure 12-6
Standard Lower Bearing (Without
Grease Line Connected). (Courtesy
Aurora Pump Division)

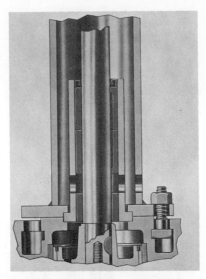

Figure 12-7
Special Relief Bearings. (Courtesy
Aurora Pump Division)

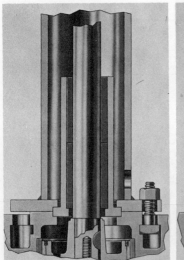 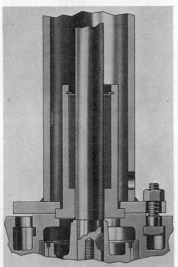 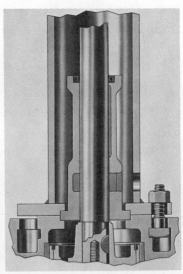

Figure 12-8. Three Types of Special Lower Bearings for Sump Pumps.
(Courtesy Aurora Pump Division)

f. Miscellaneous. Steel basin covers with a manhole are satisfactory. A high-water alarm is a good idea, but they have been known to fail. Duplex installations with mechanical alternators are recommended. Each motor base should be the elevated-head type mounted on an individual oval plate. The oval plates should, in turn, be bolted to the steel baseplate. This enables the pump unit to be removed if necessary, without disturbing the basin cover. Submerged discharge pipes invite cathodic action and are not recommended.

Flanged discharge connections are recommended. Electrical alternators may give more maintenance problems than the mechanical type. The alternator permits a different pump to operate each cycle, i.e., first one, then the other. When the water rises to an appropriate level, one pump starts to operate. The connections should be arranged so that, if the water continues to rise, both pumps operate.

At times pumps have been installed in very small sumps and the liquid waste brought in in such a way as to cause a high-velocity flow across the pump suction. This causes the impeller to be loaded unevenly, thereby increasing shaft and bearing wear and causing rough running. The best correction to such a problem is to go to a well-designed sump of sufficient size and having good inflow characteristics. If it is impossible to use an appropriate size sump, the only alternative is to put in a set of baffles to better distribute the flow and use a special bearing as shown in Fig. 12-7.

The minimum cutoff levels indicated for floats on sump pumps are usually below the minimum submergence required to prevent air entrainment. Therefore, if the sumps are pumped down to this low level, the pumps will be handling air and operating at reduced capacity.

12-4. Determining Pump Capacity. Simplified standard procedure in the selection of sewage ejectors places all types of buildings into a single classification, with the base capacity of the pump determined by the number of toilets serviced (Table 12-1). In addition, the minimum velocity required to

TABLE 12-1. Sewage Ejector Capacity

Maximum No. of Toilets	GPM
1 or 2	75
3 or 4	100
5 or 6	125
7 to 10	150
11 to 14	200
15 to 20	250
21 to 25	300
26 to 30	350

(Courtesy of Aurora Pump Division.)

keep solids moving through the impeller, volute, and pipe lines must be considered. A pump capacity of less than 50 gal/min in a 4-in. pipeline or 100 gal/min in a 6-in. pipeline would permit solids to settle out.

a. Additional Water Drainage. Under normal conditions, there will also be clear water drainage from fixtures such as sinks, urinals, lavatories, floor drains, etc. This drainage into the sewage ejector basin is highly desirable to dilute the raw sewage so that it may be more easily handled.

In establishing the capacity recommendations in Table 12-1, an allowance for fixtures up to *four times the number of toilets is included.* Where the number of fixtures exceeds that allowance, 3 gal/min for each additional fixture must be added to the pump capacity.

EXAMPLE A: Capacity of 5 toilets .... 125 GPM
No. of fixtures to be handled in addition to toilets, 25.
Deduct no. of fixtures that can be handled by pump (4 times no. of toilets) (4 × 5 toilets) ....20.
Excess fixtures .... 5 Fixtures @ 3 GPM each 15 GPM
Total capacity required .... 140 GPM

b. Water Seepage.   Water seepage in amounts up to 50% of the sewage capacity can be safely handled by sewage ejectors selected on the basis of Table 12-1. When seepage exceeds 50%, the additional amount must be added to the pump capacity to determine the required pump size.

Water seepage delivered to the sewage basin through drain tile from drainage beds will vary according to local soil conditions. As a guide to safe calculations, the following may be used:

Sandy soil—14 GPM per 1,000 ft² of bed
Clay soil—8 GPM per 1,000 ft² of bed

EXAMPLE B: Total sewage and clear water drainage from fixtures 140 GPM
Water seepage................ 80 GPM
Allowance for handling seepage up to 50% of sewage total .... 70 GPM
Additional capacity required .... 10 GPM
Total Pump capacity required .... 150 GPM

c. Unusual Additional Service Flow.   Careful consideration should be given to real or potential additional flow from unusual or abnormal additions to the service flow as determined by the method of calculation outlined above. This includes additional flow from such services as boiler drainage, air-conditioning drainage water, etc. *All of this service flow must be added to the projected pump capacity requirement.*

12-5. Determining Pump Discharge Head. The pump discharge head may be calculated by determining the distance from the lowest level of water in the basin to the street level. By using the street level in place of the sewer line as point of reference, a reasonable factor of safety is established accounting for the possibility of back pressure. To this must be added the friction loss in the discharge pipeline and fittings.

EXAMPLE C: Distance from lowest level of water to street level ...... 26 ft
Friction of discharge pipe (150 GPM 4-in. pipe 150' long) .. 2 ft
Friction of pipe fittings (elbows, check valves, gate valves, Y-branch, etc.) .... 2 ft
Total discharge head .... 30 ft

12-6. Determining Tank or Basin Size. The size of the sewage tank or basin should be carefully selected to prevent both short-cycling of the pumping operation and excessive settling out of solids. A drawn-down ratio of 3 to 1 is recommended. This means that the basin area between the low sewage level and the high sewage level should accommodate three times the capacity of the pump before the pumping cycle begins. A minimum distance of 1 ft between the lowest sewage level and the basin floor should be allowed. In addition, the distance between the top sewage level and the basin cover should be one-third the storage area between the high and low sewage levels—and not less than 2 ft. More detail is shown in Fig. 12-9. When conditions limit the

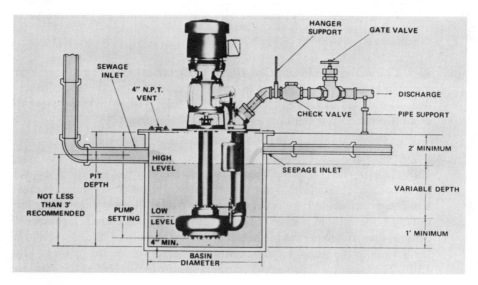

Figure 12-9. Typical Sump Pump Installation. (Courtesy Aurora Pump Division)

basin depth, the required capacity may be obtained by means of a larger basin diameter. For fast calculation:

EXAMPLE D: Total pump capacity . . . . 150 GPM

Basin to hold 3 times pump capacity (3 × 150 gal.) . . . . 450 Gal

From Table 12-2 a 48-in. diameter basin will hold 95 gal/ft of depth.

Therefore basin depth is (450 ÷ 95) . . . . 4.74 ft

Add approx. $\frac{1}{3}$ to depth for storage between high level of sewage and basin cover or minimum 2 ft . . . . 2.00 ft

Add approx. 1 ft. to depth for distance between lowest level of sewage and basin floor . . . . 1.00 ft

Total pit depth required . . . . 7.74 ft

Recommended pit depth (next standard 6-in. increment) 8.0 ft

12-7. *Calculating Round Tank or Basin Capacities.*

TABLE 13-2. Round Tank or Basin Capacities Per Foot of Depth

Simplex Basin Dia. Inches	24	30	36					
Duplex Basin Dia. Inches				42	48	54	60	72
Capacity Per Foot In Gallons	24	38	53	77	95	119	150	212

Metal Tanks are Available in Depths of 6″ Increments.
(Courtesy of Aurora Pump Division.)

12-8. *Calculating Square-Pit Capacities.* To determine the size of a square or rectangular pit commonly used in concrete construction, the same procedure as outlined in Example D may be followed. To calculate the capacity of a pit other than a round one, use the following formula:

$$\text{Cubic content in ft}^3 \times 7.48 = \text{no. of gallons}$$

12-9. *Submergence for Sump Pumps.* Submergence curves for sump pumps are difficult to obtain from pump manufacturers; however, the more progressive concerns usually have a curve for each pump. Submergence curves are usually made without a strainer being mounted on the pump. If a strainer is desired, then its height plus the height of submergence must be allowed above the bottom of the sump. Examples of submergence curves are shown in Fig. 12-10. The pumps will actually prime and operate at one half of the submer-

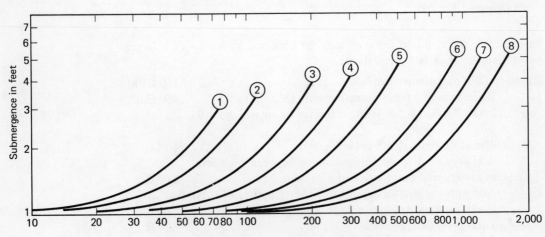

Figure 12-10. Submergence Curves for Sump Pumps.

gence shown by the curves, but they will be operating at a reduced capacity due to the entrainment of air caused by vortexing. The curves are drawn on the basis of the pump operating in a normal sump and handling cool water. The data can be used for other liquids by comparing them with water. The submergence shown is sufficient to handle these pumps up to 160°F at sea level, with the limitation that on curve no. 8 the maximum capacity would be 1,400 gal/min.

To use these curves for higher elevation it is possible to do one of two things:

1. The first method is to increase the depth of submergence 1 ft for every 800 ft of elevation up to 5,000 ft at 160°F.

   Example: If it is desired to operate a pump at 160°F and at 3200-ft elevation, divide 3,200 by 800 which gives 4 ft. If the submergence shown on the curve is 2 ft this is added to the 4 ft to give a total of 6 ft submergence necessary at 3,200-ft elevation and 160°F.

2. The second method is to drop the maximum temperature 6°F for every 1,000 ft of elevation above sea level; this is safe up to 5,000 ft.

   Example: If the sump is to be located 5,000 ft above sea level, multiply 6°F by 5 to give 30°F, which is subtracted from 160°F. Therefore these curves are safe at 5,000 ft with water temperatures up to 130°F.

The preceding discussion is based on the pump's being operated in normal sumps and handling water. There are many factors that can cause more depth of submergence to be required than is shown by these curves. Information given here is not meant to be a treatise on sump design; the examples show what factors can require deeper depth of submergence than shown on the curves.

Most sumps are designed so that the water is returned at or slightly below normal operating level and through large pipes so that the inlet velocity is small. If the liquid is returned to the sump far above normal operating level, the falling stream will definitely entrain air and cause the pump to handle an emulsion which would actually reduce the capacity. The falling water could also strike the float mechanism, causing erratic float operation, or fall in such a way as to cause wave motion on the surface and therefore erratic float operation.

12-10. Lower Bearing Life. Comments have been made several times with respect to using a special lower bearing on a wet-pit pump, usually called a sump pump. Assume, for comparative purposes that a standard bearing has a life of only 20 hours. In the same application, the special lower bearing might yield a life on the order of 40,000 hours.

12-11. Pressurization. Pressurization is a process by which clean, abrasion-free water is applied to the support column. There are several ways it may be achieved. It can be used where abrasion particles are extremely severe in a waste. There are at least seven bearing configurations that might be used together with pressurization to further extend the life of a pump. Under such circumstances, special construction materials may be required for the pump volute, impeller, and discharge piping.

Operating the pump at its selected operating point is mandatory for long pump life. For instance, a 500 GPM pump specified for a 50-ft head but operated at an actual head of 20 ft will have high radial loads and cause severe bearing and shaft wear. Similarly, if operated near shutoff head, the pump would not be recommended for continuous duty. Thus, exotic configurations cannot replace intelligent sizing of the equipment for the conditions if optimum service is desired.

12-12. Dry-Pit Pump Stations. A dry-pit pump station is shown in Figure 12-2. The dry-pit pump station requires a separate wet well. The pumps are then mounted in a dry compartment. A workman can shut suction and discharge valves in the pump piping, and pump maintenance can then theoretically be accomplished in the compartment. The workman should be able to work on the pumps in the dry pit easier than pulling a wet-pit pump. For example, a frame 182T motor weighs 70 lb. Anything heavier than this needs a hoist of some kind. The better-quality stations have appropriate lifting rings welded in the top of the station so that the weight of the entire pump can be lifted without damage to the pump-room ceiling from a structural viewpoint. Weights for the motors can easily exceed 225 lb. Thus the ability to work more comfortably in a dry-pit station is a function of the manufacturer's design.

12-13. Minimum Pump Capacity. There are hundreds of lift stations (wet or dry pit) that have pumps installed sized for 150 gal/min or less capacity, using impellers capable of passing a 3-in. sphere and operating at 1750 rpm. Many of these installations later have maintenance problems. It is immaterial who manufactures the pump. If the impeller is in the range of 9 to 12 in. in diameter and operates under the above conditions, trouble is likely to occur.

Laboratory pump curves are often made by shutting a valve partially (to throttle the flow) in the discharge line to simulate different operating heads. Quite often the pumps in the laboratory are operated under a slight suction lift. In the lab a pump may operate, for example, at 29% efficiency, 100 gal/min, 124 ft total head at 1750 rpm, and with an $10\frac{7}{16}$-inch impeller diameter. Take that same pump into the field and put it in a sewage lift station, and it may shake the impeller off the end of the shaft despite the retaining nuts, etc. It normally takes a pump a period of several weeks to incur any significant damage, but the author has observed an extreme case of a pump tearing the impeller off the shaft within a matter of minutes.

There are several factors that create these problems. In real estate developments, there is often substantial pressure to keep down the cost of sewers and pump stations. The sewers may be constructed with a 4-in. pipe size. Long 4-in. lines have a tendency set up surge conditions that alter the operating point on the pump's characteristic curve. In fact, these surges are sometimes so bad that pumps have been observed to swing from near shutoff at one end of the curve to a point at the opposite end of the curve where the impeller will cavitate.

Another problem is assuming an improper pipe friction. A designer frequently selects a friction value on the condition that the sewer is a particular age. The idea is that the pump will deliver the design load under that friction condition. The friction factor $C = 100$ is commonly used for design. Using a $4 \times 4 \times 9$ curve, shown in Fig. 12-11, assume a design capacity of 100 gal/min with a vertical lift of 32 ft and 2,500 ft of new 4-in. cast-iron force main. At $C = 100$, the friction is 1.23 ft/100 ft of force main (for capacity of 100 gal/min). The designer's calculations would be:

$$\begin{aligned}
\text{Total vertical lift} &= 32 \text{ ft} \\
\text{Friction} = 25 \times 1.23 &= \underline{31 \text{ ft}} \\
\text{Total head} &= 63 \text{ ft}
\end{aligned}$$

The pump manufacturer might cut the impeller at $8\frac{1}{16}$-inch diameter. The

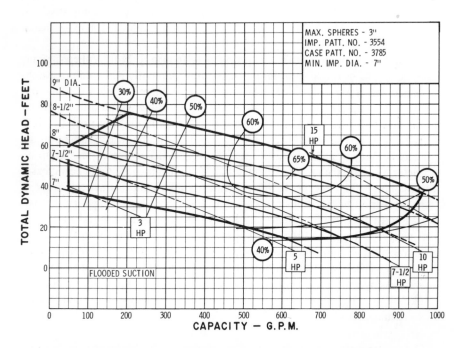

Figure 12-11. Size $4 \times 4 \times 9$ Sewage Pump Impeller Performance Curve. (Courtesy Aurora Pump Division)

actual pipe friction when the pump is installed, i.e., the pipe is new, is at least $C = 130$, and it could be as high as $C = 150$. At $C = 130$ the friction is only 0.62 as great as when $C = 100$. Examination of the pump curve indicates that the pump will, in practice, shift to the right. As it shifts, the capacity increases, and this in turn causes greater friction in the pipe.

The actual friction factor we are seeking can be determined by dividing 1.23 by 0.62 and obtaining a figure of 2.00 (approximately). If a pipe friction table is consulted, we find that, for a head loss of 2.00 ft/100 ft, the flow rate will be 130 gal/min. This is the actual flow and operating point for the $8\frac{1}{16}$-inch diameter impeller. This is one reason why pumps sometimes operate satisfactorily when new, but later give operating problems as the operating point shifts due to aging of the pipe and the corresponding increase in friction of the discharge pipe.

A convenient approximation for calculating the cost of pumping 1,000 gallons is:

$$\frac{\text{total head in ft} \times \text{cost per KWH in cents}}{\text{overall efficiency in } \% \times 3.185} = \text{cost}$$

in cents for 1,000 gallons.

In the foregoing situation, the best way to solve a low-volume pumping problem, if it is necessary to use a 4-in. force main, is to use a pneumatic sewage ejector. A 50-gal/min pneumatic ejector discharges in 30 seconds, i.e., at the 100 gal/min velocity rate. Using a duplex unit, with a compressor air reservoir system, two 5 HP 1,200 rpm compressors would complete the installation.

Another way to handle the problem is to use a $2\frac{1}{2}$-in. sphere-passing impeller. The health authorities in some states frequently will not approve the plans unless a 3-in. solid size is shown. However, there are many installations where the maintenance men have nothing but trouble with a 3-in. solid-size impeller, and they convert to a $2\frac{1}{2}$-in. solid size and end the problem. It is usually possible to put a $2\frac{1}{2}$-in. solid-size impeller in the case of a 3-in. pump. There is a loss of efficiency and head. If the impeller is not of maximum diameter, increase the impelled diameter 15%. If the impeller cannot be increased in diameter, then accept the loss in capacity and head if practical. Some of the converted installations are known to have been in service over 15 years with no further trouble. Hence the 3-in. solid-size requirement is not as valid as believed. Raggy and stringy solid material gives the most trouble.

Where high pumping heads are involved, it is frequently advisable to use either screens, comminutors, or both ahead of the pump and reduce the solid-size passing requirement of the impeller to 2 in. or even $1\frac{1}{2}$ in. in order to take advantage of increased pump operating efficiency and freedom from maintenance problems.

12-14. Determination of Pumping Rate. Sewage flow will ordinarily be somewhere between 70 and 130% of the community water consumption rate from the waterworks. Low groundwater infiltration in the sewers (due to sewers being in excellent condition) plus high water usage that does not return to the sewers (such as watering lawns) may result in the lower value. High infiltration or a combined sewer system may lead to the higher figure.

The engineer should determine the local conditions and size accordingly. One design rule sometimes used is to allow 1,000 to 2,000 gal/day infiltration per acre served, depending on the groundwater level and local condition of the sewer. An infiltration rate of 50,000 gal/day/mile of sewer is used by other engineers. It is advisable to base the sewage pump capacity on four times the maximum flow rate per day in small residential sections. This maximum flow rate consists of waste flow entering the sewers plus estimates of the sewer infiltration.

In a commercial section, the design rate might be as low as 150% of the estimated peak flow rate. These flow rates are per pump. If a duplex is used, either pump should be able to handle the demand of the station. Remember that, when two pumps are operating simultaneously, the total flow will not ordinarily double the rate of one pump. In fact, the second pump operating may contribute very little to the total flow capacity due to increase in head and similar factors.

Many state health departments recommend a design flow of 250 gal/day/capita for residential areas which have an actual water consumption per day ranging between 60 and 125 gal/capita/day. The allowance for infiltration is in addition to this value. If stormwater also goes through the lift station, the pumps must be capable of handling this flow.

12-15. Pump Characteristic Curves. Inadequate consideration of the pump characteristic curves can result in operational problems. A pump operated at a suction lift will have a different curve than one operated suction-flooded. Dry-pit and wet-pit pumps are characteristically operated suction-flooded. If the inflow into the wet well exceeds the pumping rate, it may shift the operating point. This depends on how the discharge head varies with the flow rate.

When possible, sewage pumps should have the operating point near the maximum efficiency point on the impeller. Operating points close to the cutoff point are often encountered, but this is poor practice and may result in trouble for the client later. It is difficult to formulate rules of thumb; however, at 1,750 rpm a 3-in. solid-size impeller should not be operated at less than 40% efficiency. This is true despite the fact that the pump manufacturer's curve shows satisfactory operation at a lower efficiency. More detailed explanations on characteristic curves will be given in Chapter 13.

12-16. Sewage Pump Construction. The tendency of engineers is to consider pumps as a simple machine, which is a problem of no real concern. Many an engineer unwittingly cost their clients hundreds of dollars for needless maintenance by either improperly sizing the equipment or failing to provide adequate specifications to prevent the use of inferior-quality products by the contractors. This presentation will discuss pump construction in detail.

a. Impeller Rotation Speed. In lift stations, one of the largest sources of serious trouble is attempting to operate pumps at excessive speeds. The head required on lift stations is often high enough to tempt the manufacturer to use 1,750 rpm pumps. While 1,750 rpm is a very reasonable speed for impellers having a small distance between blades, it is not a reasonable speed for a sewage pump impeller capable of passing a 3-in. sphere, when the diameter is in excess of 9 in. and the pump is operated at less than 40 to 55% efficiency. An impeller of the same conditions approaching 12-in. diameter is in a very poor design condition. When the impeller diameter is over 9 in. for the above conditions a speed of 1,150 rpm is advisable.

As impeller diameter and solid-size passage increase, the speed of rotation must decrease. In order to develop acceptable head capacity at the lower speeds, the impeller diameter increases. This results in a more expensive pump, but it is often required to insure dependable operation for sewage service.

b. Pump Frame. Closely coupled sewage pumps where the pump impeller is mounted on the extended motor shaft are not recommended. This type of unit is shown in Fig. 12-12. The most compact type of unit recommended is shown in Fig. 12-13. These are hundreds of the type of configuration in Fig. 12-12 in use, made by a number of manufacturers, and a significant percentage of them will cost the owners more money than if the above advice had been heeded. There should be a separate pump frame with two ball-bearing supports and a separate heavy-duty shaft. A standard vertical-type motor may be used. The type of frame recommended requires a slight increase in vertical space requirements and initial cost, but it is well worth the initial investment.

Figure 12-12
Close Coupled Sewage Pump.

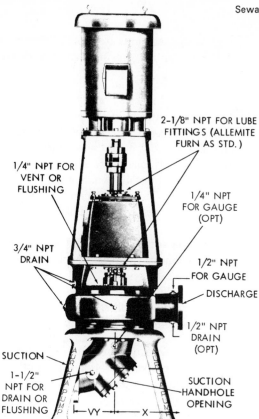

2-1/8" NPT FOR LUBE
FITTINGS (ALLEMITE
FURN AS STD.)

1/4" NPT FOR
VENT OR
FLUSHING

1/4" NPT
FOR GAUGE
(OPT)

3/4" NPT
DRAIN

1/2" NPT
FOR GAUGE

DISCHARGE

1/2" NPT
DRAIN
(OPT)

SUCTION

1-1/2"
NPT FOR
DRAIN OR
FLUSHING

SUCTION
HANDHOLE
OPENING

Figure 12-13
Recommended Type Compact
Sewage Pump. (Courtesy Aurora
Pump Division)

c. Shaft Sleeves. A shaft sleeve should be considered as a standard requirement on sewage pumps. When packing is used, a hardened stainless-steel shaft sleeve with a Brinell hardness of about 550 is recommended. Ceramic-coated sleeves at a somewhat higher Brinell are available, but the gain from their use does not compensate for the additional expense. When stainless-steel sleeves are used, it is primarily for mechanical seal applications. Bronze shaft sleeves are not recommended for sewage pumps.

d. Packing Versus Mechanical Seals. Mechanical seals are extremely popular because there is no leakage from the packing gland. However, if a mechanical seal breaks or wears out, it may take a pump completely out of service until a new seal can be obtained and installed. Since sewage pump seals are rather special, delivery is not always as rapid as would be desired. More will be said on mechanical seals in the next section.

Packing also has the disadvantage of scouring shaft sleeves if overly tightened. However, it is extremely rare that a pump with packing has to be put out of service unless the shaft sleeve is badly scoured. This normally gives warning far in advance, with adequate allowance of time to obtain parts. A lantern ring, a teflon or a bronze hollow shell with openings the size of a ring of packing, should be used, and a separate water-seal unit should be used. Water-seal units are explained in Chapter 13. This prevents the dripping of water from the packing gland from being septic.

e. Mechanical Seals. In sewage pumps, double mechanical seals are recommended if seals are to be used. A separate water-seal unit with a pressure 5 psi greater than the pump discharge pressure is recommended. It is advisable to keep two sets of mechanical seals, one shaft sleeve, and one shaft key, in stock at all times if mechanical seals are used. The major hazard of having only one spare set is that sometimes a mechanical seal will be broken during installation. Sewage, if allowed to get into a seal, may quickly ruin the seal. Thus water under higher preseure than the pump discharge, applied in the center of a double seal, will tend to keep foreign matter out of the seal.

Some factory-assembled lift stations use a filter on the discharge of the pump to remove all objectionable foreign matter. Pressure grease lubricators or pressure oiler systems can also be used. The filter is the least desirable and is not considered very acceptable by the author. The pressure grease systems are preferred to the filter, and separate water-seal clean water source is preferred over either of the above.

f. Vibration and Cavitation. Vibration and/or cavitation is usually the result of improper operation that is often directly traceable to the designer or specification writer. No matter what alloy has been used in the impeller, if cavitation is occurring, it will eventually ruin the impeller. The more severe the cavitation, the quicker it will ruin the impeller. Vibration, if sufficiently severe, may shake the pump impeller off the end of the shaft despite keyed retaining nuts that theoretically appear indestructable.

Both vibration and cavitation are often the result of using an improper operating point on the pump characteristic curve. Inadequate dynamic or hydraulic balance of an impeller can also create vibration.

g. Discharge Line Surges. Discharge-line surges can be a major problem. It is most likely to occur when very long, horizontal 4-in. pressure mains are used, and particularly when such mains connect into other lines which in turn serve pumps at one end and an elevated free discharge point some distance beyond the junction. These surges have essentially the form of an electrical standing wave on a transmission line. An instrumented pump on such a line has been observed to cycle between pump shutoff head and below minimum operating head.

h. Impeller Material. Bronze impellers or regular cast iron should not be used on sewage pumps. Iron impellers that have been properly heat-treated to make them extremely hard may be used. High-silicon cast iron is sufficiently hard that machine work is done only by grinding. Cutting an impeller is a skip-cutting type of operation that destroys carbide-tipped tools due to the shocks of alternately hitting a vane and then open air.

12-17. Sewage Ejectors. A typical sewage ejector installation is shown in Fig. 12-14. A schematic diagram of a float-controlled system is shown in Fig. 12-15. There are two common arguments against ejectors. One is that it takes more space. However it does away with the wet well, so this argument is rather inadequate. The second argument is that it requires more connected horsepower.

Like any machine, the pneumatic ejector has its place where it excels. That place is where the sewage lift station design rate is under 200 gal/min. Since an ejector is commonly sized on the basis of 30 seconds to fill, duplex ejectors can operate at double this value in a sewer.

The pneumatic ejector, by design principle, is an extremely simple and workable mechanism. Fundamentally it consists of a receiver or *pot* that allows flows to enter without obstruction. As the pot fills, compressed air is introduced to displace the contents up to a higher runoff line. The pneumatic ejector is unique as a pumping mechanism inasmuch as no mechanical parts are used to do the work on the material being pumped, and it has no practical limitations on head.

Under normal conditions, the equipment is designed to operate within a one-minute cycle. This cycle consists of two phases. First the filling of the pot, and then the discharging of its contents. Operation is fully automatic

Figure 12-14
Typical Sewage Ejector System.
(Courtesy Ralph B. Carter Co.)

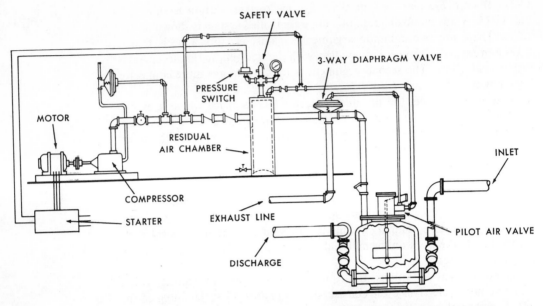

Figure 12-15. Schematic of a Float-Controlled System. (Courtesy Ralph B. Carter Co.)

with a choice of electrical or mechanical control systems. The pot does not cycle until filled. If the flow is high enough that the pot fills in less than 30 seconds, the pot will be emptied when full and operate on less than a one-minute cycle.

Figure 12-16 shows a schematic arrangement of a fully equipped duplex installation using an air reservoir system. Where reliable operation is required, this is the recommended solution. The air reservoir reduces the amount of connected horsepower required and makes more effective utilization of the compressor equipment. The air compressors are the weakest link in the pneumatic ejector system.

12-18. Discharge Check Valves. In a sewage lift, the flow is shut off at the inlet end. Hence the traditional water-hammer analysis based on flow interruption at the discharge end is not applicable. It is advisable to use spring-loaded check valves so that the valve will close more rapidly when the flow ceases. When the pump stops, the entire mass of water from the suction intake forward is free to continue forward under its own momentum. If the check valve closes before the mass reverses direction, then water hammer will be small enough not to be a problem. Thus, either a quick-closing special check valve or a spring-loaded conventional check valve is required.

12-19. Head Variations. When the pumps start, depending on the design of the discharge, the line may be filled with water. This mass of water resists

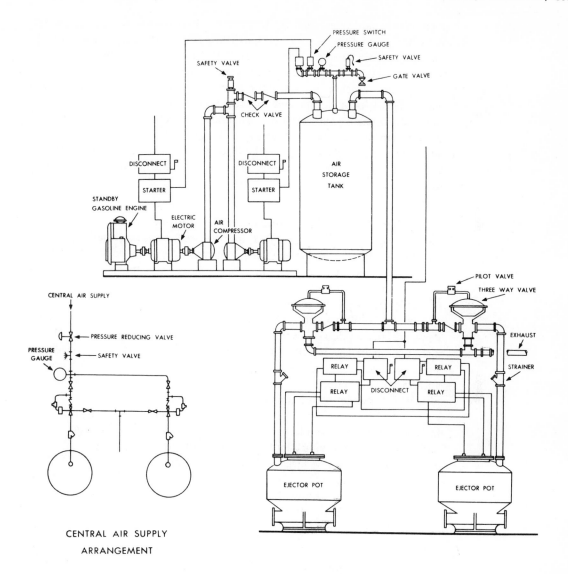

Figure 12-16. Schematic Arrangement of a Fully Equipped Duplex Eject-O-Matic Installation Utilizing an Air Reservoir System. (Courtesy Ralph B. Carter Co.)

movement, which in turn causes a shift of the operating point on the pump toward the left. Until the mass of water has stabilized, the pump may vibrate at harm to itself.

Aside from the mass of water, in some real estate developments the owners attempt to eliminate a wet well at the treatment plant. If the treatment plant is above ground, the water has to be lifted to the top of the plant. As a further complication, several lift-station force mains may enter a common force main for the ultimate part of the journey to the treatment plant. The result is that, if two or more lift stations operate simultaneously, it tends to increase the head on each station and shift the operating point toward cutoff. This is poor design. Even increase of the main size beyond the junction fails to resolve the veriations. Arrangements on the plans for only one set of pumps to operate at a time are seldom installed. If installed, they have a tendency to get disconnected. In an installation like this, pneumatic ejectors should be used or the sphere-passing size of the impeller should be reduced to $2\frac{1}{2}$ in. or smaller.

12-20. Solid-Passing Size of Impellers. Feces and toilet paper quickly become torn into very fine particles. In fact, in a wet well sewage normally looks like dirty water. Insofar as the sewage is concerned, a $1\frac{1}{4}$-in. solid-size impeller is more than adequate. The 3-in. sphere-passing capability of an impeller is based on the concept that anything that can be flushed down a toilet can be carried through the pump.

Diapers, clothing, plastic toys, sanitary napkins, and countless other items are flushed down sewers. Stand at the inlet sewer of a large treatment plant and an endless variety of objects will appear. Diapers, men's shirts, and similar articles frequently will clog up a pump. Unless prohibited by health authorities, it is better to use a coarse-screen basket or a bar screen at the wet pit and use pumps with a smaller solids-passing capability.

A $1\frac{1}{4}$-in. pump has been serving for some years in an amusement park having over 60,000 people per day on some weekends. An identical pump has for years been serving a large apartment complex. It is occasionally necessary to rake out a few large objects, such as diapers, caught on the bars. However, the pumps themselves have given no problems. There are a number of other installations known to be in service, and none of them has been a problem. Each installation does have appropriate bearings and other features discussed earlier.

12-21. Wet-Well Design. A wet well for a use in conjunction with a dry pit may be designed in a number of different ways. There is no written rule that the wet well should be circular; however, a number of them are built circular. It is possible to buy large circular sections of concrete pipe. Thus, a large hole is dug, a bottom base is cast, and the large pipe is cased into place on top of it. The bottom of the wet well is then sloped to reduce the possibility of sewage solids settling out and becoming septic. Typical wet wells are shown in Figs. 12-17 and 12-18.

A wet well should be sized to prevent the pumps from being turned back on for at least 5 minutes. At the same time, the wet well should be small enough that septic action does not take place at periods of low flow.

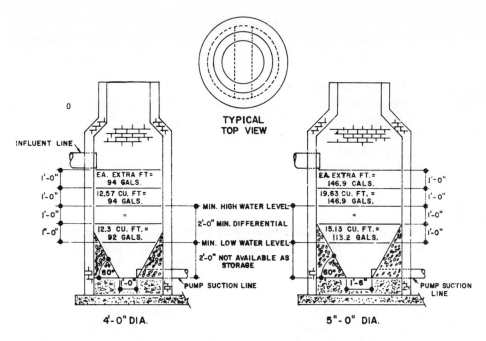

Figure 12-17. Typical Wet Well Designs. (Courtesy Can-Tex Industries)

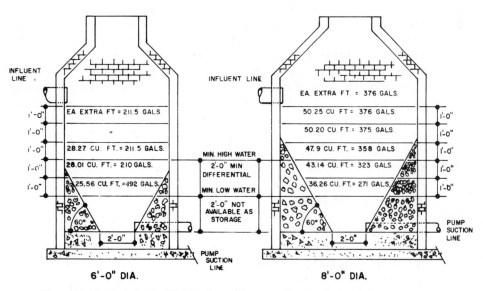

Figure 12-18. Typical Wet Well Designs. (Courtesy Can-Tex Industries)

If concrete is used, the inner surfaces of the concrete should be given a bituminous or epoxy coating as explained for construction of wells for wet-pit pumps earlier in the chapter. Brick has been used to fabricate wet wells, but labor and materials are currently so expensive that it is now seldom used. Concrete blocks are still used. They should be parged with cement mortar and then given a bituminous or epoxy coating as stated earlier.

12-22. Flotation and Sinking. It is most embarrassing, and expensive, for an engineer to have a lift station either partly float out of the ground or sink out of sight. It is therefore mandatory that the engineer determine the soil conditions at the site. If the soil is a coarse sand, gravel, or stable soil, it can be assumed that the buoyant force is equal to the weight of displaced water (62.4 pounds per cubic foot) for all parts of the structure below the highest groundwater level. The fill on top of the canned station tends to hold it down, provided it is above the groundwater table.

12-23. Self-Priming Centrifugal Trash Pump. Self-priming centrifugal trash pumps have gained substantially in acceptance in recent years. They are particularly useful in certain types of solids applications. The biggest problem encountered by the author on this type of construction has been that some of the prefabricated lift stations on the market have a quality of construction that is significantly less than others available. Unfortunately, it is sometimes difficult under present federal regulations to prepare a specification that will insure obtaining a quality lift station.

The distance of vertical lift by this type station is sometimes overly ambitious in the literature. A distance of 6 ft from the low-water cutoff level to the center of the pump suction, or a value one-half the pump discharge head, whichever is the smaller figure, is the maximum lift recommended unless a number of precautions are taken. No centrifugal pump should be expected to perform properly in suction lift unless the discharge head is more than twice the value of the equivalent lift of liquid on the suction side. In calculating suction lift, the internal friction of the piping, fittings, and pump entrance losses must be taken into consideration. A long-radius elbow should be used to change from the vertical to the horizontal plane for entry to the pump. There should be a horizontal nipple having a length of not less than 4 pipe diameters between the elbow and the pump entrance. It is desirable that the center of the pump intake be within 4 ft of the high water level in the basin when the pumps prime.

12-24. Submersible Pumps. There are some excellent submersible pumps on the market. If a submersible pump is used in a critical location, it is advisable to provide a duplex installation, provide 100% spare replacements, and provide for a means to remove the pump and replace it without having to use divers.

There is not much that can be done in the way of maintanance on the submersible units. Hence it is not unusual that the unit operates until it fails.

Many of the designs are impractical for attempting repairs locally. It is a time-consuming process to send the units back to the factory. New equipment, available out of a dealer's stock, may require piping changes or may physically fail to fit into the original pit.

When considering a submersible pump, insist on the pump characteristic curves. Do not size by a table. Determine your operating point in a conservative portion of the pump curve, and do not attempt to work on fringe area conditions. The pump and motor are both physically below the water, and it is more difficult to know when the pump is experiencing trouble than it is for some of the other types. However, the unit is still a pump, and its application cannot be abused any more than any other pumping unit.

PROBLEMS

1. Determine the total force of the water acting on one wall of a 6-ft^2 wet well when the water is 10 ft deep.

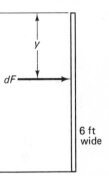

$$\sigma = \text{specific weight}$$

$$F = \int_A dF = \int_A p \, dA$$

$$F = \int_0^{10} 6\sigma \, y \, dy = 6\sigma \left. \frac{y^2}{2} \right]_0^{10}$$

$$= 6 \times 62.4 \times \frac{10^2}{2} = 18{,}720 \text{ lb}$$

2. A residential section has 200 homes. There is an elevation difference of 30 ft between the sewer inlet elevation at the lift station and the highest point in the force main, which consists of 1,400 ft. of 4-in. cast-iron pipe. Find the pump capacity and head.

The figure of 0.6 gal/min per residence can be used. The pump capacity is $0.6 \times 200 = 120$ gal/min. The friction loss in the 4-in. force main is 14 times 1.86 (the loss in 100 ft of 4 in. cast-iron pipe). This gives $14 \times 1.86 = 26.0$ ft. Add 1.3 ft. to cover friction loss through the gate valves, check valves, and elbows in the station.

$$\text{Total head} = \text{static head} + \text{friction in force main}$$
$$+ \text{friction loss in station fittings}$$

$$\text{Total head} = 30 + 14.8 + 1.3$$
$$= 57.3 \quad \text{or approximately 56 ft}$$

3. How much horsepower and what size electric motor is required to put out a circular tank 10 ft in diameter and 15 ft high when the water is 10 ft deep in the tank and is to be lifted 15 ft above the top of the tank?

$$dw = F \cdot ds$$

$$dw = 62.4 \pi r^2 \cdot ds \cdot s$$

where W = work,

F = weight of water in the differential of height $ds = 62.4\pi r^2 \cdot ds$.

Total work = W

$$= 62.4\pi r^2 \int_{15+15-10}^{15+15} s \cdot ds$$

$$= 62.4\pi r^2 \left[\frac{s^2}{2}\right]_{20\ ft}^{30\ ft}$$

$$W = 31.25\pi r^2 [S^2]_{20}^{30} = 1,227,188 \text{ ft lb}$$
$$= \text{work}$$

$$\text{Horsepower} = \frac{1,227,188 \text{ ft lb}}{33,000 \text{ ft lb/min} \times 3 \text{ min}} = 12.5 \text{ HP}$$

Therefore, a 15-HP motor is required.

4. What distances in earth and rock will give a minimum cost for excavation of a force main to run between points A and B, 600 ft horizontally and 20 ft vertically apart, with earth excavation at $12 per running ft and rock $30 per ft?

Let x = horizontal distance in rock, and $600 - x$ = distance in earth. Then

$$\sqrt{x^2 + 20^2} = \sqrt{x^2 + 400} = \text{actual distance in rock}$$

Therefore:

$$C = \text{cost} = 30\sqrt{x^2 + 400} + 12(600 - x)$$
$$= 30(x^2 + 400)^{1/2} + 7,200 - 12x$$

$$\frac{dC}{dx} = \frac{30x}{(x^2 + 400)^{1/2}} - 12 = 0 \quad \text{for a minimum value}$$

5. Determine the approximate air capacity required to operate a pneumatic sewage ejector.

The approximate empirical relation is:

$$V = \frac{Q(H + 34)}{250}$$

where V = volume of free air in ft³/min,

H = total head in ft,

Q = rate of sewage discharge in gal/min.

To allow for expansion of air in the storage tank as the sewage is displaced, the volume of air in the storage tank and the characteristics of the compressors should be selected to provide V in the above equation at a pressure at least 40% higher than that required to raise all sewage to the maximum computed lift.

REFERENCES

1. GILCHRIST, F. M. C., "Corrosion in Concrete Sewers." *Pub. Health*, 17, 10, 477 (1953).

2. POMEROY, R. D., "Protection of Concrete Sewers in the Presence of Hydrogen Sulfide." *W & S. W.*, 107, 10 (1960).

3. POMEROY, R. D., "Calcareous Pipe for Sewers." *Jour. WPCF*, 41, 8, 1491 (Aug., 1969).

4. WAKEFIELD, JOHN W., "Florida's Fringe Area Sanitation Problem." Presented at 28th Annual Meeting, Federation of Sewage and Industrial Waste Assns., Atlantic City, N.J., Oct. 10–13, 1955.

5. HASOCK, B., "Corrosion Cathodic Protection and Common Sense." Paper No. HC-3, 12-1268, presented at the National Assn. of Corrosion Engineers, Northeast Region Meeting, Pittsburgh, Pa., Nov. 13, 1957.

6. Ralph B. Carter Company, *Carter Pneumatic Ejectors*. Bulletin 5408, 1956.

RECOMMENDED SUPPLEMENTAL READING

1. Pages 141–152, 160, 165–168. Manholes, Bends, Junctions, Drop Manholes, Terminal Cleanout Structures, etc. Reference: *ASCE*—Manuals and Reports on Engineering Practice, No. 37, Design and Construction of Sanitary and Storm Sewers, *ASCE*, 1969.

2. Pages 100–119, Bearings. Reference: IGOR KARASSIK and ROY CARTER, *Centrifugal Pumps*, F. W. Dodge Corp., 1960.

3. Pages 325–351, Shaft Design for Critical Speeds. Reference: A. J. STEPANOFF, *Centrifugal and Axial Flow Pumps*, 2nd ed. John Wiley & Sons, 1957.

4. Pages 48–73. Centrifugal Pump Operation. Reference: TYLER G. HICKS, *Pump Operation and Maintenance*, 1st ed. McGraw-Hill Book Co., Inc., 1958.

5. Pages 63–98. Head on a Pump. Pages 98–127. Pump Capacity. Reference: TYLER G. HICKS. *Pump Selection and Application*, 1st ed. McGraw-Hill Book Co., Inc., 1957.

6. Pages 293–303. What is Head on a Pump? Reference: FRANK A. KRISTAL and F. A. ANNETT. *Pumps*, 2nd ed. McGraw-Hill Book Co., Inc., 1953.

7. Pages 231–254. Supporting, Restraining, and Bracing the Piping Systems. Reference: M. W. Kellogg Co. *Design of Piping Systems*, 2nd ed. John Wiley & Sons, 1956.

8. Pages 40–116. Standards for Piping. Reference: HOWARD F. RASE, *Piping Design for Process Plants*. John Wiley & Sons, 1963.

9. Pages 179–180. Pump Impellers. A. DE KOVATS and G. DESMUR, *Pumps, Fans and Compressors*. Blackie & Son, Ltd., 1958.

10. Pages 385–448. Bearings and Lubrication. Reference: RICHARD M. PHELAN, 2nd ed., *Fundamentals of Mechanical Design*. McGraw-Hill Book Co., 1962.

11. Pages 23–83. How Castings are Made. Reference: GLEN J. COOK, *Engineered Castings*. McGraw-Hill Book Co., 1961.

12. Pages 13–33. Sand Molds. Reference: D. C. EKEY and W. P. WINTER, *Introduction to Foundry Technology*. McGraw-Hill Book Co., 1958.

13 Sewage Pump Stations

Sewage treatment plant designers entirely too often fail to appreciate the hydraulic and mechanical complexity of large sewage pumps, their intakes, and piping. Some form of a coarse screen is recommended in the system design. The water volume may be adequate to warrant pumps that have rather large solids-passage capacity. Stormwater pumps may readily handle solids of 8 in. or larger diameter. Stormwater pumps usually work a limited amount of time in a given calendar year.

Sewage pumps usually have to work on a daily basis. To obtain a reasonable service life with sewage pumps, particular consideration must be given to abrasion of materials flowing through the pump, vibration, bearing lubrication, etc.

The latest issue of *Recommended Standards for Sewage Works*, Great Lakes—Upper Mississippi River Board of State Sanitary Engineers, should be used as a minimum standard in the design of all pump stations.

13-1. Larger Custom Built Pump Stations. Sewage pumping stations should not be subject to flooding. A suitable superstructure preferably located off the right-of-way of streets and alleys should be provided (1). It is important that the station be readily accessible. Sewage pumping stations should be of the dry-well type. Wet and dry wells including their superstructures shall be completely separated (1).

Insofar as materials of wet and dry pit construction are concerned, it is normally necessary to use concrete. Thus if a septic sewage problem is expected, it will be necessary to install air diffusers or mechanically aerate the sewage in the wet well adequately to maintain the sewage in the aerobic condition. It is preferable that sewage aeration take place in a baffled-off section of the wet well instead of immediately above the intakes.

   a. Wet Well. Each pump must have a separate intake. It is not desirable to have turbulence near the pump intake. In a large pump station, the wet well should be divided into two sections such that one section can be operated independently of the other in order to facilitate cleaning or repairs in the wet well. The wet-well floor should have a minimum slope of 1 to 1 to the hopper bottom. The minimum suggested dimensions for intakes at the bottom of a hoppered bottom are indicated in Fig. 13-1. The 1 to 1 slope may easily be used on three sides where it is desirable to separate the centers of the intakes by a greater distance. However, baffling should be used as necessary to prevent standing waves or unnecessary turbulence of the basin contents. The effective capacity of the wet well shall provide a holding period not to exceed 10 minutes for the design average flow (1).

   The use of a baffle at right angles to the direction of flow of the incoming sewage may be desirable if the sewage enters the chamber with a velocity of any significance.

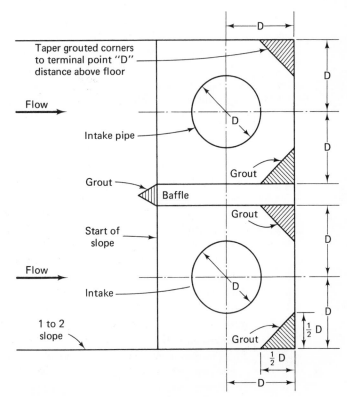

Figure 13-1
Minimum Spacing of Intakes.

b. Dry Pit. The dry pit should be accessible by ship-type ladders of the inclined type. It should be possible for a workman to comfortably and safely gain access to all components requiring routine maintenance. The dry pit should be well illuminated with explosion-proof light fixtures.

A removable grating should exist on the main pump room floor. Above this grating, at the ceiling, there should be an I beam with a movable carriage serving as a monorail. The carriage should be situated so that a hoist can be attached. Provision should exist so that the floor grating and subplatforms (if in the way) can be removed in order that the sewage pumps, valves, or other accessories in the dry pit can be hoisted to the main pump room floor level and then moved horizontally to a suitable work area or, if desired, loaded on a truck and removed from the pump house completely.

Adequate work room should exist in the pit to service all pumps and accessories as well as permit all equipment to be disconnected in the dry pit and be removed. All piping and equipment in the dry pit should be connected by means of bolted flanges. Even the straight runs of pipe should be disconnectable by means of bolted flanges.

Intake piping should be installed for extra long life. Pump stations are often required to serve far beyond their design period. The first pipe connection entering the concrete leading to the wet well should last the life of the station.

The gate valve in the intake lines should be a high-quality type that will resist corrosion for many years and still be functioning the day it is required to be shut off. Intake piping and wet-pit capacity should be provided for future additions to pump capacity. Adequate space should exist in the dry pit where additional pumping capacity can be added at a future date.

A check valve of appropriate type for use with sewage should be installed on each discharge line along with a gate valve. The check valve should be between the gate valve and the pump.

A separate sump pump must be provided in the dry pit whose sole function is dewatering the dry pit. The floor of the dry pit, and all walkway surfaces, should have adequate slope to drain to the sump pump wet well. The sump pump may discharge into the station wet well at an elevation above the station wet well overflow connection. All station wet wells should have an overflow arrangement. The sump pump is frequently slighted on quality and design detail; however, this is a mistake. The sump pump should not depend on a wire-and-weights arrangement. It should have either a sturdy rod-and-float arrangement, an electrode arrangement, or a pressure-bellows float control constructed by a reputable manufacturer of quality controls. The sump pump should have an elevated head, and the windings of the motor should be completely encapsulated in epoxy material so that accidental submersion will not ruin the motor. The sump itself should be not less than 24 in. in diameter and not less than 48 in. deep below the floor level.

Water will escape from pumps when they are being worked on, and it is desirable to have a water-pressure hose connection near the top of the dry pit where a hose can be connected to wash down the pit after the work is done.

c. Ventilation. Adequate ventilation should be provided for all pump stations. If the pump pit is below the ground surface, mechanical ventilation is required (1). The ventilation equipment should provide for at least six complete air changes per hour in the wet or dry well when the blowers are in operation.

13-2. Pump Selection. Large pumps should be so placed that under normal operating conditions they will operate under a positive suction head. Each pump should have an individual intake (1).

If only two units are used, they should have the same capacity. Each shall be capable of handling flows in excess of the expected maximum flow (1). When three or more units are provided, they should be designed to fit actual flow conditions. They should be of such capacity that, with any one unit out of service, the remaining units can handle maximum sewage flows.

In pump selection, the first considerations are current pumping demands and what the demand will be at some future design point (or points). A daily pumping cycle diagram, as shown in Fig. 13-2, made up at the present and at a future projected date, will be a useful device in sizing the pumping equipment and providing for future expansion of the station.

There is a common belief that pumps must be capable of passing 3-in. spheres and that pump suction and discharge openings must be at least 4 in. in diameter. Attempted deviation from either of these rules inevitably seems to be the reason regulatory agencies turn down plans.

References (2) through (7) contain a substantial amount of information on basic pump theory and application. The points to be stressed here are characteristic of pumps passing 3-in. or larger spheres. Pumps of this character, when operated at a value near their maximum practical operating speed, tend to vibrate if operated near shut-off head; they tend to cavitate if operated near the opposite end of the characteristic curve. In fact, cavitation may occur to a point well up on the characteristic curve. Thus, for trouble-free operation, there is a relatively restricted portion of the operating curve over which long-term, satisfactory results can be obtained.

The designer should know where the point of vibration may occur. Most pump manufacturers have a series of pump curves with a line drawn across the curves indicating the minimum capacity versus head that laboratory tests have indicated to be desirable. Laboratory tests have been run under a given condition of suction lift or submergence, and the curve represents that particular condition. The designer should never push his operating point up to that line. Figure 13-3 shows an operating curve that has the manufacturer's recommended limits at each end of the characteristic curve. Figure 13-4 indicates one reason why the designer would never press close to these

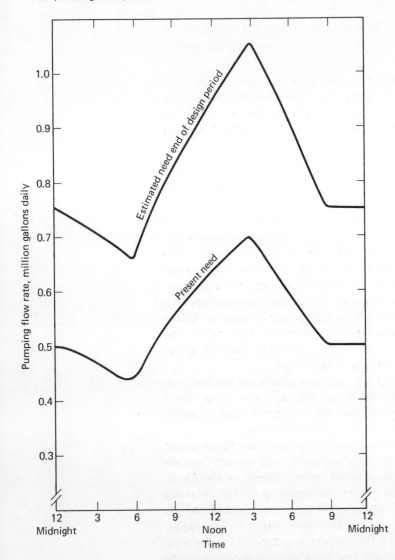

Figure 13-2
Typical Pumping Day.

design limits. In Fig. 13-4, the friction head of the pump is constantly changing over a period of time. The design point in this particular case was taken at 15 years. Certainly, the pipe size and other factors were arbitrarily selected. However, the point is that careful consideration must be given to what will occur with respect to the pump's operating point versus time.

Rules of thumb in pumps must be used with considerable judgment. However, trouble seems to occur around the 40% efficiency point on the left end of the curve on a number of pumps. A pump that will vibrate and make noise near shutoff at 1,750 rpm may cause no trouble at all operating at a similar spot on the same impeller operated at 1,150 rpm. If it is possible to achieve the head that you require with a 1,150-rpm pump, it is desirable to

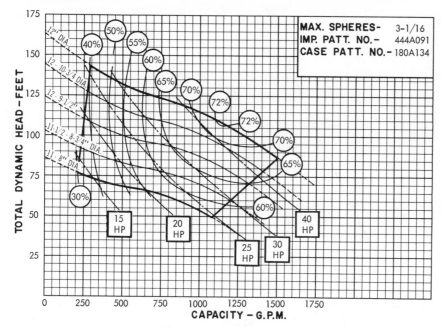

Figure 13-3. Pump Characteristic Curve for Sewage Pump. (Courtesy Aurora Pump Division)

specify in the specifications that 1,150 rpm is the maximum acceptable speed (assuming one of the lower capacity pumps).

13-3. Multiple-Pump Performance. There is more than one way to approach an explanation of multiple-pump performance. Let's approach it from the method shown in Fig. 13-5. In this particular illustration, it is assumed that two identical curves are involved. In this case, take any given head figure on one curve and double the capacity figure; this provides one point on the combined pump curve. In practice, the TDH (Total Dynamic Head) will usually increase and shift the operating point, but the ideal curve gives us a starting point to visualize what happens. If two or more curves are involved which are not identical, take some given head value common to all of them, add the capacities, and project to the right at the selected head value to the capacity point consisting of a summation of capacities. Repeating this process at several head values will define the ideal combined curve.

Next consider Fig. 13-6. The same curve used in Fig. 13-5 has been used. In this case all heads except pipe friction have been ignored. It can be seen that, for a 4-in. pipe, the increased capacity due to two pumps in parallel will be an insignificant increase over that obtained with a single pump (note the points where the pipe friction curve crosses the pump characteristic performance curves). Increasing the pipe size to 6 in. results in a significant

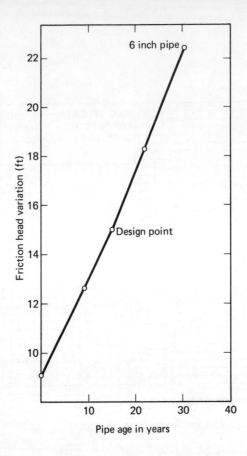

Figure 13-4
Head Variation Versus Time.

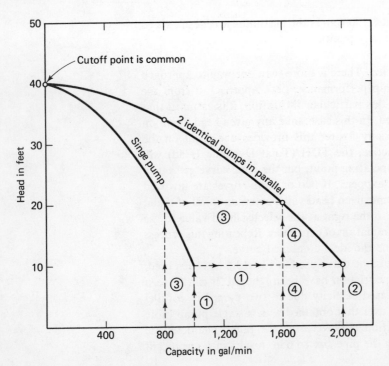

Figure 13-5
Ideal Performance Curve, Pumps in Parallel.

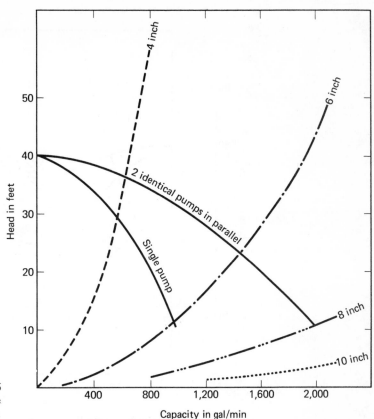

Figure 13-6
Multiple-Pump Performance
Considering Pipe Friction Only.

increase in capacity using two pumps in parallel if only pipe friction exists as a head. If only pipe friction is involved, the 8-in. pipe results in cavitation for a single pump operating, but would be usable for two pumps in parallel. The 10-in. pipe results in cavitation for either a single pump or two pumps in parallel.

Figure 13-7 shows what happens in actual practice. Here we assume that other head factors amount to 20 ft of losses. The pipe friction losses are then superimposed on top of the 20 ft of other losses. As soon as this is done, it is obvious that it is totally impractical to use the 4-in. pipe. The 6-in. pipe could be used, but it is not the best selection. Either the 8-in. or 10-in. pipe would make better use of the pumps. Plotting curves takes time, but this is frequently the best way to avoid making errors in attempting to project how the pumps are going to perform under a given set of operating conditions at present and in the future.

13-4. Characteristics. Each pump has a maximum hydrostatic test pressure, a maximum recommended-case working pressure, a maximum suction pressure, a maximum temperature at which packing or mechanical seals should

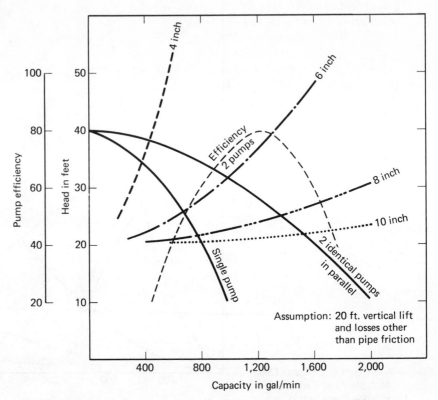

Figure 13-7. Multiple-Pump Performance Considering Actual Conditions.

function, and a maximum operating speed. It is advisable to determine these values and incorporate them in the specifications. The maximum operating speed for sewage pumps should be limited to 1,750 rpm.

The bearing life of dry-pit pumps is dependent upon a coefficient called the K factor. This factor varies with every individual pump. In general, the factor increases as the pump speed decreases for a given pump. As the K factor increases, the bearing life will increase. The same is applicable to both the outboard thrust bearing and the inboard radial bearing, although the values for the two bearings may be quite different. With proper care in pump selection, conservative speed values, and a well-designed product from a reputable manufactorer, it is often feasible to obtain bearing life in excess of 10 years, assuming that the bearings are properly lubricated.

Centrifugal pump impellers for sewage pumps should not be made from common, soft, gray cast iron. If made from cast iron, it should be a specially hardened cast iron. Hardened cast iron is extremely difficult to cut on the lathe, especially since it is a skip-cutting process where the lathe tool alternately cutting the tips of different vanes. Some manufacturers offer alloys that are extremely tough to resist the abrasive qualities of sewage.

It is desirable to keep maximum shaft deflection at the packing box below 0.002 in. to insure long packing or mechanical seal life. In large dry-pit sewage pumps, the author prefers regular packing with hardened stainless-steel shaft sleeves, interchangeable split packing inserts, gaskets to protect the shaft from pumped liquid, and a snap ring that pulls the sleeve off the shaft when the packing box insert is removed.

Grease lubrication arranged to purge old grease from bearings is preferred. The double-row thrust bearings are preferable on the pump shaft, and they should if possible have a life of 10 years or more.

The pump should have external shaft adjustment provided for renewing impeller clearance and maintaining pump efficiency. The shaft should be an extra-heavy-duty type. A slight taper facilitates removal of the impeller. The impeller key should be made of stainless steel, and the impeller should be held in place with a locknut and washer. A lifting eye tap in the end of the impeller shaft simplifies pump disassembly.

It is desirable to have $\frac{1}{2}$-in. NPT gauge taps at convenient locations at the discharge and suction openings in order to facilitate diagnosing troubles if they should occur. It is also desirable to have $\frac{3}{4}$-in. drain taps conveniently located on the pump volute. The discharge and suction flanges should be capable of 45° increment mounting to provide for eight different positions. The electric motors should be NEMA standard with P-face mounting.

13-5. Constant-Speed Pumps.

Constant-speed pumps have the advantage that they can be adjusted to the most efficient operating point on the characteristic curve if piping and other head losses are properly designed. Unfortunately, they deliver a fixed capacity throughout the period in which they are operating.

Constant-speed pumps are substantially easier to select, and there is less chance of making a mistake in calculating the operating conditions. If the pump can be selected such that the operating point requires less than the maximum diameter impeller, e.g., 25% below maximum diameter and 25% above minimum diameter, there is some room to spare if it is later discovered that the impeller operating point is incorrect.

13-6. Two-Speed Pumps.

It is easy enough to obtain two-speed electric motors even though the cost of the motors are more expensive. It is more difficult to find pumps with appropriate impeller characteristics that will operate at two speeds under a given set of operating head conditions. If care is not taken, the pumps will be in the cavitation range at the higher speed and in the vibration range at the lower speed. One compensating factor is that vibration is less likely to be a problem at the lower speed. If the static head is low, it is usually simpler to find a 700 rpm and 875 rpm two-speed unit that will perform satisfactorily than it is to find units that will perform satisfactorily at 1,150 and 1,750 rpm.

Sizing the two-speed pumps may be approached by first attempting to fix the lower-speed point near the vibration point and relatively high on the slower-speed curve. The next-higher-speed curve then must be checked by calculating the resulting TDH. Trial and error under several conditions may be required to arrive at a satisfactory combination. It is usually advisable to check dual-speed operation with the manufacturer and have the specifications with adequate provision that two-speed satisfactory operation is assured.

13-7. Variable-Speed Pumps. Figure 13-8 illustrates what would happen if a particular pump were selected and then operated at four different speeds while at all times maintaining maximum efficiency of operation on the pump characteristic curve. Careful thought will indicate that the only way such a steep head curve could be applied would be by using a pipe of comparatively small size in the discharge. This would of course require excessive horsepower to pump the higher capacity of water, and even though the pump was technically operating at its best efficiency, it would not be a practical

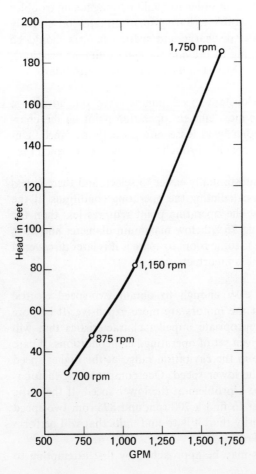

Figure 13-8
Variation of Pump Head and Capacity at Various Speeds When Operating at Maximum Efficiency.

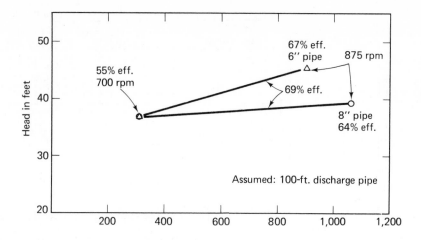

Figure 13-9. Variable-Speed Drive.

solution to municipal sewage pumping. Figure 13-9 shows two different sizes of pipes used with comparatively slow-speed and low-head pump cases. Under this circumstance, a capacity variation of 3 to 1 can be obtained for specific condition selected. This is about the practical limit of speed variation with sewage pumps in the majority of the cases, and in many applications a lesser capacity range variation will be involved.

13-8. Intakes and Submergence. Air may be entrained in the pumped liquid if the pump suction is located too close to the free liquid surface in the suction source. Pumping liquid with entrained air can cause a reduction of capacity, rough and noisy operation, vibration, loss of efficiency, and wasted power. Excessive wear of close-running parts, bearing stresses, and shaft damage are also subsequent effects.

If the capacity in gal/min and the suction inlet size or area are known, the minimum height of the suction inlet (submergence) can be determined. A properly designed suction inlet and sump can be accomplished with the help of the submergence chart illustrated in Fig. 13-10. Typical submergence requirements are shown on the chart.

EXAMPLES: 1. When the pipe size is known, the minimum submergence required for 2,000 gal/min through an 8-in. pipe is 9.6 ft.

2. When the inlet area is known, the minimum submergence required for 2,000 gal/min through a 50 in² outlet is 9.6 ft.

Minimum submergence requirements may exceed the available space requirements. When this occurs, a larger pipe size or inlet will reduce the required submergence.

EXAMPLE: 3. The minimum submergence required for 2,000 gal/min through a 10-in. pipe is 4.6 ft (5.0 ft less than required for an 8-in. pipe).

SUBMERGENCE

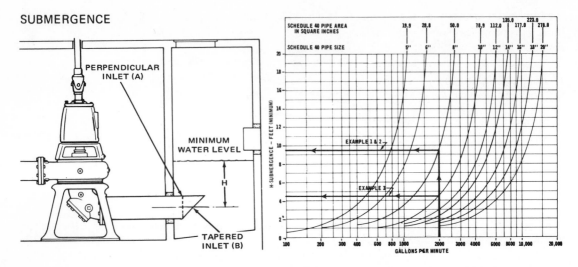

Figure 13-10. Submergence Chart. (Courtesy Aurora Pump Division)

13-9. Water-Seal Unit. It is unlawful to connect a city water supply directly to the water-seal connection on the packing box of a sewage pump. Water-seal units provide a sanitary break in the water line and literally provide an independent water system to connect to the pump packing boxes. A typical water seal unit is shown in Fig. 13-11. The float valve on the inlet water into

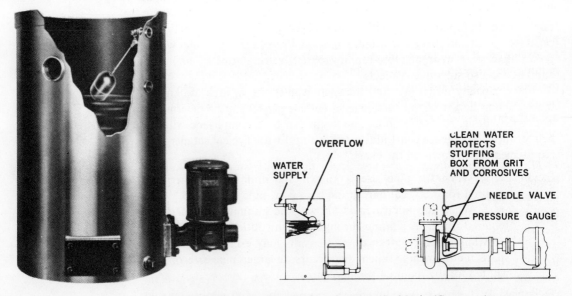

Figure 13-11. Typical Installation of a Water-Seal Unit. (Courtesy Aurora Pump Division)

the tank never comes into physical contact with the water in the seal-water tank. This provides clear-water lubricating and flushing for mechanical seals, packing, or wearing ring surfaces.

13-10. Flexible Shafting. It is customary to mount the electric motors above the floor in the main pump room. This protects the motors from any danger of flooding in the pit as well as protection from the extremely humid conditions that may prevail in the pit.

To connect the motors to the pumps below it, is necessary to use one or more sections of flexible shafting. Shafting is sized by the rpm involved, the maximum horsepower to be transmitted, and the length of the shafting required from the motor shaft to the pump shaft.

Depending on the size of the shafting, it may be necessary to use one or more so-called *steady bearings* or intermediate bearings between the electric motor and the end of the pump shaft. The coupling at the pump shaft end may be a special type. The shafting usually has to have a slight offset in order to minimize wear at one point, i.e., equalize the wear. Enclosed shafting may be obtained, but most of the installations are open shafting. A number of different arrangements can be obtained to mount the motor above the pump room floor. It is important that workmen be able to reach all steady bearings in order that they will be properly lubricated during the life of the pump.

Shafting that must operate at two speeds or at variable speeds over a speed range must be very carefully engineered. There are some speed ranges in which undesirable modes of vibration will occur that could be damaging to the shafting or associated accessories. Shafting must be provided with a slip spline which will permit removal of the pump rotating assembly without removing any section of intermediate shafting, bearings, suction, or discharge piping.

A short setting might use only an A section of shafting. A deep setting will require one A section and one or more B sections as necessary to make up the required length. A motor drive flange is required at the top end and a pump flange is required at the lower end. The details of the flange depend upon the horsepower to be transmitted and whether the shafting is standard or special tubing.

13-11. Pump Accessories. When working with dry-pit pumps connected by flexible shafting to the main pump floor, it is possible to obtain a variety of motor-mounting arrangements. The author prefers to use a steel base above the pump room floor, grouted in place, and an elevated head to support the electric motor.

It is good practice to specify a hardened stainless-steel shaft sleeve with a Brinell number of approximately 550. A split packing box, complete with rings of packing and a lantern ring, is shown in Fig. 13-12(a). Double mechanical seals, such as the Durametallic Type CRO, are usually available as an option. A double mechanical seal is shown in Fig. 13-12(b). Packing

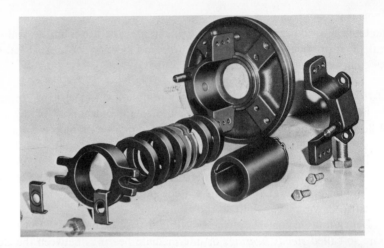

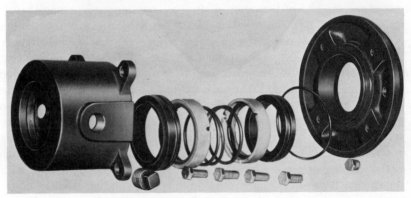

Figure 13-12. Comparison of Packing and Double Mechanical Seals. (Courtesy Aurora Pump Division) (a). Split Packing Box with Rings of Packing and Lantern Ring. (b). Double Mechanical Seal.

box options include external liquid-pressurized dead-end box (desired) or filtered pumped-liquid dead-end box (undesirable). Seal box options usually include external liquid-pressurized for flow through the box, external liquid-pressurized with a dead-end box, or filtered pumped-liquid with dead-end box (least desirable).

On vertical close-coupled dry-pit pumps, the pump should be supported by a cast-iron pedestal base and should be cast with a square footing to prevent legs from being broken during shipment. The pedestal should have openings large enough to permit access to the suction line. A handhole equal to the suction-pipe opening must be provided in the suction elbow for suction elbows of 8 in. or smaller pipe size. In suction elbows larger than 8 in. pipe size, the handhole should be 8 in. in size. It used to be the custom to make the handholes the size of the suction pipe for all sizes; however a 16-in. handhole cover weighs about 165 lb, which proved totally impractical for

Figure 13-13
Suction Nozzle with Handhole.
(Courtesy Aurora Pump Division)

workmen to handle. The pedestal must be of sufficient height so that the suction elbow will not touch the floor or foundation on which it stands.

It is sometimes necessary to use sewage pumps of the horizontal type at various locations in a plant. Suction nozzles, as shown in Fig. 13-13, are available for horizontal-type pumps. The nozzles have hand-size inspection openings to allow access to the impeller. Removing the nozzle allows the impeller and stuffing box to be removed without disturbing the suction and discharge piping or coupling alignment on some of the pumps on the market.

Figure 13-14 shows a complete sewage pump and its associated electric

Figure 13-14
Complete Sewage Pump for Sewage
Lift Stations. (Courtesy Aurora
Pump Division)

motor for the general construction found best-suited to use in sewage lift stations. Figure 13-15 illustrates a complete pump of the type used in the bottom of dry pits and coupled by flexible shafting to an electric motor mounted above the pump room floor. These can be compared with a complete wet-pit pump assembly shown in Fig. 13-16. Curb rings are not needed on dry-pit pumps.

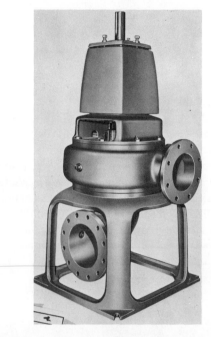

Figure 13-15
Complete Dry Pit Sewage Pump
Used with Flexible Shafting.
(Courtesy Aurora Pump Division)

13-12. Electric Motors and Motor Heaters. Sewage pump motors operated in extremely damp locations, and motors that will operate infrequently, should have heating strips included to keep out moisture.

The characteristics of sewage pumps are such that the motor horsepower may increase rather significantly as the head is reduced. It is advisable to have a reserve margin of 15% at the point of greatest horsepower demand under actual operating conditions. The margin of reserve can include any overload margin that may characteristically be built into the motor and which is so guaranteed by the manufacturer.

It is desirable in specifications to call out in detail the characteristics of the pump curve and carefully tie in the motor operating margin required. The type of motor insulation to be used should also be indicated. It is advisable to study the characteristics of available motor insulations, the operating temperatures that will result, the savings that may be realized by operating at a higher temperature, and whether the ventilation in the pump room is adequate to withstand the higher operating temperature. The author has frequently cut cost on large motors by resorting to higher-quality insulation;

Figure 13-16
Complete Wet Pit Sewage Pump
with Electric Motor and Float
Switch. (Courtesy Aurora Pump
Division)

however, it is not a step that should be taken without fully understanding all the elements involved.

13-13. *Right-Angle Drives.* It is feasible to obtain right-angle drives in order to use a gasoline, gas, or diesel engine drive or a combination electric and engine drive arrangement as shown in Fig. 13-17. The unit illustrated in Fig. 13-17 has an elevated head for optimum arrangement to the flexible shafting below. This unit is normally operated by the electric motor. The flexible shafting horizontal connect is coupled through a clutch to an engine. The engine is activated during an emergency condition when electric power is out of service.

Clutch units are, in general, hard on centrifugal pump bearings. Clutches are therefore seldom used as a direct coupling between pump and driver. Right-angle gear-drive units are normally constructed to take such loads. The clutch unit could be of the centrifugal clutch type that will not engage until the engine comes up to a specified operating speed. This is particularly desirable if the engine is to be started by automatic controls.

Right-angle gear-drive units are provided in both hollow-shaft and solid-shaft versions. Make sure that the proper combination is used.

13-14. *Engine Drives for Pumps.* For selecting electric motors, the actual horsepower of the motor is available. However, with any type of internal engine, the maximum usable horsepower under actual operating conditions is only some fraction of the engine horsepower. At sea level and standard

Figure 13-17
Combination Right Angle Gear
Drive with Electric Motor.
(Courtesy Aurora Pump Division)

conditions, the operating horsepower should not normally exceed 75% of the engine horsepower for continuous operation. Further derating is necessary with altitude, and additional derating may be necessary with a given make of engine. Higher ratings are permissible on diesel or gas engines used on stormwater pumps.

Engines will normally be water-cooled. It is frequently necessary to make arrangements for an external radiator or heat exchanger unit to cool the engine. This unit may require a separate fan. The engine will normally require a separate exhaust system and muffler, with the exhaust venting outside the building.

It is necessary to match the engine speed to that of the pump. The right-angle gear-drive unit can be matched with the proper gear ratio; however, the specifications must be specific enough that the proper engine and gear drive will be mated together. It is advisable to check on engine controls that are provided with the engine. Some controls, such as oil protection and temperature control combinations, may require a manual hold-in until the engine is running and oil pressure is at some predesignated pressure. There are controls available that are equipped with solenoids to hold the control operational for some predesignated period of time to allow it to come up to the proper operating condition.

The engine battery must receive a trickle charge when not in use to keep it ready at all time for emergency service. It is necessary to designate this charging equipment in the specifications or on the plans. A fuel tank and other accessories may be required separate from the engine. All these details must be determined and clearly set forth in the specifications or designated on the plans. Natural gas or digester gas is an important source of power in some sewage treatment plants.

13-15. Emergency Start Panels. There are several types of controls that may prove useful in meeting a given set of conditions. In an important pump station, it is usually desirable to have two sources of electric power if available. In this case, automatic load-transfer switches are available to shift power automatically from a preferred source to an auxiliary source. These switches can also be used to switch back to the original source of power when it is restored.

Controls are available that can be arranged so that, when regular power fails, the electric power is disengaged and a set of cranking controls is activated to automatically start the engine. Engines should be operated at periodic intervals to insure that they will function when required. A manually operated engine setup is of course much simpler than the automatically operated units just mentioned.

13-16. Useful Calculations. Various useful pump calculations can readily be made with the following equations.

$$HP = \frac{GPM \times ft\ head}{3,960 \times pump\ efficiency} \tag{13-1}$$

$$HP = \frac{GPM \times 8.3\ lb \times ft\ head}{33,000} \tag{13-2}$$

$$Efficiency = \frac{water\ horsepower}{brake\ horsepower} \tag{13-3}$$

$$KWH\ per\ 1,000\ gal\ pumped = \frac{head\ in\ ft \times 0.00315}{pump\ efficiency \times motor\ efficiency} \tag{13-4}$$

$$Cost\ to\ pump\ 1,000\ gal = \frac{TDH\ in\ ft \times cost\ per\ KWH\ in\ cents}{overall\ efficiency\ in\ percent \times 3.185} \tag{13-5}$$

where GPM = gallons per minute,
　　　TDH = total dynamic head.

REFERENCES

1. *Recommended Standards for Sewage Works*, rev. ed. Great Lakes—Upper Mississippi River Board of State Sanitary Engineers, Health Education Service, Albany, N.Y., 1971.

2. KRASSIK, IGOR, and CARTER, ROY, *Centrifugal Pumps*. F. W. Dodge Corp., New York, 1960.

3. KRISTAL, FRANK A., and ANNETT, F. A., *Pumps*. McGraw-Hill Book Company, Inc., 1953.

4. HICKS, TYLER G., *Pump Selection and Application*, 1st ed. McGraw-Hill Book Company, Inc. New York, 1957.

5. STEPANOFF, A. J., *Centrifugal and Axial Flow Pumps*, 2nd ed. John Wiley & Sons, Inc., New York, 1957.

6. KOVATS, A. DE, and DESMUR, G., *Pumps, Fans and Compressors*. Blackie & Son, Ltd., London, 1958.

7. CSANADY, G. T., *Theory of Turbomachines*. McGraw-Hill Book Company, New York, 1964.

Disinfection and Pathogen Removal   **14**

This chapter contains the basic information needed to select, design, size, and calculate the consumption of chlorine using a true vacuum-type chlorination system. It also contains other useful information on the general subject of disinfection and pathogen removal (1, 2, 3, 4, 5, 6).

14-1. Disinfection and Pathogen Removal. The objective for the disinfection of water or sewage is the prevention of the transmission of disease through the agency of water. A precise evaluation of the effectiveness of disinfection in the prevention of disease cannot be made because of many complex and interrelated factors which must be considered. The effectiveness of disinfection must therefore be based on statistics of death rates and cases of illness per given number of people before disinfection was employed and the results after disinfection practices were developed. This presentation will not attempt to discuss the various statistics contained in other sources. Its treatment will be in terms of the effectiveness of a particular disinfectant concentration in killing certain types of pathogens in a given set of conditions and contact period. It will also compare various disinfectants in terms of problems, advantages, or disadvantages encountered.

14-2. Coliform Bacteria. The intestinal tract of man is inhabited by large numbers of bacteria having certain common characteristics that are easy to identify. These bacteria have about the same resistance as some of the common enteric disease bacteria to the killing action of the commonly used disinfectants.

The coliform group, as specified in the U.S. Public Health Service Drinking Water Standards, is defined in the eleventh edition of *Standard Methods* (7): "The coliform group includes all of the aerobic and facultative anaerobic, Gramnegative, nonsporeforming, rod-shaped bacilli which ferment lactose with gas formulation within 48 hours at 35°C." This definition is the basis for much of the discussion in references cited within this paper.

The use of the coliform index is contingent upon two assumptions. First, disease-producing organisms are equally or more susceptible to natural or artificial destructive powers (such as disinfection). Second, the presence of coliform organisms in small amounts in water suggests the possible presence of excretal pollution. Similarly, the absence of coliform is considered as a criterion of probable safety (8).

E. typhosa, the causative agent of typhoid fever, is usually assumed in chlorination practice to be as vulnerable to chlorine as coliform organisms, and this organism is also assumed to be representative of the vulnerability of other intestinal *Salmonella* and *Shigella* pathogens. Kehr (8) has reported that coliforms die off naturally more rapidly than some enteric pathogens and that some strains of *Salmonella, Shigella, M. tuberculosis*, and other organisms are more resistant to chlorine under certain conditions.

14-3. Control of Public Water Supplies. Except for water used on interstate carriers, the authority for control of public water supplies is vested in the various states and their political subdivisions (10). The U.S. Public Health Service water quality standards do not specify minimum treatment, only limiting criteria of chemical and bacteriologic safety. The quality of U.S. public water supplies has in general been judged in terms of the U.S. Public Health drinking water standards (11, 12).

14-4. Chlorination of Water Supplies. The first use of chlorine gas to treat a public water occurred on November 22, 1912, at Niagara Falls, New York (13).

Butterfield (14) identified the primary factors governing the bactericidal efficiency of both free chlorine and chloramine as:

1. The time of contact of organism and bactericidal agent—the longer the time, the more effective the sterilization.
2. The temperature of the water in which the contact is made—the lower the temperature, the less effective the sterilization.
3. The pH of the water in which contact is made—the higher the pH, the less effective the sterilization. Thus, when the combination of high pH and low temperature is encountered, the poorest results are anticipated.

Wyss (15) states that the germicidal action of chlorine involves a complex series of events, and while interest may center in the destruction of organisms in a natural environment, the reactions involved may be so complex that it may be impossible to arrive at any understanding from the study. He further indicates that, of the factors affecting the death of microbes through the action of chlorine, the one which has ruined more good experiments is that which concerns the disinfectant-wasting side reactions. When someone reports that it takes 20 mg/l of chlorine about 20 minutes to kill a reasonable number of bacterial vegetative cells, most people who work in the field now understand that during this period probably 19 mg/l of chlorine have been destroyed by some disinfectant-wasting side reaction and that a fraction of a milligram per liter has succeeded in killing the organisms during the period.

Faber (16) indicates that the presence of any organic compound will decrease the bactericidal effect of the chlorine. The intensity with which any oxidizing agent enters into chemical reaction is measured by its oxidation potential. When chlorine combines with other substances, its oxidation potential is reduced or may be completely neutralized.

When chlorine is applied, the chlorine residual will increase as the dosage is increased up to a certain dosage. Addition of extra dosage causes the residual to commence to decrease until it reaches a point called the *break point*. Increase of dosage beyond the break point results again in increased residual.

Chlorine, as normally applied, may not kill all pathogenic organisms (17). It will not kill the cysts of *E. histolytica,* and it is ineffective against certain microscopic organisms. It is, however, an effective barrier against waterborne pathogenic bacteria under ordinary conditions.

The time of exposure and concentration are the primary factors influencing the extent of kill with free chlorine (18, 19). In a given time interval, it requires 40 times as much chloramine as free chlorine to obtain a 100% kill of *Escherichia coli* at normal pH (20). For *E. typhosa,* this ratio is approximately 25 to 1.

Hays (21) indicates that the bactericidal efficiency of hypochlorite solutions is significantly increased if the pH of the water is less than 6.0. There was little loss of available chlorine, at pH levels from 4.0 to 9.0, within 24 hours.

Chang (22) found that the cysts of the protozoan *Endamoeba histolytica* may be destroyed by superchlorination with a contact period of 30 minutes or more. Chang (23) has stated that some amoebas (in the spore stage) and nematodes are not killed by 10 ppm of residual chlorine, although exposure to prechlorination of this concentration may impair their ability to pass through a rapid sand filter.

Popp (24) conducted a number of tests using various chlorine sources with chlorine contents up to 0.6 ppm. He worked with different pH levels and bacterial loadings and found that reduction of organisms did not follow exponential law but occurred in two steps. The first step, lasting about 12 seconds, was accompanied by rapid reduction of organisms. The second step was characterized by a slow reduction.

Clarke and Kabler (25) reported that it took approximately 7 to 46 times as much free chlorine to obtain comparable kills of Coxsackie virus as was required to kill *E. coli* cells.

Neefe et al. (26) concluded that treatment of contaminated water with sufficient chlorine to provide a chlorine residual of 1 ppm after 30 minutes' contact did not inactivate or attenuate the causative agent of infectious hepatitis. Increasing the chlorine residual to 15 ppm after a 30-minute contact period results in a definite attenuation of the hepatitis agent.

Kelly and Sanderson (27) concluded that complete inactivation of enteric viruses (beyond the limit of detection) was not achieved by the usual conditions of bacterial disinfection of water supplies, i.e., free residual chlorine concentrations of 0.2 ppm for 10-minute contact time at pH 7. Concentrations of free residual chlorine of from 0.2 to 0.3 ppm inactivated viruses after contact periods of 30 minutes. Factors which affected the inactivation of viruses were strain, pH level, free residual chlorine concentration, exposure time, and temperature. Rises in the pH value of one unit above pH 7 more than doubled the contact period necessary for inactivation of several strains.

Hallinana (28) recognized that forms of active residuals of chlorine, other than free chlorine, existed. The term *combined available chlorine* developed as a sort of empiric definition to cover this condition. Kjellander and Lund (29) studied the effects of the different forms of combined chlorine on *Escherichia coli* and poliovirus. The study was carried out using residuals of combined chlorine obtained by addition of free chlorine below the break point, or NH_2Cl, or chloramine T. It was found that unlike forms of combined chlorine containing identical equivalents of residual combined chlorine possess different chlorination efficiencies. It was therefore suggested that the term *combined chlorine* may not always be an adequate parameter for chlorination.

A Public Health Service publication (30) stated:

Chlorine, the disinfectant used almost universally in safeguarding water, is not effective against all microorganisms in the concentrations of chlorine normally used in waterworks systems. Certain bacteria are resistant or invulnerable to chlorine in the amounts generally applied. Research is required on new and more effective disinfectants and should be supported.

Morris (30) commented with respect to the above statement: "What is needed in the near future is an increase in standards and control of proven chlorination more than research on hypothetic new or more effective disinfectants than seem unlikely ever to exist."

Snow (31) indicated, in general, that six major variables have been shown to affect the efficiency of chlorine disinfection:

1. The types and concentrations of the chlorine present.

2. The equilibrium relationships between coexistent chlorine forms (governed largely by the pH of the water).
3. The type and density of organisms (virus, bacteria, protozoa, helminth, and other forms) and their species resistivity to chlorine.
4. The duration of contact of the organisms with the chlorine.
5. The temperature of the water insofar as it determines the rate of reaction of chlorine compounds and the rate of kill of organisms.
6. The concentration of substances whose rapid oxidation by chlorine is manifested as chlorine demand.

Lovtsevich (32) found enteroviruses to be more resistant to chlorine than colon bacteria. He concluded that the coliform index cannot be used as a reliable indicator in relation to viruses.

14-5. Field Investigation. A field investigation was conducted by the author to determine the simplest, safest, most economical, and most practical method of disinfection of water and sewage at roadside rest and recreation areas. The findings are equally applicable to any small- or medium-size wastewater treatment plant.

Initially it was assumed that some form of liquid-disinfectant pumping scheme would be the most practical. The author had previously used units, with various manufacturers' metering pumps, to pump a wide range of chemicals including both calcium hypochlorite and sodium hypochlorite. Both powdered and liquid sodium hypochlorites are available. The hypochlorites are easy to package and are available in small and large quantities.

Insofar as the mechanical capability of metering pumps was concerned, there was no problem. Good-quality, reliable, and satisfactory units were available. The number of times operators were caught improperly metering the liquid was more frequent than desirable. With respect to economy, the cost per pound of hypochlorites is approximately three times more than for liquid chlorine. Another disadvantage is that hypochlorites lose strength with age.

The factor that finally resulted in rejecting hypochlorites was the matter of safety. Handling hypochlorite can be hazardous since the gas released is pure chlorine. The oxidizing power of hypochlorites is strong enough that there is also a potential explosion hazard. A typical documentation of the case against hypochlorites is F. P. File M65, American Insurance Association, Engineering and Safety Department, 85 John Street, New York, N.Y. 10038, Special Interest Bulletin No. 214, February, 1965, on calcium hypochlorite. Simple economics excludes hypochlorites from larger plants.

Gas chlorination was investigated. Chlorine is supplied commercially in cylinders that vary in size from 1 to 2,000 lb. It is also available in tank cars. The 100-lb and 150-lb cylinders seem to predominate. Chlorine gas compressed in a cylinder forms liquid chlorine with chlorine gas at the top of the cylinder. The flow of gaseous chlorine from the container depends on the internal pressure, which in turn depends on the temperature of the liquid chlorine. The liquid becomes gas when it absorbs its heat of vaporization from the surrounding air. This has a cooling effect upon the cylinder and the air around the cylinder. When the cylinder is cooled sufficiently, moisture in the air condenses on the metal surface, and if the cooling continues, the moisture becomes frost. A frost buildup insulates the cylinder and inhibits heat absorption of the liquid.

The chlorine in the cylinder is both liquid and gas. When gas is withdrawn from the cylinder, the pressure in the cylinder is slightly lowered and the liquid "boils" to form more gas. Pressure in the cylinder depends on the temperature of the liquid, and therefore the pressure does not actually change as the gas is used. The pressure will remain essentially the same whether the cylinder is full or almost empty, as long as there is liquid in the cylinder and the temperature is the same. Pressure versus temperature values are shown in Table 14-1.

TABLE 14-1. Chlorine Pressure Versus Temperature

Temperature °F	Pressure in Psi
100°F	140
70°F	85
32°F	37
30°F	0
−150°F	chlorine freezes

Various types of chlorination equipment were investigated. Systems that had pressure in any part were rejected. In many small plants, there is an attendant present for only a short period of time each day. Many of the small- and medium-size treatment plants have children that roam in the vicinity. It was considered undesirable for any hazard to be unnecessarily injected by the presence of the chlorination system. One type of so-called vacuum system has a pressure line from the cylinder to the chlorinator mounted on the wall. If the pressure line broke or became disconnected, chlorine would escape freely.

It was concluded that only a fail-safe device on the cylinder itself would be a practical means of chlorination under the conditions of the small- and medium-size plant. It was finally concluded that a cylinder-mounted true vacuum system chlorinator containing all regulation, adjustment, and safety components would meet the criteria. Over 150 units of this type were investigated under field conditions and found to function satisfactorily. This unit is illustrated in Fig. 14-1. In this type of system, the line can be broken at any point, and the unit merely shuts itself off.

The primary precautions were to use a new lead gasket with each connection change and chain all cylinders in a vertical position to keep them from being knocked over and damaging the chlorinator.

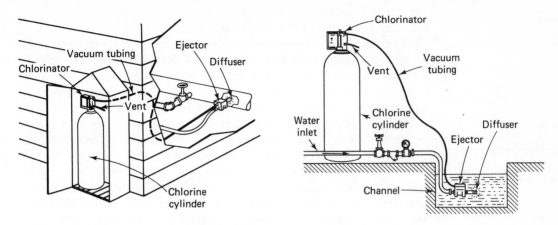

Figure 14-1. Typical Vacuum Units. (Courtesy Capital Controls Co.)

Most chlorinators are oversized by a factor of four or five times the normal withdrawal rate. One factor observed was that the true vacuum chlorinator apparently functioned at temperatures below freezing. Installations were visited in which the winter temperature was at least as low as 20°F and probably lower. Yet the operators indicated the system functioned satisfactorily for their needs. Chlorine test kits were employed in a number of instances to verify this opinion.

It is feasible to use a cylinder-mounted regulator and vacuum valve on the gas cylinder and have the metering tube and rate valve mounted separately. In this system, no part of the system is under chlorine pressure. However, regardless of the system, it is essential that the parts used inside the vacuum valve and regulator be resistant to the corrosive properties of chlorine in contact with moisture. At the time of the investigation, only one company's product met what was considered desirable the criteria of necessary corrosion-resistant materials.

Chlorine cylinders are constructed of steel, and the outlet valves are constructed of bronze. But chlorine in the cylinder is completely dry; therefore, it is essentially noncorrosive to many common metals in the absence of moisture. It is equally true that moisture does frequently get into the interconnecting piping and forms acid, and immediately corrosion starts. One unit that was inspected had five corrodible metal springs as well as a corrodible metal screen on the intake filter. The manufacturer may have subsequently replaced these parts with noncorrodible materials, but it is advisable to check such details. It is advisable to secure corrosion-resistant materials even if it increases the cost per unit.

The equipment in Fig. 14-1 had silver surfaces that could be disassembled for inspection and cleaning, whereas another manufacturer's unit apparently is not serviceable and must be replaced.

A feature of concern on one product inspected was the vent on the control unit. If a malfunction occurs, gas escapes into the left-hand section of the chamber and vents to the atmosphere. The inlet-safety-valve force was about 3 lb compared to the 12-lb force in the equipment shown in Fig. 14-1.

Since the original investigation, additional equipment has entered the market that is quite similar to that in Fig. 14-1. However, no information has been accumulated on this equipment. The potential deficiencies of concern have been noted. The usual history of any new device from any manufacturer is a series of configurations, each superior to its predecessor.

All chlorination equipment is subject to maintenance problems. There are small amounts of particles or impurities in chlorine. These small particles, called "gunk" or other such names, may accumulate and eventually interfere with the operation of the regulator. Sealing compounds from the connections, as well as impurities, can be a source of gunk. One distinctive feature observed in the true vacuum-type unit was that it had fewer problems than other types. Atmospheric moisture, for one thing, is not as apt to bother it. In some systems, if precautions are not taken, enough moisture from the atmosphere gets into the system to put the equipment out of operation.

14-6. *Vacuum Chlorination Systems.* The previously mentioned field investigation and subsequent additional research on the subject led to the conclusion that small- and medium-size plants should have a true vacuum-type gas chlorination system. No part of the system should at any time be under chlorine pressure once the gas leaves the cylinder. It is also essential that the device on the cylinder be so designed that, if the vacuum is lost, the valve is automatically in a fail-safe condition without any vents or other devices that could permit chlorine to escape. It must be possible to leave the system unattended without danger to persons in the vicinity of the treatment plant.

Virtually all of the chlorination equipment in the United States, with the exception of a few swimming pool chlorinators, is made by five companies. The five firms all have the capability of making fine, quality equipment. At any given time, one firm may hold patents that provide a definite advantage on certain features. The factor advocated here is that of a principle and not of the specific product. The specification should be set forth and not the brand name alone. It is usually only a question of time before several of the manufacturers will manage to meet the specification.

Figure 14-2 shows a typical arrangement for chlorinating a deep well pump water supply. If the line from the chlorinator to the ejector is broken or the connection yanked loose, nothing will happen because a valve in the chlorinator closes as soon as the vacuum is broken. Figure 14-3 shows three different ways that the chlorine can be ejected into water. Figure 14-4 shows a working installation that is buried in snow.

The *Chlorine Manual* states that a dependable, continuous discharge rate of chlorine gas from a single 100-lb or 150-lb cylinder is 42 lb/24 hr under the following conditions:

1. Without sweating (the condensation of moisture on the cylinder).

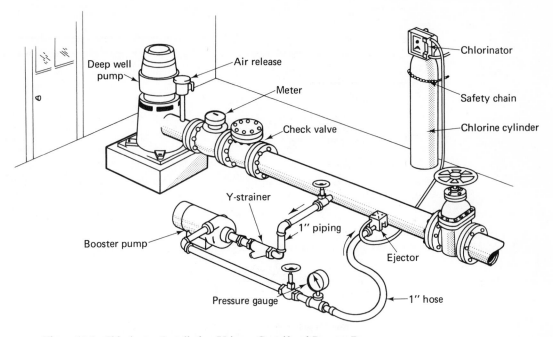

Figure 14-2. Chlorinator Installation Using a Centrifugal Booster Pump.
(Courtesy Capital Controls Co.)

2. At 70°F and normal air circulation conditions.
3. Discharging against 35 psi back pressure.

The vacuum-type chlorinator used in the author's research project was found capable of a gas feed rate of at least 100 lb/24 hr in an enclosure maintained at 20°F. The unit has been tested with the gas being withdrawn against a negative pressure of 2 lb.

Insofar as the true-vacuum chlorination equipment is concerned, a protective housing is not required unless the ambient temperature is below 20°F. However, to prevent vandalism and ensure the safety of children, it is suggested that a small protective housing be used. The interior should be small enough that a person cannot enter and only the equipment is contained inside the housing. If the housing is small enough that a person cannot enter, there is no need for a fan unless state regulations demand one. A typical housing is shown in Fig. 14-5.

Since chlorination systems in many small plants must function a large portion of the time without an operator present, it is essential that an automatic switchover system be used. The recommended dual system by one manufacturer is shown in Fig. 14-6. The dimensions of most components are given in Fig. 14-7. A water pressure differential should be developed through a pressure-reducing valve (Fig. 14-8) when a booster pump is not available, provided the flow and pressure loss across the valve can be tolerated. If

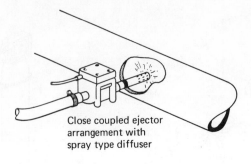

Close coupled ejector
arrangement with
spray type diffuser

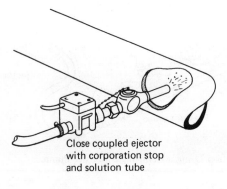

Close coupled ejector
with corporation stop
and solution tube

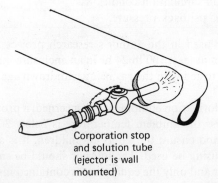

Corporation stop
and solution tube
(ejector is wall
mounted)

Figure 14-3
Ejector Installations. (Courtesy
Capital Controls Co.)

booster pumps are used, it is advisable to use two separate pumps. The reducing-valve arrangement may not be suitable for many installations due to the pressure loss entailed.

It should be pointed out that the configuration required is two completely separate chlorinators mounted on two separate chlorine cylinders. Do not become confused in your specifications with a chlorinator with a dual range. Figure 14-9 shows a dual-range chlorinator for chlorinating two separate places at the same time from the same cylinder. No protection is given here if the cylinder becomes exhausted or a piece of equipment fails.

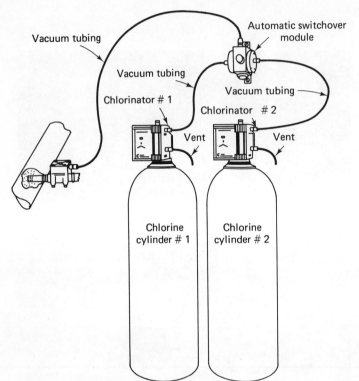

Figure 14-6
Automatic Switchover System.
(Courtesy Capital Controls Co.)

It is not recommended to have multiple cylinders hooked up to a common header so as to turn off old cylinders and turn on new ones. This can be very hazardous since slight temperature differences can cause chlorine to transfer from one cylinder to the other, and cylinder valves do not always close tightly after being used. The greatest hazard is caused by the fact that operating personnel change from time to time and are not always completely familiar with the techniques of changing or the hazards involved.

14-7. Calculations for Water Treatment. For water treatment, a dosage rate of 1–10 ppm is typical, although well water may be adequately treated in most cases by 1–5 ppm. The dosage is the amount of chlorine required for disinfection of the water.

$$\text{PPD} = \text{GPM} \times 0.012 \times \text{dosage (ppm)} \qquad (14\text{-}1)$$

where PPD = pounds per day (lb/24 hr) chlorine feed rate, GPM = gallons per minute, and ppm = parts per million (pounds of chlorine per million pounds of water). One gallon of water weighs 8.34 lb.

Chlorinator size should be based on the maximum expected flow rate at any time. Use GPM rather than daily total or average flow rates.

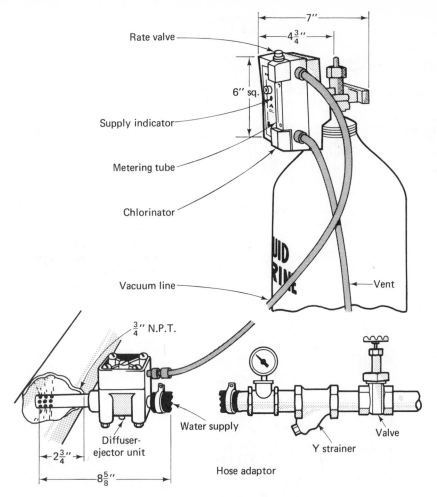

Figure 14-7. Dimensional Data. (Courtesy Capital Controls Co.)

14-8. Chlorinator Sizing Guide. The following procedure will enable you to size a chlorinator for any application.

Many factors determine the exact amount of chlorine to be fed in a given application. Gas chlorinators operate over a wide range of flow rates and usually can be converted to higher or lower rates quite simply. The maximum flow rate of a gas chlorinator is at least 20 times its minimum rate, with any given capacity metering tube, using vacuum-type chlorinators. For this reason, most chlorinators are oversized to have a safety factor of 3 to 5 times the required dosage of the specific application.

The capacity of the chlorinator is often dictated by federal, state, and

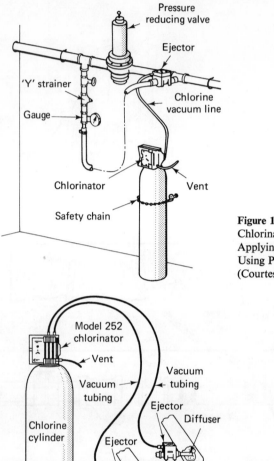

Figure 14-8
Chlorinator Installation for
Applying Chlorine into Pipeline
Using Pressure-Reducing Valve.
(Courtesy Capital Controls Co.)

Figure 14-9
Dual-Range Chlorinator.
(Courtesy Capital Controls Co.)

local authorities. Therefore, the values shown in Table 14-2 are to be used only as a guide under average conditions.

14-9. Sewage Chlorination. When chlorine is added to sewage, the organic solids in suspension and solution, as well as any reducing substances present, absorb chlorine until their capacity is satisfied. A degree of bacterial kill does occur with partial chlorination, but no free chlorine and chloramines will be present until the complete demand is satisfied. Efficient chlorine disinfection of the sewage requires sufficient chlorine to meet the complete chlorine demand plus an additional amount to produce a chlorine residual. The amount of residual required depends upon the contact time, character of the effluent, temperature, pH, original count of bacteria in the unchlorinated

TABLE 14-2. Chlorinator Sizing Table

Chlorination Treatment for	Typical Dosage Rates (ppm)
Algae	1–10
Bacteria	1–10
BOD reduction	5–12
Color	1–10
Cyanide:	
Reduction to cyanate	2 times cyanide content
Complete destruction	8.5 times cyanide content
Hydrogen sulphide	
Taste and odor control	2 times H_2S content
Destruction	8.4 times H_2S content
Iron bacteria	1–10
Iron precipitation	0.64 times Fe content
Manganese precipitation	1.3 times Mn content
Odor	1–10
Sewage	
Fresh raw sewage	6–12
Septic raw sewage	12–25
Trickling filter effluent	3–10
Activated sludge effluent	2–8
Sand filter effluent	1–5
Slime	1–10
Sterilization	25–50
Sulfur Bacteria	1–10
Swimming pool	1–10
Taste	1–10
Water:	
Cooling	1–10
Chiling	5–25
Washdown	25–50
Well	1–5
Surface	1–10

(Courtesy of Capital Controls Co.)

sewage or effluent, and similar factors. Increasing residual would, in turn, decrease the contact time for a given kill, other factors remaining constant.

Wilson (33) indicates that methods used in water chlorination are not usually applicable to sewage treatment. A review of the published information reveals discrepancies in the effects of sewage chlorination. The chemistry involved in sewage and industrial waste chlorination is complex (34). This is due primarily to the fact that numerous germicidal chlorine derivatives of varying effectiveness are formed. The chemistries of sewage and water chlorination are basically the same and, in general, differ only in the quantities present of materials which affect chlorination efficiency (35). This similarity has led to a false confidence in chlorination, as applied in sewage treatment, and to some fallacies in the control of processes.

In the chlorination of sewage and wastes, the economics of the problem usually precludes the addition of sufficient chlorine to satisfy all the chlorine demand and to convert all ammonia and nitrogen compounds to chloramines (34). Unless this is done, the existence of HOCl in chlorinated sewage or waste is one that lasts a short period of time, and the chlorine compounds which remain are unfortunately those of relatively low disinfecting efficiency.

The relationship of the number of bacteria surviving at any time, to the elapsed time, is given by Chick's law (36, 37) as:

$$\log B = \log B_0 - kt \qquad (14\text{-}2)$$

where B is the final number of bacteria, B_0 is the initial number of bacteria, t is the elapsed time, and k is the velocity constant.

Since this is a constant-rate process, the absolute number killed per unit of time decreases with the passage of time for the simple reason there are less bacteria available to be killed.

Walton and Clup (38), in a literature review, discussed fifty-seven references. They found that there have been many attempts to correlate the chlorine residuals used for control purposes with the resulting bacteria kill. In the discussion of effectiveness, it was noted that the relationship between the residual chlorine and residual coliform density varies greatly from plant to plant or with change in the characteristics of the sewage received at a given plant.

Another source (39) states: "It does not seem reasonable to set up rigid coliform standards and at the same time to permit the control of chlorination on the basis of chlorine residuals, which do not necessarily produce the desired coliform numbers in the effluent." The chlorine residual is only a collateral and secondary test to the ultimate objective to be achieved (maintenance of coliform numbers not in excess of standards).

Laboratory studies (40) indicate that chlorine dosage cannot be used as the basis of achievement of stipulated residual coliform organisms. In settled sewage, the amount of chlorine required for a 30-minute contact time is less than one-half that required for a 5-minute contact time. Dilution of sewage with water in the ratio of 1:4 decreases the chlorine requirement by one-half with a 5-minute contact time and by one-fourth with a 30-minute contact time.

The ideal situation is to have equipment that could adjust the rate of chlorine feed in accordance with changes in the chlorine demand of the waste. Equipment suitable for this purpose has been developed. Previously the rate of chlorine application has been governed by the following general methods (41):

1. Manual adjustment.
2. Semiautomatic control: chlorinators start and stop in conjunction with the sewage pumps.

3. Step control: in multipump installations, the feed may be automatically synchronized to deliver a certain dosage for a particular
   combination of sewage pumps in operation.
4. Automatic control: chlorine application rate is proportional to
   the flow.

Grune (42) concluded that the effectiveness and efficiency of chlorination
is limited, and it cannot be employed alone where any degree of treatment is
required, but chlorination supplements many other processes and has found
its proper place among them.

Thomas (43) concluded that chlorination and comminution were not
sufficient to protect a bay having a sea outfall 4 ft below the surface. The
result was oil slicks, floating fat, and fecal particles on the bay's surface and
washed upon the beaches.

Ammon and Wieselsberger (44) reported that it is technically and economically feasible to disinfect digested domestic sewage sludge by chlorination.
They found that actively gasifying sludge requires materially higher chlorine
dosages than thoroughly digested sludge.

Bergsman and Vahlne (45) found that tubercle bacilli could be removed
by using 2 ppm on raw sewage and 15 ppm on septic tank effluent. Jensen
reported a 10 ppm dose required to destroy tubercle bacilli in activated-
sludge effluent. It has been reported by Miller and Anderson (47) that tubercle
bacilli can be recovered from sewage effluents that have been chlorinated.

Chlorine is more effective in disinfecting seawater than freshwater (34).
This fact could be of particular interest for recreational usage where saltwater
swimming pools are built near the sea.

Kelly and Sanderson (49) found that, for resistant strains of viruses,
an increase in contact period was a better disinfecting procedure than an
increase in concentration of chlorine. They found that a high pH value
enhanced the rate of inactivation. It was also indicated that present-day
standards of disinfection were inadequate.

In addition to variations in the character of normal domestic sewage and
waste volume variations, other factors within or outside a sewage plant may
influence the chlorine dosage required (50). Among these are:

1. Changes in sewage temperature.
2. Cesspool and septic-tank discharges.
3. Chlorine-demand industrial wastes such as sodium thiosulphate
   (hypo), cyanides, phenols, and sulphur compounds or dyes.
4. Saltwater infiltration.
5. Discharge of high volumes of certain detergents.
6. Digestion-tank supernatant liquor.
7. Elutriates (overflow liquor) from digested sludge elutriation
   (sludge washing).

8. Vacuum-filter filtrates high in suspended solids.

It should be pointed out that, in recreation area sanitation engineering, dumping the contents of trailer or boat waste-holding tanks into the system involve the same principle as cesspool and septic-tank discharges. Chemicals and detergents used in cleaning may be on sufficient scale to compare with items 3 and 5 in the list above.

Heukelekian and Faust (51) point out that chlorination of effluents as presently practiced destroys only some of the pathogenic organisms (*Salmonella* and *Shigella*) that may be present, leaving the viruses, mycobacteria, and parasitic worms unaffected. Higher chlorine dosages and residuals must be maintained in order to destroy them.

14-10. Sewage Chlorination Design. In chlorinating sewage, it must be emphasized and understood that various organic pollutants will be oxidized by the chlorine. Chlorine is an extremely active disinfectant under acidic conditions, but it has a much lower effectiveness if the sewage is alkaline. The alkalinity and pH operate to cause the chlorine to function at a slow kill status.

Unless chlorine comes in contact with a pathogen, it will not effect a kill. The pathogens and organisms inside chunks and flocs are protected. This is why the final effluent must be chlorinated regardless of whether chlorine was also applied earlier in the process.

In northern climates during the winter, it may take over five times longer for the chlorine to kill the pathogens. The bacteria such as *E. coli* are not a good indication of kill in sewage. The cysts of *E. histolytica*, spore formers such as *B. anthracis* and tubercle bacilli, enteric virus, and bacilli are much more difficult to kill, yet they must be killed before the water is safe for consumption.

Figure 14-10 indicates one means of injecting chlorine into a chlorination tank. It would be more desirable to have the injector mounted inside the incoming pipe where it would be flowing full.

Table 14-3 indicates vacuum line sizes, Table 14-4 provides ejector pressure data, and Table 14-5 provides typical booster pump data from one manufacturer.

TABLE 14-3. Vacuum Line Size

	Total Vacuum Tube Length (ft)		
lb/24 hr	100	200	500
50	$\frac{3}{8}$	$\frac{3}{8}$	$\frac{1}{2}$
100	$\frac{3}{8}$	$\frac{1}{2}$	$\frac{1}{2}$
200	$\frac{1}{2}$	$\frac{5}{8}$	$\frac{3}{4}$
500	$\frac{5}{8}$	$\frac{3}{4}$	1

(Courtesy of Capital Controls Co.)

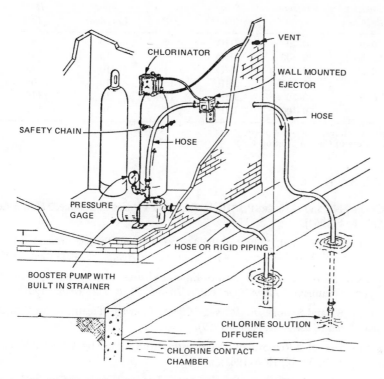

Figure 14-10. Sewage Chlorination. (Courtesy Capital Controls Co.)

TABLE 14-4. Ejector Pressure Data

Pressure at Point of Injection		Required Water Supply Pressure—psig & kg/cm²								
		Chlorine 0.6, 1.5 lb/day 11, 28 g/hr		Chlorine 4, 10 lb/day 75,200 g/hr		Chlorine 25 lb/day 500 g/hr		Chlorine 50 lb/day 900 g/hr		Chlorine 100 lb/day 1900 g/hr	
psig	kg/cm²										
0		10	0.7	10	0.7	15	1.1	20	1.4	30	2.1
5	0.4	15	1.1	15	1.1	20	1.4	25	1.8	35	2.5
10	0.7	25	1.8	25	1.8	30	2.1	35	2.5	45	3.2
20	1.4	40	2.8	40	2.8	45	3.2	50	3.5	65	4.6
30	2.1	50	3.5	50	3.5	55	3.9	70	4.9	80	5.6
40	2.8	65	4.6	65	4.6	70	4.9	85	6.0	95	6.7
50	3.5	75	5.3	75	5.3	80	5.6	95	6.7	105	7.4
60	4.2	90	6.3	90	6.3	95	6.7	105	7.4	115	8.1
80	5.6	110	7.7	110	7.7	115	8.1	130	9.1	140	9.8
100	7.0	135	9.5	135	9.5	140	9.8	150	10.5	160	11.2

(Courtesy of Capital Controls Co.)

TABLE 14-5. Booster Pump Data

Chlorinator Capacity (lb/24 hr)	Main Pressure (psi)	Pump Model*	HP	Booster Pump Flow (gal/min)	Booster Pump Head (ft)
0–10	0–70	JHD	$\frac{3}{4}$	10	95
(or less)	70–100	JHD	$\frac{3}{4}$	10	95
0–25	0–70	JHD	$\frac{3}{4}$	10	95
	70–100	JHE	1	10	105
0–50	0–70	JHF	$1\frac{1}{2}$	10	115
	70–100	JHF	$1\frac{1}{2}$	10	115

*Specify voltage, single or 3 phase.
(Courtesy of Capital Controls Co.)

REFERENCES

1. CLARK, N., and KABLER, P., "Human Enteric Viruses in Sewage." *Health Lab. Sci.*, 1, 4 (1964).

2. CARLSON, H. J., et al., "Effect of the Activated Sludge Process of Sewage Treatment on Poliomyelitis Virus." *Amer. Jour. Public Health*, 33, 1083 (Sept., 1943).

3. CLARKE, N. A., et al., "Removal of Enteric Viruses from Sewage by Activated Sludge Treatment." *Amer. Jour. Public Health*, 51, 8, 1118 (Aug., 1961).

4. KELLY, S., and SANDERSON, W. W., "Effect of Sewage Treatment on Viruses." *Sew. & Ind. Wastes*, 31, 6, 683 (1959).

5. KELLY, S., et al., "Removal of Enteroviruses from Sewage by Activated Sludge." *Jour. WPCF*, 33, 10, 1056 (1961).

6. KELLY, S., et al., "Poliomyelitis and Other Enteric Viruses in Sewage." *Amer. Jour. Public Health*, 47, 72 (1957).

7. *Standard Methods for Examination of Water and Wastewater*, 11th ed. Amer. Public Health Assn., AWWA, and WPCF, New York, 1960.

8. *Water Quality and Treatment*, 2nd ed. American Water Works Assn., New York, 1950.

9. KEHR, ROBERT W., and BUTTERFIELD, CHESTER T., "Notes on the Relation Between Coliforms and Enteric Pathogens." *Public Health Reports*, 58, 15, 589 (1943).

10. LAUBUSCH, EDMUND J., "Water Disinfection Practices in the United States." *Jour. AWWA*, 52, 11, 1416 (1960).

11. *Drinking Water Standards*. Public Health Service, Publ. No. 956, 1962.

12. *Interstate Quarantine Drinking Water Standards—Miscellaneous Amendments.* U.S. Public Health Service, Federal Register, March 1, 1957, pp. 1271-2.

13. FABER, H. A., "How Modern Chlorination Started." *W & S. W.*, 99, 11, 455 (Nov., 1952).

14. BUTTERFIELD, C. T., "Bactericidal Properties of Chloramines and Free Chlorine in Water." *Public Health Reports*, 63 (July, 1948).

15. WYSS, O., "Disinfection by Chlorine—Theoretical Aspects." *W & S. W.*, 109, R-155 (1962).

16. FABER, H. A., "Contemporary Chlorination Practice." *Jour. AWWA*, 39, 200 (March, 1947).

17. BABBITT, H. E., et al., *Water Supply Engineering*, 6th ed. McGraw-Hill Book Co., Inc., New York, 1962.

18. BUTTERFIELD, C. T., et al., "Influence of pH and Temperature on the Survival of Coliforms and Enteric Pathogens when Exposed to Free Chlorine." *Public Health Reports*, 58, 51, 1837 (1943).

19. WATTIE, E., and BUTTERFIELD, C. T., "Relative Resistance of *Escherichia coli* and *Eberthella typosa* to Chlorine and Chloramine." *Public Health Reports*, 59, 1661 (1944).

20. ARTHUR, G. B., "Compact Activated Sludge Plant Designed for Flexible Operation." *Public Works*, 86, 3, 106 (March, 1955).

21. HAYS, H., et al., "Effect of Acidification on Stability and Bactericidal Activity of Added Chlorine in Water Supplies." *Jour. Milk & Food Tech.*, 26, 147 (1963).

22. CHANG, S. L., and FAIR, GORDON M., "Viability and Destruction of the Cysts of Endameba Histolytica." *Jour. AWWA*, 33, 1705 (Oct., 1941).

23. CHANG, S. L., "Viruses, Amebas and Nematodes and Public Water Supplies." *Jour. AWWA*, 53, 288 (1961).

24. POPP, L., "Bacteriological Studies on the Effect of Chlorine for the Disinfection of Water." *Gas-u Wasserfach.* (Germany), 95, 100 (1054).

25. CLARKE, N. A., and KABLER, P. W., "The Inactivation of Purified Coxsackie Virus in Water by Chlorine." *Amer. Jour. Hygiene*, 59, 119 (Jan., 1954).

26. NEEFE, JOHN R., et al., "Disinfection of Water Containing Causative Agent of Infectious (Epidemic) Hepatitis." *Jour. Amer. Med. Assn.*, 128, 1076 (1945).

27. KELLY, S., and SANDERSON, W. W., "Effect of Chlorine in Water on Enteric Viruses." *Amer. Jour. Public Health,* 48, 10, 1323 (1958).

28. HALLINANA, F. J., "Tests for Active Residual Chlorine and Chloramine." *Jour. AWWA*, 36, 296 (1944).

29. KJELLANDER, JAN, and LUND, EVVA, "Sensitivity of *Esch. Coli* and Poliovirus to Different Forms of Combined Chlorine." *Jour. AWWA*, 57, 893 (1965).

30. MORRIS, J., "Future of Chlorination." *Jour. AWWA*, 58, 11, 1475 (1966).

31. SNOW, W. B., "Recommended Chlorine Residuals for Military Water Supplies." *Jour. AWWA*, 48, 12, 1510 (1956).

32. LOVTSEVICH, E. L., "Sanitary Significance of Coliform Tests in Chlorination of Water With Relation to Enteroviruses." *Gigiena i. Sanitariya* (USSR), 8, 12 (1962).

33. WILSON, H., "Some Problems in the Chlorination of Sewage." *Jour. and Proc. Inst. Sewage Purif.*, part 4 (1950), p. 533.

34. MOORE, E. W., "Fundamentals of Chlorination of Sewage and Industrial Wastes." *W & S. W.*, 100, 5, R-197 (1953).

35. NUSBAUM, I., "Sewage Chlorination Mechanism." *W & S. W.*, 99, 7, 297 (1952).

36. CHICK, H., *Jour. of Hygiene*, (Cambridge), 8, 92 (1908).

37. CHICK, H., *Jour. of Hygiene*, 10, 1051 (1910).

38. WALTON, G., and CLUP, G. L., "Chlorine Disinfection in Primary Sewage Treatment—A Review of the Literature Parts 1 through 4. *W & S. W.*, 110, 415, 438 (1963); 111, 37, 80 (1964).

39. HEUKELEKIAN, H., and FAUST, S. D., "Water Pollution by Organic Pesticides." *Jour. AWWA*, 56, 267 (1964).

40. HEUKELEKIAN, H., and SMITH, M. B., "Disinfection of Sewage with Chlorine. I. Laboratory Experiments on the Effect of Chlorine Dosage on Residual Coliform Organisms." *Sew. & Ind. Wastes*, 22, 12, 1509 (1950).

41. "Solubility of Atmospheric Oxygen in Water." *Proc. ASCE*, 87, SA5, 73 (Sept., 1961).

42. GRUNE, WERNER N., "Sewage Chlorination in Review." *W & S. W.*, 102, 8, 351 (Aug., 1955).

43. THOMAS, R. H., "Marine Dilution and Inactivation of Sewage." *Internatl. Jour. Air and Water Poll.*, 7, 845 (1963).

44. AMMON, F. V., and WIESELSBERGER, F., "Chlorination of Digested Municipal Sewage Sludge." *Gesundh. Ing.*, 73, 1 & 2, 15 (1952).

45. BERGSMAN, A., and VAHLNE, G., "Chlorination of Sewage from Tuberculosis Sanitariums." *Nord. Hyg. Tidskr.* (Sweden), 32, 49 (1951).

46. JENSEN, K. E., "Presence and Destruction of Tubercle Bacilli in Sewage." *Bull. World Health Organ.*, 10, 171 (1954).

47. MILLER, F. J. W., and ANDERSON, J. P., "Two Cases of Primary Tuberculosis After Immersion in Sewage Contaminated Water." *Archives of Disease in Childhood*, 29, 152 (1954).

48. ALFIMOV, N. N., and YAGOVOI, P. N., "Bactericidal Effect of Chlorine in Sea Water and Fresh Water." *Gigiena i Sanit.* (Moscow), 25, 85 (1960).

49. KELLY, S., and SANDERSON, W. W., "Effect of Chlorine in Water on Enteric Viruses. II. The Effect of Combined Chlorine on Poliomyelitis and Coxsackie Viruses." *Amer. Jour. Public Health*, 50, 1, 14 (1960).

50. TYLER, R. G., ORLOB, G. T., and WILLIAMS, F. W., "Chlorination of Raw and Digested Sludge." *Sew. & Ind. Wastes*, 22, 7, 875 (1950).

51. HEUKELEKIAN, H., and FAUST, S. D., "Compatibility of Wastewater Disinfection by Chlorination." *Jour. WPCF*, 33, 9, 932 (1961).

Index